Tribology of Additively Manufactured Materials

The Elsevier Series on Tribology and Surface Engineering

The **Elsevier Series on Tribology and Surface Engineering** is a series of books that summarize foundational knowledge and cutting-edge applications, while also outlining recent developments in the fields of tribology and surface engineering. The series aims to provide systematically organized references for researchers and engineers whose objectives are to develop optimal tribological systems or to establish new theories of tribology, ultimately advancing the field. Books in the series will contain foundational knowledge, applied examples and best practices, as well as the latest in groundbreaking research.

Key Features:

- Displays an international editorial board and roster of volume editors and authors
- Presents an interdisciplinary mix of topics from classical theory to cutting-edge industrial applications of new research
- Demystifies the complex scientific and technological issues of tribology and surface engineering
- Empowers researchers and engineers to develop optimal tribological systems, to establish new theories of tribology and to facilitate the advancement of the field

Elsevier Series on Tribology and Surface Engineering

Tribology of Additively Manufactured Materials

Fundamentals, Modeling, and Applications

Edited by

Pankaj Kumar
Department of Mechanical Engineering, University of New Mexico, Albuquerque, NM, United States

Manoranjan Misra
Department of Chemical and Materials Engineering, University of Nevada, Reno, NV, United States

Pradeep L. Menezes
Department of Mechanical Engineering, University of Nevada, Reno, NV, United States

ELSEVIER

Elsevier
Radarweg 29, PO Box 211, 1000 AE Amsterdam, Netherlands
The Boulevard, Langford Lane, Kidlington, Oxford OX5 1GB, United Kingdom
50 Hampshire Street, 5th Floor, Cambridge, MA 02139, United States

ISBN: 978-0-12-821328-5

For information on all Elsevier publications visit our website at
https://www.elsevier.com/books-and-journals

Publisher: Matthew Deans
Acquisitions Editor: Dennis McGonagle
Editorial Project Manager: Emily Thomson
Production Project Manager: Prem Kumar Kaliamoorthi
Cover Designer: Vicky Pearson Esser

Typeset by TNQ Technologies

Contents

List of contributors

Stephen A. Akinlabi Department of Mechanical Engineering, Butterworth Campus, Walter Sisulu University, Butterworth, Eastern Cape, South Africa

Esther T. Akinlabi Department of Mechanical Engineering Science, University of Johannesburg, Auckland Park Kingsway Campus, Johannesburg, South Africa; The Directorate, Pan Africa University for Life and Earth Sciences Institute, Ibadan, Nigeria

Ethan Austin Department of Mechanical Engineering, University of Nevada, Reno, NV, United States

H.R. Bakhsheshi-Rad School of Mechanical Engineering, Faculty of Engineering, Universiti Teknologi Malaysia, Johor Bahru, Johor, Malaysia; Advanced Materials Research Center, Department of Materials Engineering, Najafabad Branch, Azad University, Najafabad, Iran

Mohammadreza Daroonparvar Department of Mechanical Engineering, University of Nevada, Reno, NV, United States; Research and Development Department, ASB Industries Inc., Barberton, OH, United States

Christopher Evangelista Department of Mechanical Engineering, University of Nevada, Reno, NV, United States

Dallas Evans Department of Mechanical Engineering, University of Nevada, Reno, NV, United States

Carlos Flores Department of Mechanical Engineering, University of Nevada, Reno, NV, United States

Keith Hall Department of Mechanical Engineering, University of Nevada, Reno, NV, United States

David Hart Department of Mechanical Engineering, University of Nevada, Reno, NV, United States

Sunil Hassan Department of Mechanical Engineering, Butterworth Campus, Walter Sisulu University, Butterworth, Eastern Cape, South Africa

Mathew Heintzen Department of Mechanical Engineering, University of Nevada, Reno, NV, United States

Branden Hughes Department of Mechanical Engineering, University of Nevada, Reno, NV, United States

Tien-Chien Jen Department of Mechanical Engineering Science, University of Johannesburg, Auckland Park Kingsway Campus, Johannesburg, South Africa

Charles M. Kay Research and Development Department, ASB Industries Inc., Barberton, OH, United States

Thomas Kotowski Department of Mechanical Engineering, University of Nevada, Reno, NV, United States

Pankaj Kumar Department of Mechanical Engineering, University of New Mexico, Albuquerque, NM, United States; Department of Chemical and Materials Engineering, University of Nevada, Reno, NV, United States

Amanendra K. Kushwaha Department of Mechanical Engineering, University of Nevada, Reno, NV, United States

Cody Lee Department of Mechanical Engineering, University of Nevada, Reno, NV, United States

Joel Leslie Department of Mechanical Engineering, University of Nevada, Reno, NV, United States

Rasheedat M. Mahamood Department of Material and Metallurgical Engineering, University of Ilorin, Ilorin, Nigeria; Department of Mechanical Engineering Science, University of Johannesburg, Auckland Park Kingsway Campus, Johannesburg, South Africa

Connor Marshall Department of Mechanical Engineering, University of Nevada, Reno, NV, United States

M.A. Mat Yajid School of Mechanical Engineering, Faculty of Engineering, Universiti Teknologi Malaysia, Johor Bahru, Johor, Malaysia

Angus McCarroll Department of Mechanical Engineering, University of Nevada, Reno, NV, United States

Pradeep L. Menezes Department of Mechanical Engineering, University of Nevada, Reno, NV, United States

Matthew Montez Department of Mechanical Engineering, University of Nevada, Reno, NV, United States

Leslie T. Mushongera Department of Chemical & Materials Engineering, University of Nevada, Reno, NV, United States

Chandramohan Palanisamy Department of Mechanical Engineering, Sri Ramakrishna Engineering College, Coimbatore, India; Department of Metallurgy, University of Johannesburg, Johannesburg, South Africa

Radul Patel Department of Mechanical Engineering, University of Nevada, Reno, NV, United States

Md Hafizur Rahman Department of Mechanical Engineering, University of Nevada, Reno, NV, United States

Dipen Kumar Rajak Department of Mechanical Engineering, Sandip Institute of Technology & Research Centre, Nashik, Maharashtra, India; Department of Mechanical Engineering, G. H. Raisoni Institute of Business Management, Jalgaon, Maharashtra, India

Alessandro M. Ralls Department of Mechanical Engineering, University of Nevada, Reno, NV, United States

Raghu Raman Department of Mechanical Engineering, Sri Ramakrishna Engineering College, Coimbatore, India

M. Razzaghi Advanced Materials Research Center, Department of Materials Engineering, Najafabad Branch, Azad University, Najafabad, Iran

Henry Rendler Department of Mechanical Engineering, University of Nevada, Reno, NV, United States

Enrique Rubio Department of Mechanical Engineering, University of Nevada, Reno, NV, United States

Diego Armando Saldana Department of Mechanical Engineering, University of Nevada, Reno, NV, United States

Ethan Slater Department of Mechanical Engineering, University of Nevada, Reno, NV, United States

Eric Tompkins Department of Mechanical Engineering, University of Nevada, Reno, NV, United States

Jacob Welty Department of Mechanical Engineering, University of Nevada, Reno, NV, United States

Keegan Willis Department of Mechanical Engineering, University of Nevada, Reno, NV, United States

Carson Zollars Department of Mechanical Engineering, University of Nevada, Reno, NV, United States

About the editors

Dr. Pankaj Kumar is an Assistant Professor in the Mechanical Engineering Department at the University of New Mexico, USA. Dr. Kumar's research is in the broader areas of advanced manufacturing and materials design for novel structural and functional applications. His research includes material concepts with specific emphasis on advanced manufacturing, materials processing, and physical and mechanical metallurgy. He teaches mechanical engineering design and advanced manufacturing courses to undergraduate and graduate students at the University of New Mexico.

Manoranjan Misra is a Foundation Professor at the University of Nevada, Reno, USA. He has published over 200 journal articles and his research interests are metallurgy, and chemical and material processing.

Pradeep L. Menezes (corresponding editor) is an Associate Professor in the Department of Mechanical Engineering at the University of Nevada, Reno, USA. Before joining this university, he worked as an Adjunct Assistant Professor at the University of Wisconsin–Milwaukee (UWM), and as a Research Assistant Professor at the University of Pittsburgh. Dr. Menezes's research career has produced more than 150 peer-reviewed journal publications (citations more than 7000; h-index: 40), 30 book chapters, six books related to tribology, over 100 conference papers/presentations, and a patent.

About the editors

Preface

Additive Manufacturing (AM) provides the unique advantages of reducing the cost over conventional manufacturing methods by reducing the overall raw materials and energy waste in manufacturing engineering components for aerospace, medical, and automobile applications. During operation, the engineering AM components can be subjected to tribological conditions, such as sliding or rolling contacts with counter materials along with corrosive environments. These conditions can lead to an early failure and/or reduce the performance of components. Understanding the tribological mechanisms could be key to determining the reliable applications of these components in critical engineering applications. This book emphasizes the fundamentals of various AM techniques and the tribology and corrosion of AM materials.

Chapter 1 discusses the commonly used AM techniques for manufacturing materials ranging from polymers to metals to composites. In this chapter, the basic principle, operations, and applications of each technique have been covered. The choice of manufacturing methods, the materials, AM parameters, and postprocess treatment has been discussed to customize properties and performance.

Chapter 2 details the application of the cold spray (CS)-based AM technique in manufacturing the engineering components. A comparison of solid-state manufacturing has been made with the fusion-based thermal spray processes. The basic principle of CS-based AM operation and utilization has been discussed, and the fundamental manufacturing mechanism is covered. The application of CS-based AM for the manufacturing of different materials such as titanium alloys, aluminum alloys, nickel alloys, and metal materials composites has been included in this chapter. This chapter intends to provide the fundamentals of an emerging solid-state AM technique for manufacturing the structural components.

High accuracy manufacturing is critical for many industries such as medical, automotive, aerospace, and manufacturing for prototyping and modeling. Prototyping and modeling are very important for product design and development, allowing preoperative planning for manufacturing at a large scale and respective diagnosis. Stereolithography (SLA) is one of the most emerging 3D printing technologies that have precise control over manufacturing. This printing technique employs a precise laser to cure a liquid polymer, layer by layer, onto an inverted built platform. Chapter 3 gives an overview of SLA processing and its applications in various industries. The chapter also covers the mechanical, tribological, corrosion, and tribocorrosion properties of SLA manufacturing components.

Due to their unique gradient structure and properties, the functional graded metallic materials (FGMs) proved to be alternative materials systems where the operating conditions change abruptly or slowly in given applications such as boilers and jet engines. Chapter 4 details the feasibility of using AM techniques to manufacture FGMs. This chapter presents both the conventional and AM techniques for manufacturing. Further, the mechanical, tribological, and corrosive properties of FGMs have been discussed, and their application in industries has been portrayed.

Fusion Deposition Modeling (FDM) is an inexpensive AM technique for a range of fabrication applications from prototyping to mass production. While FDM is considered the rapid prototyping technique, recent advancements enable FDM to manufacture end-use products. In Chapter 5, the basic principle and operations of FDM are discussed in detail, from the selection of the material to the materials process. The properties, including the mechanical properties, tribological properties, corrosion, and tribo-corrosion properties of the FDM components, have been presented. In addition, the application of FDM in industries has been included.

Chapter 6 specifically focuses on the fundamental of process and microstructure development in fusion-based AM techniques by considering an example of laser powder bed fusion. This chapter discusses the impact of laser parameters on the microstructure and the properties evaluation. A phase field modeling study is included to understand the underlying mechanisms involved in the microstructure development in fusion-based AM techniques.

Chapter 7 focuses on surface roughness development in AM components. Large surface roughness has been a significant challenge in AM components. This chapter details surface roughness evolution as affected by the AM process parameters. A comprehensive assessment of surface roughness development in various AM techniques is given. The surface roughness assessment has been made with respect to the process conditions and the common materials. The postprocessing techniques to control the surface roughness of the AM components are also highlighted in this chapter.

Chapter 8 addresses the tribology of AM metallic materials, emphasizing the influence of AM factors involved in AM techniques such as microstructural features, posttreatment, surface, and modification on the tribological behavior. The chapter also discusses various mechanisms of wear performance in AM components. Tribology performance of various materials are considered in this chapter, such as titanium alloy and composite, Inconel alloy, and stainless steel manufactured by different AM techniques.

Manufacturing medical implants using AM techniques has been considered important because AM can manufacture patient-specific implants. The significance of AM in medical implants has been considered in Chapter 9. This chapter focuses on the tribology behavior of AM metallic components specific to medical implants. In this, the properties of titanium and titanium alloys, their biomedical applications, the need for surface modification, methods used in surface modification of titanium and its alloys for biomedical applications, the use of AM for surface modification, and for improved corrosion and wear resistance properties have been discussed. It is assessed in this chapter that AM significantly influences the production and the economy of the fabrication of medical titanium-based implants as well as other biomedical materials with tailored mechanical, biological, and the needed surface properties.

The corrosion behavior of solid-state CS AM components has been discussed in Chapter 10, along with the recent activities to increase the corrosion resistance of the materials. This chapter provides an understanding of the corrosion behavior of CS materials as affected by the processing and postprocessing parameters considering different materials.

This book is projected for professionals as well as university students to provide the fundamental understanding and the recent developments in the AM manufacturing field. A large data set and the review of the recent work are included that are relevant to different fields of interest such as materials and manufacturing, mechanical, and chemical. Professionals associated with the development and applications of materials and manufacturing will find this book very insightful for their professional reference. We have also added an extensive list of references at the end of each chapter, and it makes this book an excellent source of references in the field of AM and materials.

Comprehensive knowledge of tribology, AM, and their interrelationship through this book has been possible with the collective efforts of various research groups around the world.

Powder bed fusion–based additive manufacturing: SLS, SLM, SHS, and DMLS

1

Amanendra K. Kushwaha [1], Md Hafizur Rahman [1], Ethan Slater [1], Radul Patel [1], Christopher Evangelista [1], Ethan Austin [1], Eric Tompkins [1], Angus McCarroll [1], Dipen Kumar Rajak [2] and Pradeep L. Menezes [1]

[1]Department of Mechanical Engineering, University of Nevada, Reno, NV, United States; [2]Department of Mechanical Engineering, G. H. Raisoni Institute of Business Management, Jalgaon, Maharashtra, India

1.1 Introduction

Powder bed fusion processes are a form of additive manufacturing (AM) that starts with a computer-aided design (CAD) model. A computer program is used to slice the model into thin layers. The layer thickness can be adjusted to achieve the desired precision. The computer can then create a path for the heat source (typically a laser) to follow for each layer to bond the desired material [1–3]. The powder is then spread across the surface and bonded by the heat source. This process is repeated for each layer until the part is complete. The process parameters can be varied across the different layers during printing, making this manufacturing process very discrete and unique [4–8].

There is a lot of material wastage in traditional subtractive manufacturing processes, such as machining, drilling, milling, etc. Since the powder bed fusion process involves the addition of material to create a product, there is virtually no wastage [4–11]. This allows commercial industries to create complex 3D functional geometries in minimal time with no wastage, thus maximizing their profit by saving material, time, and special tooling costs. In this chapter, we are going to discuss four major powder bed processes, namely, selective laser sintering (SLS), selective laser melting (SLM), selective heat sintering (SHS), and direct metal laser sintering (DMLS). SLS was one of the first AM processes developed in 3D printing, initially created in the 1980s by Dr. Carl Deckard and Dr. Joe Beaman [12]. This technology used lasers to heat plastic powder and melt it into a desired shape layer by layer. However, this process took extraordinary amounts of time and money [1,2,13–16].

SHS is another lesser-known technology that uses a heat source instead of a laser for 3D printing. SHS is typically used to print specialized and complex geometries, which find its applications in the automobile, biomedical, and aerospace sectors, whereas SLM is mostly used in manufacturing industries. Unlike SHS, SLM uses a

Tribology of Additively Manufactured Materials. https://doi.org/10.1016/B978-0-12-821328-5.00001-9

laser beam to melt the metal powders to weld them together in one layer as well as weld them to the previously solidified layer. DMLS process is very similar to SHS as both methods use powdered metal and lasers to 3D print metal parts, starting with a 3D CAD model. Like SLM, DMLS prints the part layer by layer by dispensing a thin layer of metal powder over the build platform and then using the laser to heat the selected areas to make a solid part. Since AM works by adding layers of melted material, no specialized tooling tip is required for different parts of a complex geometry anymore. Thus, it is very useful for projects involving complex geometries, where now the only requirement is to make a 3D CAD model and slice it into thin layers using a slicing software. The structural integrity of the parts produced using each manufacturing process is a crucial factor that makes the process reliable and safe to be implemented in critical applications.

SLS is an ideal AM process for aerospace applications due to the rigid structure of parts generated during the printing process [4–7]. During SLS, the internal temperature of the printing volume remains close to the melting point of the material being sintered. The controlled and regulated environment coupled with the print volume temperature results in the printed parts having a much more controllable and uniform grain boundary structure [17]. SLS printing technology received attention from the pharmaceutical industry as it allows pharmaceutical companies to print pills [18,19]. SLS recently has seen an increase in potential materials being printed, due to more accurate systems incorporating closer heat regulation. Overall, the SLS process produces high-quality parts having good granular composition requiring lower postprocessing. However, it has a few drawbacks, including the time needed to reach appropriate temperatures, the rough surface texture of products, and the relatively high porosity of products [17]. While SLS offers a good granular composition, typically better than other AM processes like DMLS, SLS is restricted to lower melting temperature materials due to a lower manufacturing energy input. SLS can be used to process a wider selection of materials including glass, ceramics, and plastics as compared to the DMLS process which can only be used for metals alloys [16,20–24].

On the other hand, SHS is a consumer-grade practical prototyping device mimicking SLS in process execution and final product quality, however, using a thermal heated printing head rather than a laser. SHS with controlled printing environments allows the plastic powder to be near melting temperature before sintering. The structure of parts produced by SHS has similar results to that of SLS, showing a high-quality surface finish and dense grain structures [13–16,20]. Grain boundary, grain structure, and inclusions from the process are key factors that determine the reliability of the part produced [25]. The grain boundaries found in parts printed using DLS show high integrity due to the low temperature experienced by the metal powder during the sintering process. The grain structures for these DMLS-printed parts are much finer than wrought alloy counterparts, showing high densities. Due to pool melting, DMLS-printed parts have porous layers that can become yielding points under strain. However, high-porosity DMLS alloy components can be advantageous for medical applications, as discussed in Section 4.3.

1.2 Types of powder bed sintering processes

There are two fundamental steps involved in creating a 3D CAD model for any AM process. The first step is to create a 3D model of the part to be printed using CAD software. The second step would be to take this CAD file and slice it either using a third-party slicing software or the printer's in-built slicing feature. During this step, the model is sliced up to desired layer thickness resolution of the printer. The printer software reads this file and generates layers of powder that the laser can traverse to sinter them together. Once the layer is printed, the next layer of powder is spread over the previous layer, and the process continues till all the layers are printed [26]. Based on the powder bed fusion technique, there are four primary types of 3D printing processes as discussed below.

1.2.1 Selective laser sintering

SLS is an AM process developed around the mid-1980s through the Defense Advanced Research Projects Agency (DARPA). The lead designers of the SLS process were Dr. Carl Deckard and Dr. Joe Beaman from the University of Texas at Austin. SLS uses a high-power laser to fuse the material powders layer by layer to a predetermined shape designed using a CAD program and sliced using a slicing software. SLS has evolved over the years to become a complex sequence of operations based on fundamental functions [12].

The main parts of an SLS printer include a laser, a roller, the printing bed, a fabrication piston, and occasionally a powder delivery piston (Fig. 1.1). Once the sliced CAD file is loaded into the printer software, the first task is to generate a thin layer of polymer or metal particulate on the printing bed. The printer then uses a laser to

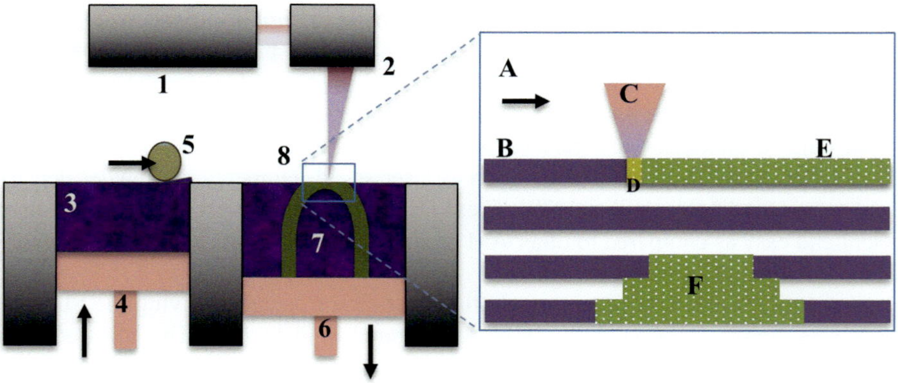

Figure 1.1 A simplified figure of an SLS printer showing the two-piston system operation. SLS process: 1, laser; 2, scanner system; 3, powder delivery system; 4, powder delivery piston; 5, roller; 6, fabrication piston; 7, fabrication powder bed; 8, object being fabricated (see inset); A, laser scanning direction; B, sintered powder particles (brown state); C, laser beam; D, laser sintering; E, preplaced powder bed (green state); and F, unsintered material in previous layers.

trace a 2D shape of the layer of the 3D object, melting/sintering a thin lamina of nylon or metal. The process is repeated to form multiple layers of 2D shapes, which keep fusing due to the laser intensity of every consecutive layer, forming a 3D object [17].

Compared to other AM processes, the SLS technique is a relatively simple process with four fundamental tasks, namely:

1. Dispensing powder from the reservoir,
2. Generating a flat surface,
3. Sintering a 2D parameter, and
4. Receding the fabrication piston.

The first step of the process involves the powder piston pushing powder out, effectively making an area where a mound of powder protrudes out from the printing bed surface. In the second step, this mound is dispersed uniformly by dragging a bi-axial gantry over the powder surface, making the entire printing surface fractionally thicker by one layer as defined in the slicing software. The gantry that flows over the surface has a roller on it that simultaneously pushes and rolls over the powder, ensuring a consistent coating throughout the surface. The gantry height can vary based on the layer thickness selected, allowing the varying amount of powder to be laid down as the gantry passes over the printing plane. Once the initial layer is spread uniformly, the third step is sintering in which the laser begins to move in a line path shaping a 2D outline of the layer of the 3D part. The laser process parameters can be varied based on the type of material and the required output. For the last step of the printing process, the powder delivery piston is raised to have a mound of powders, and the fabrication piston is lowered by one layer thickness. Now, the gantry spreads and flattens another layer of powder over the surface, which can then be sintered. This process is repeated until all the layers of the 3D model are printed [2]. Fig. 1.1 shows the breakdown of the SLS system and the integration of the two chambers utilized by the printer.

In the SLS and other powder bed fusion processes, the powder bed acts as a support structure for the built part, thus eliminating the need to build a support structure. The powders can directly be sintered together, to make final parts and functional prototypes. The sintering process ensures stronger molecular bonding for these functional parts. Apart from this, the SLS process also offers a high production rate and a wide range of materials that can be processed using this technique [27]. These factors are further dependent on variables such as laser speed, temperature, powder density, and matrix binding capabilities [28]. These features make these processes more convenient and easy as compared to other AM processes such as fused deposition modeling (FDM) and stereolithography (SLA).

According to the US patent search, the most recent patent involving the SLS process was filed on January 28th, 1997, that expired in 2011 [29]. This allowed companies to build, create, and sell at-home SLS printers. One such example is the benchtop SLS. The benchtop SLS printer uses a diode or fiber laser instead of the traditional CO_2 laser, which is widely used in industries. The main reason for this is the power consumption being significantly reduced compared to a CO_2 laser. The benchtop SLS starts at a price of about \$10,000 and with low running and maintenance costs. This is extremely affordable as compared to its industrial counterparts, which start at a minimum of $200,000 [30]. The benchtop SLS system has two main downfalls, (1) size and (2) material. Since this is a home-use system, the overall size of the unit is small; therefore, it cannot print large products. Another issue with this system is the laser intensity, which is not powerful enough for all kinds of materials. For example, this printer cannot be used to work with high melting point materials such as aluminum and silica-infused ceramics. Generally, the usage of benchtop printers is limited to plastic polymer materials [31].

Compared to home systems, industrial SLS systems are robust, and significant research has been carried out in recent years to improve the systems even more. Thijs et al. [1] critically analyzed the current progress of SLS to explain its future capabilities. They specifically highlighted the structural pitfalls and manufacturing benefits of SLS using aluminum alloy powders. Over the literature, SLS has increasingly been targeted as a primary process for additively manufacturing aluminum components. Pure Al, Al–Mg, and Al–Si powders are some of the most studied materials due to their high yield strength properties [17]. The scanning speed (v_s) and the power of the laser are the key process parameters that determine the feed rate of the laser [7]. SLS is widely used for metal alloy powders as it offers improved grain boundary stability and surface integrity of the finished product [32]. These process parameters and the desired outcomes are determined by the sintered metal's specific heat and melting temperatures [33]. As the laser power (P_L) increases and scan rate decreases, pools of melted metal form, which on cooling, become "balled" creating an area of poor distribution during the next layer formation. This issue propagates itself through the consecutive layers that are being fused, causing epitaxial growth. Epitaxial growth occurs when residual heat from the fusion boundary affects the layer's thermal distribution and thermal expansion. Due to this thermal boundary, residual heat gets collected on the boundary of the last layer resulting in the pooling of heat, which can cause the metal to melt rather than sinter [9]. The achieved grain structure affects the overall density of the finished product; the lower the density, the lower the strength of the part [10,13,14].

The SLS process allows the unused powder to be reused for printing other parts if mixed with 30%–50% fresh powder, which offers great savings in material cost [34]. However, there are several drawbacks of the SLS process as compared to other AM processes [35]. One drawback of the SLS process is that it cannot be used to print a hollow, fully enclosed product. Since the entire system is sitting in powder, hollow parts without any holes will remain filled with unsintered powder [36]. Another downside to SLS printing is that the temperature of the powder and the part must be held within 2°C for the entire process, including preheating, sintering, and storage before removing from the powder. This is essential to reduce the warping and distortion of the printed part due to thermal stresses [37].

The density of the printed part from the SLS process relies on the peak power of the laser instead of the laser beam duration. Because of this, a pulsed laser is typically used in the SLS process. Usually, the SLS printer heats the material powder close to its melting point. This allows the laser to be on for less time and only heats the material high enough for the sintering to occur [38]. This also allows for a better final surface finish across the part as a result of uniform heating. One of the most significant differences that separate this process from others is the lack of support structures to create the object. In the SLS process, the object is always surrounded by unsintered material that supports the object as it is built [39]. This gives flexibility to the SLS process to print intricate designs without modeling and caring for support structures.

1.2.2 Selective heat sintering

SHS is yet another AM process used in the 3D printing of complex geometries. This process was invented and exclusively patented by a Danish company, Blueprinter. Thus, it is not being used as widely as other AM processes; however, it is used for high-quality, high-precision applications such as for aerospace material parts. The overall process is very similar to SLS. The basic difference between SHS and SLS is that SHS uses a thermal-based printer head, while SLS uses a laser to melt the powder [30]. Since the thermal print head apparatus requires less space as compared to the laser, the 3D printer size for SHS is generally much smaller and, therefore, cheaper than that of SLS. In the SHS process, a thin layer of thermoplastic powder is spread over the surface. The thermal printhead then moves over the surface and melts layers of the thermoplastic powder into a solid layer by layer. Once a layer is printed, the powder bed piston moves down to lower the print bed, and an automated roller rolls onto the surface to add a new layer of material. The thermal printhead now sinters this layer. This process is repeated until all the layers are printed to form the final 3D shape. Similar to SLS, the SHS process also does not require the building of any support structure. This gives the user complete freedom to build complex shapes. However, the SHS process is generally limited to using thermoplastic polymeric powders due to a thermal printhead with a limited heating capability as compared to the laser beams in other laser printing processes such as SLS, SLM, and DMLS [30].

1.2.3 Selective laser melting

SLM originated at the Fraunhofer Institute in Aachen, Germany, in 1995, to improve the SLS process [40]. While SLS heats the powder grains to the point where they bind together, SLM fully melts the metal powder so that the material becomes liquefied to form a homogeneous component [15]. Although this difference typically creates stronger and more durable components, it usually takes longer for the printed part to cool down [41]. This extended cooling duration makes SLM process slower than SLS. Classification of SLS versus SLM techniques is carried out by the binding mechanism used, which is unusual because the conventional method separates processes by the materials they use. This chapter discusses SLM because the American Society for Testing and Materials (ASTM) International has grouped SLM into the "laser sintering" category. However, it is not a true sintering process considering the liquefaction aspect [16].

SLM works by dispersing a thin layer of powder in a build chamber, which is then heated up to just below the melting point. The laser then traverses over the surfaces and raises the temperature further to create a melt pool, which solidifies as it cools. This process happens in layers; once the first layer is finished, another layer of powder is spread onto the first layer. This is followed by heating the powder to the melting point, which then melts and fuses the powders in the current layer as well as adheres this layer to the previously printed layers [42]. As the layers are being added one by one, the platform is lowered, creating room for more layers, until the final part is obtained. Parts printed using SLM generally have a rougher surface texture than SHS parts, thus requiring postprocessing operations such as polishing.

The SLM process offers plenty of benefits over other AM manufacturing processes such as high print speed and short cycle time for functional prototype applications [43]. Before the advent of 3D printing, creating a prototype would take weeks or even months to complete. Revising the prototype design post failure would take another few weeks. Additionally, the cost and effort required to prepare the special tooling and machine configurations were very high. SLM process can be used to print functional prototypes that can be manufactured and tested within a short cycle time. Any design changes can be readily made in the model, printed, and tested within a short period [44]. Thus, this process is very significant to small-scale manufacturing units, which no longer require industrial machinery to create prototypes. Design and prototyping of functional parts often lose additional time due to outsourcing-specific machining tasks, which can now be easily achieved using these AM processes [44]. With the AM process, everything can be carried out with a single 3D printer machine, thus much smaller warehouse space is needed to store and prepare functional components as compared to industrial machinery requirements for similar tasks in an industrial setup. SLM and other sintering processes can create intricate internal features and cavities that would otherwise not be possible with conventional casting or machining [44]. Other than machine costs and time saved, SLM also offers cost savings on experienced machine operators and maintenance technicians who would otherwise be needed in an industrial setup to create complex parts [45].

The accessibility of CAD software designing and automated SLM opens up design opportunities to individuals who would not normally create the same parts with traditional methods. When using AM methods, it is typical for only 1 or 2 machines to be needed for the entire span of the production cycle. Significant cost savings are also intrinsic to AM process due to minimal wastage of material. When using subtractive manufacturing, a large amount of waste is created because the desired part is shaped out of simply shaped stock material. While this waste could be recycled and remade into more parts, this process would incur additional costs. The workspace on typical SLM machines can be up to one cubic meter such that any conceivable part can be created that fits within the volume of the workspace. The configuration of the printer rarely changes throughout its use other than varying the process parameters based on the material and application of the parts produced. The original CAD software can be used to make changes to the design or change the layer thickness using the slicing software. Other than these software changes, there are no equipment or tooling changes required to the SLM machine [46]. Another interesting benefit to the SLM process is the possibility of using multiple materials within the same part. This mixing of materials enables unique mechanical properties including a gradient in properties that are typically difficult to achieve using traditional methods. Traditional methods also inevitably produce a certain percentage of defective parts within a batch of parts. Due to the shorter cycle time, these design issues can be easily identified and rectified quickly with SLM [47].

SLM capabilities are a huge competitive advantage in the business world. The fast production of functional prototypes reduces the likelihood of miscommunication with customers, as they can witness testing and functionality sooner, leading to higher quality control of manufactured parts in a shorter cycle time [48]. Additionally, market testing can begin earlier in the design process, and customers can provide valuable feedback on the design and functionality. This higher product confidence creates significant investments, which strengthen the resources available for product designers. In addition, due to reduced outsourcing and minimized processing and postprocessing steps, the overall environmental impact of the manufacturing process goes down. These advantages provide a significant advantage over companies using traditional methods.

1.2.4 Direct metal laser sintering

DMLS is an AM process used in the 3D printing of metallic parts. In this technique, the metallic powder substrate is dispersed on a build platform and selectively heated by a laser beam to form the part [23]. Once a layer is completely sintered, a recoater arm applies another layer of powder over the build zone, and the next layer is sintered. This process is repeated in a layer-by-layer fashion until the part is completed. The main difference between DMLS and SLM is the temperature that the metal powder is subjected to. While SLM heats the powder until the metal is fully melted, DMLS uses a lower heat and only heats the surface enough to sinter the particles together [49]. Since it does not melt the material, DMLS is generally preferred when working with metal alloys, whereas SLM is used mainly for the 3D printing of pure metals. The major advantage of DMLS is that it can print a metal part out of an alloy or pure metal powder without impacting its material properties. Since DMLS uses a lower intensity laser, there is less energy consumed during the DMLS process; thus, it is more efficient than the SLM process, which uses higher intensity laser to melt the metal powder [49].

However, there are some drawbacks of the DMLS process that includes high initial setup cost. In addition to this, the DMLS process generally produces relatively porous parts compared to SLM since it does not completely melt the metal powder. In some cases, DMLS-printed parts may have significantly lower fatigue and yield strength than cast parts due to higher surface defects and micro-cracks [24,50–52]. However, there are numerous forms of postprocessing techniques such as heat treatment and optimization methods to achieve stronger parts. Postprocessing DMLS parts with hot isostatic pressing (HIP) is one of the most effective ways to improve the high cycle fatigue strength. It increases part density by removing voids that might cause micro-cracks [53–55]. Parts printed with the DMLS process using 316L stainless steel powders that underwent HIP, machining, and polishing are shown to have comparable fatigue properties as compared to the wrought 316L stainless steel. Also, the specimens generally exhibited higher fatigue strengths when the cyclic stresses were applied parallel to the build plane [56,57]. Therefore, it is important to consider the direction of the stresses the part is intended to handle when establishing the printing orientation of the part.

Material properties of DMLS-printed parts can also be optimized by controlling specific process parameters. A set of criteria designed to optimize DMLS-printed products has been proposed by Verma et al. [23]. Usage of the genetic algorithm is recommended to ensure that the part is sliced in an optimal way to reduce surface inaccuracy and improve computational efficiency [58—60]. Genetic algorithms are typically used to simulate the process of biological evolution to determine the optimal outcome [61—63]. This could be used to optimize the basic DMLS process parameters, including P_L, hatching distance (h_d), and v_s [64].

P_L is defined as the amount of power supplied to the laser to sinter the powder. It is important to use a high laser power to sinter the parts correctly but not too high that the material melts completely [65]. h_d refers to the distance between laser scan lines. Increasing the h_d will decrease the production time but will increase the porosity [64]. v_s refers to the speed at which the laser passes over the build platform. Increasing the v_s naturally leads to a reduction in production time while effectively reducing the width of the scan line. P_L, h_d, and v_s can be used to compute energy density (E_d) using Eq. (1.1):

$$E_d = \frac{P_L}{h_d v_s} \tag{1.1}$$

Power density (P_d) is the amount of energy consumed by the machine to sinter an area of metal in J/mm^2. The 3D form of E_d can be found by including the powder layer thickness (t) in the above equation. Therefore, the power density could be calculated by Eq. (1.2):

$$P_d = \frac{P_L}{h_d v_s t} \tag{1.2}$$

The power density of the laser impacts certain material properties of sintered parts, such as density, and surface finish [66]. However, this does not take into account other important factors, such as hatching pattern, h_d, slicing technique, laser diameter, and many other parameters that can affect the final part [67]. Therefore, while the power density has been proven to be correlated with some mechanical properties, it is not an ideal indicator of part performance. The effects of P_L, h_d, and v_s on DMLS parts made of AlSi10Mg alloy were examined by Krishnan et al. [68]. Their experiment concluded that h_d is the most influential parameter on the mechanical properties as it increases the porosity of the part, thereby decreasing its density and hardness [64,68]. Increasing the v_s has also been shown to have a similar effect on the sintered parts. Due to the reduction in melt pool size due to the smaller laser width associated with a faster v_s; the density and hardness decrease as the v_s is increased [64].

The microstructures of a metal part produced by DMLS may vary between different planes [69]. Ghasri-Khouzani examined the microstructures and hardness in the AISI 316L stainless steel test specimens for different planes using the DMLS process [70].

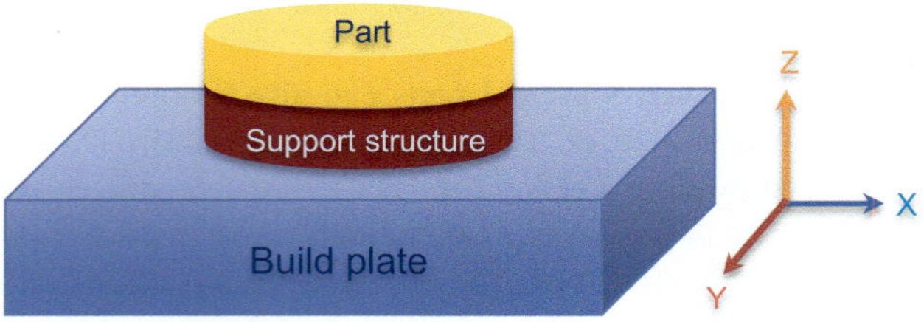

Figure 1.2 Axes definition of the orientation of the built part.

The specimens tested were disks with a 45-mm diameter and 20-mm height. The process parameters suggested by Electro-Optical Systems included a powder layer thickness of 20 μm, P_L of 195 W, and 80 μm hatch spacing [71]. The process parameters such as preheating temperature and postprocessing heat treatments also influence the mechanical properties of the built part [72]. These mechanical properties of the printed part are directional. Thus, build direction considerations are taken into account while designing the part to be printed. Fig. 1.2 illustrates the directions of the axes used to define the planes of the built part. For microscopy, the specimens were cross-sectioned both parallel and perpendicular to the build plane, ground, and polished down to 0.05 μm finish [70]. The specimens were then studied using an optical microscope, electron back-scattered diffraction, and X-ray diffraction (XRD) analysis. XRD is a comprehensive tool used to identify complex phases and compounds in a specimen [73]. The study found that the vertical section (XZ-plane) exhibited mostly columnar structures, with fewer equiaxed structures, whereas the horizontal section (XY-plane) consisted mainly of equiaxed structures, with fewer columnar structures [70].

The experiments discussed above demonstrate that if the production process is optimized, DMLS has the capability to create functional, high strength, and complex parts from a 3D model, making it a very useful tool for industries that require small or unique, high-performance metal parts.

1.3 Materials properties

The basic operating principle of each powder bed fusion process is nearly the same with few differences in the heating methodology, peak temperatures, and type of materials that can be processed using these techniques. The process parameters for the powder bed fusion process can be altered based on the 3D printing technique employed and the type of material. Table 1.1 presents a few examples of different processing methods and the impact of process parameters on the mechanical properties of printed parts. The table also presents how different postprocessing techniques can enhance the material properties of 3D-printed components. The mechanical properties obtained by each of the powder bed fusion manufacturing processes are discussed below.

Table 1.1 Process parameters and obtained materials properties for different powder bed processing techniques.

Processing method	Materials	Process parameters	Material properties	Remarks	Ref.
SLS	Starch cellulose, cellulose acetate	• Particle size for both powders = 106–125 μm For starch cellulose • Laser power = 3 W • Energy density = 0.301 J/mm^2 For Cellulose acetate • Laser power = 1.70 W • Energy density = 0.1711 J/mm^2	Starch cellulose • Density = 0.522 g/cm^3 • Tensile strength = 2.277 MPa • Failure deformation (%) = 4.167 Cellulose acetate • Density = 0.460 g/cm^3 • Tensile strength = 3.729 MPa • Elongation at failure = 4.167%	• Optimized the process parameters such as laser power and scan speed to produce biopolymer scaffold structures. • Specimens with smaller particle sizes showed better mechanical properties.	[74]
SLS	Polyethylene powder	• Particle size = 231 μm • Laser power = 5 W • Powder bed temperature = 101°C • Laser scan speed = 2500 mm/s • Laser scan spacing = 0.25 mm	• Tensile strength = 5.98 MPa • Tensile modulus = 170.67 MPa • Elongation at failure = 42.9%	• Double-laser scanning increases the energy input and provides strength to the sintered parts.	[75]

Continued

Table 1.1 Process parameters and obtained materials properties for different powder bed processing techniques.—cont'd

Processing method	Materials	Process parameters	Material properties	Remarks	Ref.
SLS	Alumide, polyamide PA2200	• Laser power = 25 W • Laser velocity = 1500 mm/s • Scan spacing = 0.25 mm • Energy Density = 0.066 J/mm^2 • Beam offset = 0.15 mm • Building chamber temperature = 170°C • Removal chamber temperature = 153°C • Layer thickness = 0.1 mm • Scaling factors = 2.2%	Alumide • Modulus = 1871 MPa • Quasi-static limit = 7.39 MPa • Yield strength = 13.16 MPa • Tensile strength = 17.17 MPa • Elongation at failure = 6.01% • Energy absorption = 0.90 MJ/m^3 • Poisson's ratio = 0.35 PA2200 • Modulus = 1321.90 MPa • Quasi-static limit = 13.21 MPa • Yield strength = 21.12 MPa • Tensile strength = 33.24 MPa • Elongation at failure = 15.62% • Energy absorption = 4.45 MJ/m^3 • Poisson's ratio = 0.42	• The microscopy reveals that the Alumide samples undergo brittle fracture, while the PA2200 undergoes ductile fracture.	[76]

SLS	Hydroxyapatite (HA), silica sol, and sodium tripolyphosphate (STPP) slurry	• Laser energy = 7–16 W • Scan speed = 120–550 mm/s • Layer thickness = 50–100 μm • Scan hatch = 0.1–0.15 mm • Spot size = 0.3 mm	• Pore size = 5–25 μm • Porosity = 17–20% • Surface roughness = 12–14 μm • Compression strength = 41–43 MPa • Bending strength = 1.2–1.4 MPa • Dimensional shrinkage = 14–15%	• Fabricated bio-composite scaffolds for bone implants. • Heat treatment at 1200–1400°C produces denser scaffolds with enhanced strength and lower surface roughness.	[77]
SHS	High-density polyethylene (HDPE)	• Heat energy = 22.16–28.48 J/mm^2 • Layer thickness = 0.1–0.2 mm • Heater feed rate = 3–3.5 mm/s • Printer feed rate = 100–120 mm/min	Loss modulus • Predicted = 6.05 • Experimental = 6.162 • Error = 1.81% Storage modulus • Predicted = 8.703 • Experimental = 8.514 • Error = 2.17% Damping parameter • Predicted = 0.142 • Experimental = 0.149 • Error = 4.69%	• Optimum parameters for improved viscoelastic properties were: (1) Heat energy = 26.32 J/mm^2 (2) Layer thickness = 0.1 mm (3) Heater feed rate = 3.5 mm/s, and (4) Printer feed rate = 116.38 mm/min.	[78]

Continued

Table 1.1 Process parameters and obtained materials properties for different powder bed processing techniques.—cont'd

Processing method	Materials	Process parameters	Material properties	Remarks	Ref.
SLM followed by heat treatment	Inconel 718	• Laser power = 170 W • Scan speed = 25 m/min • Overlap = 30% • Layer thickness = 20 μm	SLM'ed • Yield strength = 889–907 MPa • Tensile strength = 1137–1148 MPa • elongation = 19.2–25.9% • modulus = 204 GPa • Microhardness = 365 HV SLM'ed and Heat treated • Yield strength = 1102–1161 MPa • Tensile strength = 1280–1358 MPa • elongation = 10–22% • modulus = 201 GPa • Microhardness = 470 HV	• During the melting process, rapid cyclic heating and cooling result in excellent metallurgical bonding between the layers with grain refinement.	[79]

SLM followed by heat treatment	AlSi7Mg	• Laser power = 350 W • Scan speed = 1000–2000 mm/s • Hatch spacing = 0.10–0.15 mm • Layer thickness = 30 μm • Annealing temperature = 250, 300, and 350°C.	SLM'ed • Microhardness = 110.52 HV • Relative density = 99.8% Postannealing (250°C) • Microhardness = 100.94 HV • Tensile strength = 304.28 MPa • Yield strength = 208.53 MPa • elongation = 14.45% Postannealing (350°C) • Microhardness = 100.94 HV • Tensile strength = 210.35 MPa • Yield strength = 152.01 MPa • elongation = 30.83%	• The SLM'ed samples displayed a mix of brittle and ductile fracture; however, the heat-treated samples showed the ductile mode of fracture.	[80]
SLM	AlSi10Mg	• Laser power = 200 W • Scan speed = 1400 mm/s • Scan spacing = 105 μm	• Modulus = 68 GPa • Tensile strength = 391 MPa • elongation at failure = 5.55% • Microhardness = 127 HV • Impact energy (Charpy test) = 3.94 J	• SLM specimens show anisotropic behavior during elongation due to excessive borderline porosity formed during the laser scanning.	[2]

Continued

Table 1.1 Process parameters and obtained materials properties for different powder bed processing techniques.—cont'd

Processing method	Materials	Process parameters	Material properties	Remarks	Ref.
DMLS followed by HIP, T6 heat treatment, and aging	AlSi10Mg	• Laser power = 370 W • Scan speed = 1300 mm/s • Hatch distance = 0.19 μm • Layer thickness = 30 μm	As stress-relieved • Yield strength = 210 MPa • Tensile strength = 302 MPa • Strain = 10.7% • Hardness = 84.7 HB HIP'ed • Yield strength = 108 MPa • Tensile strength = 176 MPa • Strain = 25% • Hardness = 50.2 HB HIP'ed and 4h aged • Yield strength = 308 MPa • Tensile strength = 345 MPa • Strain = 5.8% • Hardness = 112 HB T6 heat treatment and 4h aged) • Yield strength = 234 MPa • Tensile strength = 279 MPa • Strain = 5% • Hardness = 90 HB	• The as stress-relieved sample shows a ductile fracture mode. • HIP'ed samples show even more ductile fracture having dimpled fracture surface. • Post-HIP and T6 treatment the samples show a cleavage-type fracture.	[78]

DMLS followed by solution treatment, water quenching, and isothermal heat treatment	Co-Cr-Mo alloy	• Laser power = 200 W • Laser spot diameter = 100 μm • Build platform temperature = 80°C Solution treatment • Temperature = 1150°C • Time = 1 h Heat treatment • Temperature = 800°C • Time = 4 h • The samples were water quenched after each of the processes.	Solution treated • Yield stress = 612 MPa • Tensile strength = 1030 MPa • Fracture Elongation = 20%. Solution and heat treated • Yield stress = 712 MPa • Tensile strength = 982 MPa • Fracture elongation = 2.3%.	• A better combination of mechanical properties was achieved for solution-treated samples. • The tensile strength and ductility are reduced post the heat treatment.	[81]

Continued

Table 1.1 Process parameters and obtained materials properties for different powder bed processing techniques.—cont'd

Processing method	Materials	Process parameters	Material properties	Remarks	Ref.
DMLS followed by direct aging, solution heat treatment, and subsequent aging	Maraging steel	• Laser power = 285 W • scan speed = 960 mm/s • Hatch distance = 0.11 mm • Layer thickness = 0.04 mm • Energy density = 67.47 J/mm^3 Aging treatment • Temperature = 490°C • Time = 6 h • Air cooling Subsequent heat treatment • Temperature = 490°C • Time = 2 h • air cooling • Followed by aging treatment (490°C for 2 h) and allowed to cool in air.	As built • Hardness = 36 HRC • Corrosion rate = 0.0016 mm/year Aged • Hardness = 56 HRC • Corrosion rate = 0.0032 mm/year Subsequent heat treated • Hardness = 55 HRC • Corrosion rate = 0.0437 mm/year	• The rate of corrosion increases after heat treatment. • The aging and heat treatment increase the hardness of DMLS-printed samples.	[82]

1.3.1 Selective laser sintering

SLS manufacturing technology includes a wide variety of materials that can be printed to the manufacturer's design requirements. Some of the most common materials used in the SLS process include plastics, glass/ceramics, and metals. This wide selection of materials for SLS allows for building prototypes and final products with low production numbers for varied applications. The most important factor in choosing the material is that the material's melting point should be well within the specification of the laser used for the system [83]. If the material being used has too high of a melting point, then the process will fail and not create the desired outcome of the CAD file.

The most popular material used in the SLS process is nylon. Nylon is a polymer that could be obtained both from petroleum and biological sources; however, they are typically nonbiodegradable [84]. Nylon-based materials, such as Nylon 12, can be 3D printed using SLS to create products that are heat and fire resistant [43]. Flex thermoplastic elastomer (TPE) is a polyurethane that is very commonly used in SLS due to the ease of manufacturing and has good elastic properties. TPE is a relatively economical option for the printing process. It is generally used in gaskets and rubber boots as a protection from harsh and adverse conditions [85]. Since the TPE is very malleable, the printing process tends to be very slow, allowing for layer thickness of 0.1–0.2 mm with a good accuracy. Nylon 12 GF is a glass-infused polymer with high tensile strength and very high resistance to extreme temperatures. Thus, this material requires a high-temperature printing system along with a strong laser [86]. The glass infusion also allows the material to be used in high temperature resistance and high load areas of the printed part. Some of the most common applications of nylon GF 12 are sporting good products as well as housings and enclosures [84]. Nylon 12 AF is an aluminum-filled polymer that has extremely good wear resistance and surface finish characteristics [87]. The aluminum filling gives the polymer a wide range of melting temperatures that need to be accounted for while selecting the laser intensity and system temperature [88]. This material is commonly used in automotive and aerospace applications due to its excellent surface properties [89].

The SLS process can be adapted for a wide range of materials and intricate designs [90]. Some of the final printed products require postprocessing. For example, the aluminum-infused polymer needs to be polished for a better surface finish and reduced surface roughness.

1.3.2 Selective heat sintering

Polyamide 12 (PA 12) and its composites are still the most widely commercialized materials in SHS/SLS, making up more than 80% of the current SLS/SHS material market [91]. This is due to their excellent laser sintering processability, including a wide sintering temperature range, good powder flowability, and high sintering rate. However, the intrinsic characteristics of the PA 12 matrix, such as relatively low mechanical properties and melting temperatures, limit the range of applications for these materials [92].

Besides polymers, SHS is also used to print ceramic materials. They have a wide range of applications in industries, such as aerospace, defense, automotive, and machining sectors due to their low thermal conductivity, high mechanical strength, high wear resistance, and excellent corrosion resistance. The starting point for manufacturing ceramic parts is usually powder mixed with binders and stabilizers. Ceramic components are conventionally shaped from powder to form a green compact, followed by machining and sintering steps to achieve a high densification functionality. Examples of conventional ceramic forming technologies include slip casting, tape casting, pressing, direct consolidation, and injection molding. Limitations of these conventional techniques include high wear rates for machining tools, high labor requirements, and higher costs. There is also a need for tight process control to avoid undesirable shrinkages from sintering green parts [93]. Recent research also indicates the development of biocompatible ceramics [94], which can be used for dental, body prosthesis, and tissue engineering scaffolds [95]. However, more investigation is needed to establish ceramic as viable input material for SHS.

One of the disadvantages of SHS is that it is a fairly new technology, so there are still limited materials that can be used to make parts. The most commonly used material in SHS is M-Flex, a white powder similar to nylon. It has a tensile strength of 7.5 MPa, making it strong and durable. It can be used for rapid prototyping to create low-cost concept figures and testable parts. It is also useful to create prototypes that require moving parts with high precision [96].

1.3.3 Selective laser melting

The most commonly used materials in the SLM process are copper, aluminum, stainless steel, tool steel, cobalt chrome, and titanium [97]. Parts created using SLM tend to be less porous and have a more unified crystalline structure, making them stronger than the similar part created using SLS [98]. SLM produces parts with near full density when using metals, but the results are still porous when using polymers. Therefore, when dealing with polymer manufacturing, SLS is preferred. Additionally, SLM is used when working with pure metals, while DMLS is used when working with metal alloys [20–23].

For a material to be used in SLM, it must first be manufactured into a powder. These powders are usually gas atomized prealloys because of their relative ease of production on an industrial scale. These gas atomized powders can be preprocessed using various techniques to produce nanocrystalline powders. Bulk materials produced with nanocrystalline grain structures are known to provide superior properties [99]. Goll et al. [100] used the SLM technique to work with nanocrystalline powders to produce bulk FeNdB-based permanent magnets. Their studies show the feasibility to produce nanocrystalline bulk materials using the powder bed fusion process. The survival of the nanocrystals depends on a melt pool depth, which was maintained at 20–40 μm. The reason for keeping the melt pool depth low is that the total liquefaction of the metal powder may induce the formation of secondary phases and grain growth. It may also introduce problems like balling and running that causes residual stresses and deformation [101].

Both SLS and SLM are often referred to as rapid manufacturing or rapid prototyping and are relatively newer technologies. Due to this, there is a lack of fully formed international standards or set procedures to outline large-scale production using the 3D printing process. Therefore, most manufacturers are still using conventional methods such as casting, forging, and machining for bulk manufacturing. The usage of these processes is expected to grow over the next few decades.

1.3.4 Direct metal laser sintering

The DMLS process uses a powder bed to generate near-perfect dense materials layer by layer using a powerful laser. The heat generated by the laser and its interaction with the powder bed determines the mechanical and chemical properties of the printed components. Some of the other factors that can affect the outcome of the DMLS process are scan speed, P_L, and h_d. The corrosion resistance of DMLS-printed parts is a varying trait from material to material; however, aerospace parts have seen an increase in interest due to its ability to make fully dense parts that have complex structures that cannot be created in a single piece using conventional milling and computer numerical control (CNC) machines. These complex structural parts printed in one piece allows for significant weight reduction due to the absence of rivets and bolts and the presence of cooling channels that would otherwise not be possible with traditional methods.

The specimens printed by DMLS are commonly printed with varying amounts of silicon, copper, magnesium, zinc, and titanium with aluminum as a bulk material. Aluminum alloy is a common material used in DMLS and other sintering processes due to its low density and high relative strength. It is chosen due to its innate eutectic composition allowing for high weldability. Regardless of material composition, large irregular pores occurred at melting pool boundaries; these areas suffered higher grain decomposition due to the gases being trapped. Caligano et al. [102] isolated the surface finish and granular composition as the primary determining factor of corrosion resistance. By reducing surface imperfections and granular gaps, the specimen can be effectively sealed from potential corrosion. The tests utilized four differently treated specimens. The specimens that performed the highest were treated with 10 g/L $Ce(NO_3)_3$ or Ce (III). The treatment reduced grain boundaries throughout the part, therefore, prolonging the part life in the corrosive testing environment. Once treated, the specimens are polished to reduce porosity and then tested in a corrosive environment [103]. A direct correlation can be made between a material's porosity and its compressive strength. The mechanical and the corrosion properties of the DMLS printed part are both affected by high porosity and low surface quality. Both of these problems can be mitigated by polishing the final product or using chemical polish to prevent surface wear. Therefore, while issues with surface finish need to be addressed, the solution is simple and well established [72].

Heat treatments are another primary factor that adversely affects a material's ability to handle corrosion. Heat treatments tend to increase grain boundaries causing a high aptitude for pitting and inclusions in the metal, which negatively impacts the material's ability to survive corrosive conditions [104]. Studies were conducted on the heat treatment and electrochemical behavior of DMLS-printed nickel-based superalloys. The results show that DMLS-printed parts had a higher corrosion resistance for untreated specimens due to tighter grain boundaries; however, they lacked strength due to a reduction in density. The results further showed that DMLS-printed parts performed far better after heat treatment than normal heat-treated metals [103].

DMLS has numerous benefits that include fast production and freedom of design. Owing to these, DMLS-printed parts have been used in the aerospace industry for years. SpaceX 3D-printed the main oxidizer valve body in one of the nine merlin 1D engines and, on January 6, 2014, successfully launched the rocket into space [105]. Before the actual DMLS process, the metal powder is atomized in argon gas atomizers to get the correct alloy chemistry and particle size. Parts manufactured with DMLS have the potential to reach up to 99% density [106]. This is great for parts required in oil and gas industries, aerospace parts, functional prototypes, custom medical implants, and tools. In the dentistry industry, DMLS is used to make prosthetics, bridges, crowns, and partial dentures as they are durable and last a long time [107−109].

A popular material for DMLS manufacturing is Monel K500. This material is composed of nickel−copper alloy, titanium, and aluminum. It is a material that has been tested with liquid rocket engines and is valued by the aerospace industry for its oxygen compatibility at high pressures [106]. Parts manufactured using this material with the DMLS process include heat sink chamber spools, nozzle spools, oil pipelines, and manifolds on rocket engines. This material is nonreactive to liquid oxygen and is tested to be noncorrosive. DMLS manufacturing is also used throughout aerospace, tooling, and medical sectors where having a corrosion-resistant material is crucial.

DMLS process eliminates tool-path generation making it faster to manufacture complex tools and attachments with cooling channels. These parts after posttreatment have high dimensional accuracy and excellent surface quality [110]. There are several different materials used in DMLS, some of which include stainless steel 17−4 PH, which is corrosion resistant and strong; cobalt chrome, which is used in dental and custom medical implants; and nickel alloy 625 and 718, which are used for their high tensile strength and resistance to rupture [111−113].

Although DMLS printing is revolutionary and very useful, there are numerous drawbacks to consider. A downside to using DMLS processes is the high initial setup cost. Industrial DMLS machines typically cost an average of $500,000 or more depending on the desired applications [114]. Additionally, the time required for printing the parts is much longer for this process than for traditional methods. It is also required for DMLS to manufacture several iterations of a new product before it is ready for mass production due to surface imperfections, which are common when working with certain materials. Another important consideration for the DMLS process is the optimal balance of each process parameter variable such as scan speed, P_L, and h_d. Among these, the scan speed causes the greatest amount of variation in properties for completed parts [115]. This can be considered a benefit because it allows for greater control over precise features in parts. On the other hand, it can be the cause of significant error if not addressed and identified properly during production. DMLS part designs must be altered and modified to conform to the relatively small build volume that most industrial DMLS printers are capable of. Although the build areas vary based on the machine manufacturer, the most common dimensions are around 250 × 250 × 300 mm^3. Small machines have built platforms of roughly 100 × 100 × 80 mm^3, and large machine platforms are roughly 400 × 400 × 380 mm^3 [58]. Alternative technologies such as FDM can accommodate volumes of over 2,475,000 cm^3, which is several folds higher than the DMLS process [116]. Processing parameters for DMLS typically depend on the property of the material being processed and its potential application [117].

DMLS printing is often accompanied by postprocessing steps to provide strength, durability, and required surface finish. Parts that are too small or complex may not be able to undergo these postprocessing procedures, which would make DMLS an invalid option in that specific application. Finished DMLS parts are very porous compared to a completely melted part. Although this may seem like a downside, it can be controlled if not outright eliminated. Most parts manufactured with DMLS printers are small due to the limited printer size. Even though it is such an expensive manufacturing process, it is still widely used across many industries [117]. Postprocessing of DMLS-printed parts results in strong functional parts [117]. There is almost no waste when manufacturing with DMLS as any powder that is not used is being reused to print another part.

Parts made from DMLS are used throughout the aerospace industry, for example, rocket engine manifolds, injectors, and combustor liners. Many energy companies use DMLS to develop rotors, stators, and prototype turbines. With DMLS being able to manufacture with biocompatible materials, we can manufacture dental devices, surgical tools, and implants [94].

1.4 Applications

1.4.1 *Selective laser sintering*

SLS has many applications in pharmaceutical practices and could restructure the existing standards of manufacturing. SLS, with its flexibility with individual fabrications, could change the way medicine is prepared. This technology could push the sector away from mass production toward units that more accurately address patients' individual needs. When medicine is mass-produced in the way that it is currently, companies are limited to fixed-dose units that may not be as effective as personally tailored medicine [118]. Easy modifications to existing parts by altering the CAD model are now possible, which was hardly doable with traditional methods requiring tool and die changes. However, there are still doubts as to its safety and effectiveness. Pharmaceutical companies are wary of the degradation and contamination caused by the laser beam used during the sintering process, in addition to the lack of study and approval for ingested thermoplastic polymers. The required drug that will be encapsulated must be carefully considered due to its exposure to severe light and heat during the sintering process around it. An example of a remedy to this issue is to combine the drug with a polymer that can withstand these conditions [19]. One obvious concern of the new products in the pharmaceutical industry is their biocompatibility. In the case of 3D-printed medications, the pills must be biodegradable with the approval of the Food and Drug Administration, which commercial SLS materials do not comply with [118]. Besides this, there are some design constraints that include: the diffusion/delivery characteristics of the design, that is, orally disintegrating, immediate, or sustained profile, allergic reactions, and interactions with other medications.

It is important to note that printing the required design dimensions is challenging to some extent, as some polymers are incapable of printing on a microscopic scale. In recent years, several novel materials have been successfully printed with the SLS process, which include polycaprolactone, high-density polyethylene (HDPE), Kollicoat IR (e.g., polyvinyl alcohol and polyethylene glycol copolymer), Eudragit (e.g., methacrylic acid and ethyl acrylate copolymer), hydroxypropyl methylcellulose, kollidon VA64 (e.g., vinylpyrrolidone-vinyl acetate copolymer, polyethylene oxide, cellulose acetate, and ethyl cellulose [118]. The capsule products commonly used in the pharmaceutical sector are composed of a drug and biodegradable thermoplastic polymer. This is due to the human body being more receptive to synthetic polymers rather than metals. It has long been considered that polymers have a significant advantage over metals in the context of medical applications because the isotonic saline solution that makes up the body's extracellular fluid is extremely hostile to metals, but it is not generally associated with the degradation of many synthetic high-molecular-weight polymers [119].

The feedstock material used in SLS has the most similarities with materials used in traditional manufacturing methods, making it more compliant than other 3D printing methods for manufacturing medical devices. Also, SLS is a solvent-free process that eliminates the need for additional drying steps to evaporate any residual binder, which is a significant advantage over other 3D-printing processes such as binder jetting [118]. Moreover, SLS does not need preprocessing, unlike other 3D printing methods, nor does it need additional materials that often cause unwanted toxicity. All of these factors lead to the conclusion that SLS is the safest and most effective 3D printing technology for pharmaceutical applications in the coming decades.

1.4.2 Selective heat sintering

SHS is a very similar process compared to SLS. The core differences revolve around how the powdered material is heated. Heat sintering is a lot more efficient than SLS. The laser in the SLS process requires more energy and is more intense than the SHS process. This means that the SHS process is more cost-efficient and can be reduced to small printing size machines. As discussed in the SLS section, the printer's smallest size is larger than what can be put on a desk in a home environment. A few examples of SLS printers are small enough to be put onto desks, but the printing size is small relative to the same size printer for an SHS printer. The SHS process is generally more cost-efficient and seen in small businesses and home applications for creating prototypes and single-unit components.

SHS is used in the automotive industry to build automotive part prototypes. Since prototype builds for automobiles are not mass-produced, manufacturers can print and test parts quickly and cheaply. Automotive manufacturers like Koenigsegg utilize many different types of printing to build their automobiles [120,121]. In 2011, Koenigsegg could assemble 15 high-spec vehicles per year, and now, the number has increased to several hundred, which is still much lower compared to other manufacturers [121].

Another large industry that utilizes the SHS process is the aerospace industry. Just like SLS, the process can use infused materials like glass and aluminum components. This means the materials involved in the process are capable of withstanding aerospace standards and demands. SHS process is ideal for the aerospace industry where the parts and prototypes can be quickly printed and tested before actual deployment. This ensures the safety of people or parts onboard spacecraft by early detection of flaws.

3D-printed prosthetics and orthodontic parts are also very common in the biomedical sector that utilizes the SHS process. Since it is much cheaper than the SLS, this is ideal for small businesses creating special orthodontic parts for individual customers. The different materials available for the process also allow prosthetics companies to meet customer-specific needs. Since this process is cheaper than other processes, prosthetics and orthodontic parts could be made available at affordable prices. SHS is often used as a desktop printer due to its low cost of manufacturing and small printer size, compared to other sintering processes. This process does not compare in terms of cost to desktop FDM printers available at much lower costs relative to the powder bed sintering printers.

Engineering parts made in aluminum alloys have been fabricated by traditional manufacturing processes such as casting, forging, extrusion, and powder metallurgy. However, these traditional manufacturing processes have resulted in parts having coarse grain structures and poor mechanical properties as a consequence of the low cooling rates associated with these processes [122]. Also, the adoption of tooling for making parts through these traditional manufacturing routes increases the cost of production and the lead time [123]. Therefore, SLS and SHS have become a promising route for manufacturing businesses engaging in the fabrication of aluminum parts and aiming to deliver their new customized products quickly and gain more consumer markets for their products. This is because SHS allows parts to be fabricated without the requirement of any part-specific tools, shortening the design and production cycle, and ensuring significant time and cost savings [17]. With the expected rapid development and improvement in process capabilities of SHS in the next decade, it has the potential to take over the manufacturing, transportation, medicine, sports, and electronics sectors [124].

The main drawbacks of the SHS printing process are poor surface quality, low dimensional accuracy, and material properties that do not meet the prerequisite for industrial applications. Postprocessing treatments like polishing, painting, heat treatment, and furnace infiltration have been employed [125]. Again, these postprocessing treatments also introduce the burden of elongating the production cycle and increasing the production cost. Meanwhile, an investigation for appropriate processing parameters and materials required to improve surface finish, dimensional accuracy, and mechanical strength of SHS-fabricated parts is being undertaken. This will in turn eliminate the postprocessing steps, shortening the lead-time, and reducing the production costs.

1.4.3 Direct metal laser sintering

DMLS is a frequently used 3D printing technology that finds its applications in the aerospace and medical industries. In the aerospace industry, DMLS is used for parts with complicated geometries or features that would be difficult or otherwise impossible to create through traditional manufacturing techniques [125]. DMLS can be used to create custom, high-strength prosthetics to replace damaged bones in the medical field. In this case, the porosity of the metal is an advantage as it allows the existing bone to grow into the prosthetic structure [49].

DMLS is widely used in the medical industry to create custom prosthetics, implants, and supports. Ciocca et al. [126] studied the use of DMLS to create a customized titanium mesh to guide bone regeneration. In this case, computerized tomography (CT) scans were used to develop a tomographic image of the maxilla, and the desired bone augmentation was designed by examining the cross-sectional slices of the bone around the implants. The mesh was designed by creating a 3D model of the maxillary arch using the tomographic images obtained from the CT scans. A thickness was applied, and an array of holes was added to complete the mesh. The mesh was printed out of titanium alloy, which is advantageous for biomedical applications because of its high strength-to-weight ratio, excellent corrosion resistance, and biocompatibility [126]. Since the part was directly printed from the model of the mandible, it fitted securely and did not require the use of bone screws. The mesh was positioned in a place over the maxilla, and a combination of vital bone and alloplastic mixed with grafting material was added inside the mesh to support bone regeneration. After 8 months, the regenerated bone was sufficient to install the implants. Jardini et al. [127] demonstrated the improvements in cranioplasty surgery using the DMLS technology. In this medical case, a head injury caused the loss of the majority of the cranial frontal bone because of a posttrauma defect, which had to be reconstructed to restore the structural integrity of the cranium. Again, the CT scans were used to generate the cranial geometry and design the specific implant. For preoperative planning, the cranium model of the patient was printed using SLS with PA 2200 material, while the actual implant was printed by DMLS using Ti6Al4V alloy. This reduces the duration of surgery due to precise planning of all the anatomical and geometrical complexities. Fig. 1.3 shows the 3D-printed model of the cranium and the specific titanium alloy implant in place. After precise planning, the implant was surgically put into place using biocompatible screws made up of the same material. Fig. 1.4 shows the DMLS-printed implant fitted into the cranium during the operation. This implant is meant to provide structural integrity to the skull and improve the overall aesthetic of the forehead.

These case studies demonstrate the effectiveness of DMLS technology for custom medical applications. The ability to design a part from a 3D model of a specific bone structure and directly print it out of strong, corrosion-resistant metal is highly advantageous. These models are crucial for preoperative planning, training, diagnosis, and designing of implants and specialized tools for biomedical applications. The accuracy of the parts produced by DMLS allows doctors to design and print parts for specialized medical cases. These support structures provide precise containment and better support to the soft tissues and bones during and after the operation as compared to the traditional methods [128]. As

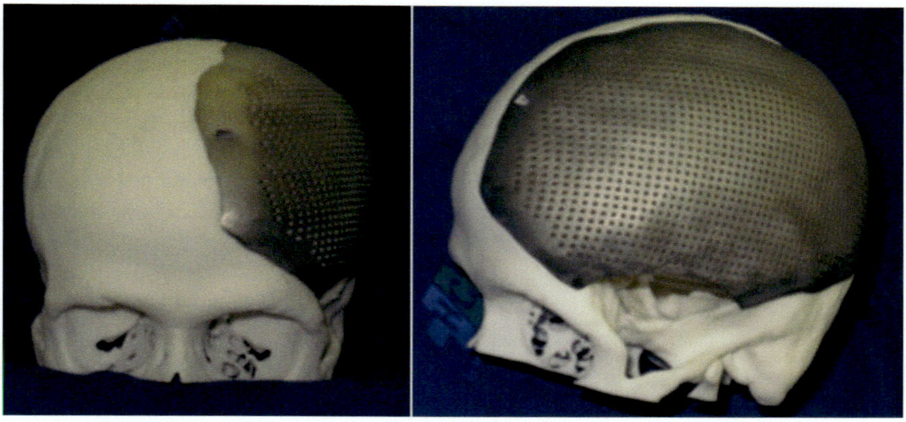

Figure 1.3 3D-printed model of the cranium and the implant for preoperative planning.
Reprinted with permission from A. Jardini, et al., Improvement in cranioplasty: advanced prosthesis biomanufacturing, Procedia Cirp 49 (2016) 203–208. Copyright Elsevier, 2016.

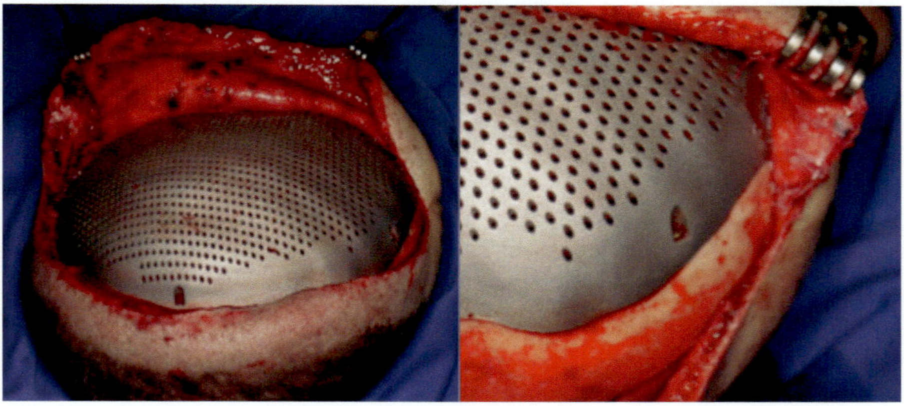

Figure 1.4 DMLS-printed implant fitted into the cranium during the surgical operative procedure.
Reprinted with permission from A. Jardini, et al., Improvement in cranioplasty: advanced prosthesis biomanufacturing, Procedia Cirp 49 (2016) 203–208. Copyright Elsevier, 2016.

technology continues to improve, additively manufactured parts will likely become even more prevalent in the medical field due to the relative ease of producing customized products compared to traditional manufacturing methods.

DMLS is also used in the aerospace industry to produce specialized parts with good corrosion resistance and mechanical properties. AM methods allow manufacturers to work with specialized alloys that are difficult to manufacture using traditional metal forming processes. IN-718 is a nickel-based superalloy with high strength, high hardness, and low thermal conductivity [17]. However, its high hardness makes it difficult to machine and IN-718 tools produced by machining wear out quickly [129]. The case study by B.

Anush Raj et al. [130] discusses developing an IN-718 alloy for turbine components using a DMLS process. The study compares the microstructures, electrochemical properties, and corrosion resistance of DMLS produced test specimens with those of a test specimen of commercially available IN-718 [130]. The DMLS samples were fabricated horizontally for optimal fatigue strength [65], and heat treatment was used to optimize the microstructure and mechanical properties [104]. Electromechanical and corrosion tests were then conducted per ASTM standards. It was found that heat-treated DMLS parts developed higher amounts of γ, γ', and δ precipitates than the commercially available samples. These precipitates have been shown to improve the mechanical and corrosion properties due to the protective film of oxides that develop over the surface [131−134]. High current intensity during the printing process has been linked to higher corrosion rates in the parts produced [118,135,136]. Therefore, the DMLS-printed parts are expected to have better corrosion resistance properties due to lower laser intensity in comparison to other laser-based processes. This is proven by scanning electron microscopy (SEM) and energy dispersive X-ray analysis (EDS) corrosion morphology testing. Corrosion of additively manufactured metals generally originates in the metal pool boundary [73,137,138]. EDS was used in order to identify the segregation of elements in the commercial and DMLS samples. It was observed that the higher segregation of nickel alloy detected in the DMLS samples as compared to the commercial samples led to better corrosion resistance properties in the DMLS samples.

This case study demonstrates some of the benefits of using AM techniques in the aerospace industry. Moreover, AM allows for the use of superalloys in the production of complex parts, which is difficult to achieve using traditional methods. Because DMLS does not melt the metal completely, it is preferred over SLM for applications where alloys are involved and phase transformation of constituents needs to be avoided. In addition, the corrosion resistance properties of the DMLS materials are better than other AM processes. Due to all these reasons, DMLS is a viable technique that shows immense potential to be applied in the coming decades.

1.5 Conclusions

Powder bed fusion−based AM processes allow manufacturers to print custom, complex, and functional parts directly out of raw powders. This offers a lot of savings in terms of costs associated with the design, tooling, machinery, skilled labor, and material loss. Due to the simplification of the overall manufacturing process, any design changes can be readily accommodated and tested in a short cycle time. This chapter focused on the four powder bed fusion processes: SLS, SLM, SHS, and DMLS.

SLS is known to be a rapid process, producing parts that require postprocessing. These parts are not as strong as a part created from SLM. SLS heats the powder to the point where it binds together, while SLM fully melts the powder so that it becomes a homogeneous component, which takes a longer time to cool down.

SHS is yet another energy-efficient powder bed fusion process. The typical material used in SHS is M-flex, which does not have a high melting point, so it uses less energy than both DMLS and SLS. Generally, larger parts can be manufactured by SHS as compared to SLS for the same sized machines due to a smaller printhead in the SHS machine. SHS printers are typically cheaper and more compact than SLS printers.

SLM is a slightly different process as compared to the powder bed sintering processes. In SLM, the powder is melted by the laser creating a melt pool, which ensures minimal porosity, high strength, and better interlayer bonding in the printed parts. This process cannot be used to work with alloys where the phase transformation for the constituent materials must be avoided.

DMLS works similarly to the SLS, but it can be used to work with metal alloys in powder form. It uses low heat to heat the powder surface just hot enough to sinter the particles. Since there is no melt pool formation, the alloy powders are sintered uniformly. Like other processes, DMLS has no waste material and is cost-efficient as it uses less heat to sinter the parts together. Despite these benefits, parts manufactured using DMLS generally have lower fatigue and low yield strength due to increased porosity and micro-cracks. Since the metal does not completely melt, the parts created from DMLS are relatively porous as compared to parts produced by SLS or SHS.

Owing to these features, powder bed fusion processes are used in varied industries and applications. SHS is best known for creating low-cost testable prototypes for moving parts. SLS could be applied in the pharmaceutical industry and replace the current capsules manufacturing process we use. SLS's speed and cost efficiency can be used to produce drugs in many different doses rather than mass-producing a single fixed-dose. Another strong application of these AM processes is rapid prototyping. Prototyping is a major part of the engineering process before the product reaches the consumer. SLS and SHS are used to prototype parts that can be tested quickly. Overall, powder bed fusion processes have undergone substantial improvements in their potential to be used for various engineering and medical applications over the last decade. As composites have begun dominating markets and lasers have become increasingly efficient, SLS, SLM, and DMLS have become potential candidates for prototyping and manufacturing processes reducing the need for expensive tooling.

Acknowledgments

The authors would like to acknowledge the help from Raven Maccione and Harmony Werth in finding some articles relevant to the additive manufacturing processes and proofreading the manuscript.

References

[1] L. Thijs, et al., Fine-structured aluminium products with controllable texture by selective laser melting of pre-alloyed AlSi10Mg powder, Acta Materialia 61 (5) (2013) 1809–1819.

[2] K. Kempen, et al., Mechanical properties of AlSi10Mg produced by selective laser melting, Physics Procedia 39 (2012) 439–446.

[3] B. Zhang, et al., CAD-based design and pre-processing tools for additive manufacturing, Journal of Manufacturing Systems 52 (2019) 227–241.
[4] L.F. Mondolfo, 4 - Mechanical properties, in: L.F. Mondolfo (Ed.), Aluminum Alloys, Butterworth-Heinemann, Oxford, 1976, pp. 68–95.
[5] L.F. Mondolfo, Al–Mg aluminum–magnesium system, in: L.F. Mondolfo (Ed.), Aluminum Alloys, Butterworth-Heinemann, Oxford, 1976, pp. 311–323.
[6] L.F. Mondolfo, Al–Si aluminum–silicon system, in: L.F. Mondolfo (Ed.), Aluminum Alloys, Butterworth-Heinemann, 1976, pp. 368–376.
[7] M.M. Dewidar, K.W. Dalgarno, C.S. Wright, Processing conditions and mechanical properties of high-speed steel parts fabricated using direct selective laser sintering, Proceedings of the Institution of Mechanical Engineers, Part B: Journal of Engineering Manufacture 217 (12) (2003) 1651–1663.
[8] E.O. Olakanmi, Selective laser sintering/melting (SLS/SLM) of pure Al, Al–Mg, and Al–Si powders: effect of processing conditions and powder properties, Journal of Materials Processing Technology 213 (8) (2013) 1387–1405.
[9] S. Das, Physical aspects of process control in selective laser sintering of metals, Advanced Engineering Materials 5 (10) (2003) 701–711.
[10] J.C. Lippold, W.A.T. Clark, M. Tumuluru, An investigation of weld metal interfaces, The Metal Science of Joining (1992) 141–146.
[11] D.D. Singh, T. Mahender, A.R. Reddy, Powder bed fusion process: a brief review, Materials Today: Proceedings 46 (2021) 350–355.
[12] D.M. Yu, D. Wang, Mechanical performance study for rapid prototyping of selective laser sintering, Trans Tech Publications Ltd. (2014).
[13] I. Yadroitsev, et al., Factor analysis of selective laser melting process parameters and geometrical characteristics of synthesized single tracks, Rapid Prototyping Journal 18 (2012).
[14] O. Rehme, C. Emmelmann. Reproducibility for Properties of Selective Laser Melting Products. Proceedings of the Third International WLT-Conference on Lasers in Manufacturing, pp. 227–232, Munich, 2005.
[15] D. Manfredi, et al., Direct metal laser sintering: an additive manufacturing technology ready to produce lightweight structural parts for robotic applications, La Metallurgia Italiana 105 (2013).
[16] Technologies, A.C.F.o.A.M. and A.C.F.o.A.M.T.S.F.o. Terminology, Standard Terminology for Additive Manufacturing Technologies, Astm International, 2012.
[17] E.O. Olakanmi, R.F. Cochrane, K.W. Dalgarno, A review on selective laser sintering/melting (SLS/SLM) of aluminium alloy powders: processing, microstructure, and properties, Progress in Materials Science 74 (2015) 401–477.
[18] R. Thakkar, et al., Impact of laser speed and drug particle size on selective laser sintering 3D printing of amorphous solid dispersions, Pharmaceutics 13 (8) (2021) 1149.
[19] F. Fina, et al., Selective laser sintering (SLS) 3D printing of medicines, International Journal of Pharmaceutics 529 (1–2) (2017) 285–293.
[20] J.P. Kruth, et al., Consolidation phenomena in laser and powder-bed based layered manufacturing, CIRP Annals 56 (2) (2007) 730–759.
[21] J.P. Kruth, et al., Benchmarking of Different SLS/SLM Processes as Rapid Manufacturing Techniques. Proceedings of the International Conference Polymers & Moulds Innovations PMI (2005).
[22] D.D. Gu, et al., Laser additive manufacturing of metallic components: materials, processes and mechanisms, International Materials Reviews 57 (3) (2012) 133–164.

[23] A. Verma, S. Tyagi, K. Yang, Modeling and optimization of direct metal laser sintering process, The International Journal of Advanced Manufacturing Technology 77 (5) (2015) 847–860.
[24] E. Yasa, J.-P. Kruth, Microstructural investigation of selective laser melting 316L stainless steel parts exposed to laser re-melting, Procedia Engineering 19 (2011) 389–395.
[25] N. Sanaei, A. Fatemi, Defects in additive manufactured metals and their effect on fatigue performance: a state-of-the-art review, Progress in Materials Science 117 (2021) 100724.
[26] F. Chen, G. Mac, N. Gupta, Security features embedded in computer aided design (CAD) solid models for additive manufacturing, Materials & Design 128 (2017) 182–194.
[27] R. Singh, et al., Powder bed fusion process in additive manufacturing: an overview, Materials Today: Proceedings 26 (2020) 3058–3070.
[28] Y. Zhang, A. Bernard, Generic build time estimation model for parts produced by SLS, in: High Value Manufacturing: Advanced Research in Virtual and Rapid Prototyping. Proceedings of the 6th International Conference on Advanced Research in Virtual and Rapid Prototyping, 2013.
[29] C.R. Deckard, Method and Apparatus for Producing Parts by Selective Sintering, Google Patents, 1989.
[30] M. Baumers, C. Tuck, R. Hague, Selective heat sintering versus laser sintering: comparison of deposition rate, process energy consumption and cost performance, in: 2014 International Solid Freeform Fabrication Symposium, University of Texas at Austin, 2015.
[31] A. Amado, et al., Advances in SLS powder characterization, in: 2011 International Solid Freeform Fabrication Symposium, University of Texas at Austin, 2011.
[32] E. Olakanmi, R. Cochrane, K. Dalgarno, Densification mechanism and microstructural evolution in selective laser sintering of Al–12Si powders, Journal of Materials Processing Technology 211 (1) (2011) 113–121.
[33] I. Gibson, D. Shi, Material properties and fabrication parameters in selective laser sintering process, Rapid Prototyping Journal 3 (1997).
[34] K. Kellens, et al., Environmental assessment of selective laser melting and selective laser sintering, Methodology 4 (5) (2010).
[35] Sinterit, PA12 Smooth v2 with Lower Refresh Ratio: New Standards in SLS Powders Refreshing, Sinterit, Poland, 2021.
[36] Materialise, Design Guidelines – PA 12 – Laser Sintering (SLS), Materialise, Belgium, 2021. Available from: https://www.materialise.com/en/manufacturing/materials/pa-12-sls/design-guidelines.
[37] Formlabs, Guide to Selective Laser Sintering (SLS) 3D Printing, Formlabs, Somerville, 2021. Available from: https://formlabs.com/blog/what-is-selective-laser-sintering/.
[38] S. Kumar, Selective laser sintering: a qualitative and objective approach, JOM 55 (10) (2003) 43–47.
[39] A. Lindberg, et al., Mechanical performance of polymer powder bed fused objects–FEM simulation and verification, Additive Manufacturing 24 (2018) 577–586.
[40] L. Kashapov, et al., Plasma electrolytic treatment of products after selective laser melting, Journal of Physics: Conference Series 669 (2016) (IOP Publishing).
[41] M. Masoomi, et al., An experimental-numerical investigation of heat transfer during selective laser melting, in: 2014 International Solid Freeform Fabrication Symposium, University of Texas at Austin, 2015.
[42] S. Bremen, W. Meiners, A. Diatlov, Selective laser melting: a manufacturing technology for the future? Laser Technik Journal 9 (2) (2012) 33–38.

[43] P.K. Gokuldoss, S. Kolla, J. Eckert, Additive manufacturing processes: selective laser melting, electron beam melting and binder jetting—selection guidelines, Materials 10 (6) (2017) 672.
[44] Twi-global, What Are the Advantages and Disadvantages of 3D Pringting, The Welding Institute, Cambridge, 2021. Available from: https://www.twi-global.com/technical-knowledge/faqs/what-is-3d-printing/pros-and-cons.
[45] D.S. Nguyen, H.S. Park, C.M. Lee, Optimization of selective laser melting process parameters for Ti-6Al-4V alloy manufacturing using deep learning, Journal of Manufacturing Processes 55 (2020) 230−235.
[46] B. Previtali, et al., Comparative costs of additive manufacturing vs. machining: the case study of the production of forming dies for tube bending, in: 28th Annual International Solid Freeform Fabrication Symposium−An Additive Manufacturing Conference. Austin, TX, USA 7−9 August, 2017.
[47] H. Yang, et al., Six-sigma quality management of additive manufacturing, Proceedings of the IEEE 109 (4) (2020) 347−376.
[48] T. Rayna, L. Striukova, From rapid prototyping to home fabrication: how 3D printing is changing business model innovation, Technological Forecasting and Social Change 102 (2016) 214−224.
[49] G. Jones, Direct Metal Laser Sintering (DMLS)−Simply Explained, All3DP GmbH, Munich, 2021.
[50] A. Riemer, et al., On the fatigue crack growth behavior in 316L stainless steel manufactured by selective laser melting, Engineering Fracture Mechanics 120 (2014) 15−25.
[51] H.K. Rafi, T.L. Starr, B.E. Stucker, A comparison of the tensile, fatigue, and fracture behavior of Ti−6Al−4V and 15-5 PH stainless steel parts made by selective laser melting, The International Journal of Advanced Manufacturing Technology 69 (5) (2013) 1299−1309.
[52] A.B. Spierings, T.L. Starr, K. Wegener, Fatigue performance of additive manufactured metallic parts, Rapid Prototyping Journal 19 (2013).
[53] M. Shellabear, O. Nyrhilä, DMLS-Development history and state of the art, in: Laser Assisted Netshape Engineering 4, Proceedings of the 4th LANE, Erlangen, 2004, pp. 21−24.
[54] B. Farber, et al., Correlation of mechanical properties to microstructure in Inconel 718 fabricated by direct metal laser sintering, Materials Science and Engineering: A 712 (2018) 539−547.
[55] D.H. Smith, et al., Microstructure and mechanical behavior of direct metal laser sintered Inconel alloy 718, Materials Characterization 113 (2016) 1−9.
[56] T.M. Mower, M.J. Long, Mechanical behavior of additive manufactured, powder-bed laser-fused materials, Materials Science and Engineering: A 651 (2016) 198−213.
[57] S.N. Dwivedi, S.K. Tyagi, M.K. Tiwari, Optimal part orientation in fused deposition modeling: an approach based on continuous domain ant colony optimization, The International Journal of Advanced Manufacturing Systems 10 (2) (2007) 95−110.
[58] A. Verma, M.K. Tiwari, Role of corporate memory in the global supply chain environment, International Journal of Production Research 47 (19) (2009) 5311−5342.
[59] S.H. Choi, S. Samavedam, Visualisation of rapid prototyping, Rapid Prototyping Journal 7 (2001).
[60] S.K. Tyagi, K. Yang, A. Verma, Non-discrete ant colony optimisation (NdACO) to optimise the development cycle time and cost in overlapped product development, International Journal of Production Research 51 (2) (2013) 346−361.

[61] D.E. Goldberg, Genetic Algorithms in Search, Optimization, and Machine Learning, Addison-Wesley Longman Publishing Co., Inc, Boston, MA, 1989, 75 Arlington Street, Suite 300.
[62] S.K. Tyagi, et al., Development of a fuzzy goal programming model for optimization of lead time and cost in an overlapped product development project using a Gaussian Adaptive Particle Swarm optimization-based approach, Engineering Applications of Artificial Intelligence 24 (5) (2011) 866–879.
[63] A. Verma, et al., Stochastic modelling and optimisation of multi-plant capacity planning problem, International Journal of Intelligent Engineering Informatics 2 (2–3) (2014) 139–165.
[64] M. Krishnan, et al., On the effect of process parameters on properties of AlSi10Mg parts produced by DMLS, Rapid Prototyping Journal 20 (2014).
[65] G. Nicoletto, et al., Surface roughness and directional fatigue behavior of as-built EBM and DMLS Ti6Al4V, International Journal of Fatigue 116 (2018) 140–148.
[66] A.B. Spierings, M.U. Schneider, R. Eggenberger, Comparison of density measurement techniques for additive manufactured metallic parts, Rapid Prototyping Journal 17 (2011).
[67] S. Greco, et al., Selective laser melting (SLM) of AISI 316L—impact of laser power, layer thickness, and hatch spacing on roughness, density, and microhardness at constant input energy density, The International Journal of Advanced Manufacturing Technology 108 (2020) 1551–1562.
[68] R. Li, et al., 316L stainless steel with gradient porosity fabricated by selective laser melting, Journal of Materials Engineering and Performance 19 (5) (2010) 666–671.
[69] M. Badrossamay, T.H.C. Childs, Layer Formation Studies in Selective Laser Melting of Steel Powders, International Solid Freeform Fabrication Symposium, University of Texas at Austin, 2006.
[70] M. Ghasri-Khouzani, et al., Direct metal laser-sintered stainless steel: comparison of microstructure and hardness between different planes, The International Journal of Advanced Manufacturing Technology 95 (9) (2018) 4031–4037.
[71] H. Krauss, C. Eschey, M.F. Zaeh, Thermography for monitoring the selective laser melting process, in: 2012 International Solid Freeform Fabrication Symposium. University of Texas at Austin, 2012.
[72] A. Aversa, et al., Effect of process and post-process conditions on the mechanical properties of an A357 alloy produced via laser powder bed fusion, Metals 7 (2) (2017) 68.
[73] H. Stanjek, W. Häusler, Basics of X-ray diffraction, Hyperfine Interactions 154 (1) (2004) 107–119.
[74] G.V. Salmoria, et al., Structure and mechanical properties of cellulose based scaffolds fabricated by selective laser sintering, Polymer Testing 28 (6) (2009) 648–652.
[75] J. Bai, et al., The effect of processing conditions on the mechanical properties of polyethylene produced by selective laser sintering, Polymer Testing 52 (2016) 89–93.
[76] D.I. Stoia, E. Linul, L. Marsavina, Influence of manufacturing parameters on mechanical properties of porous materials by selective laser sintering, Materials 12 (6) (2019) 871.
[77] F.-H. Liu, Synthesis of biomedical composite scaffolds by laser sintering: mechanical properties and in vitro bioactivity evaluation, Applied Surface Science 297 (2014) 1–8.
[78] D. Rajamani, E. Balasubramanian, Investigation of sintering parameters on viscoelastic behaviour of selective heat sintered HDPE parts, Journal of Applied Science and Engineering 22 (3) (2019) 391–402.
[79] Z. Wang, et al., The microstructure and mechanical properties of deposited-IN718 by selective laser melting, Journal of Alloys and Compounds 513 (2012) 518–523.

[80] T. Zou, et al., Effect of heat treatments on microstructure and mechanical properties of AlSi7Mg fabricated by selective laser melting, Journal of Materials Engineering and Performance (2021) 1−12.
[81] M. Béreš, et al., Mechanical and phase transformation behaviour of biomedical Co-Cr-Mo alloy fabricated by direct metal laser sintering, Materials Science and Engineering: A 714 (2018) 36−42.
[82] G. Özer, A. Karaaslan, A study on the effects of different heat-treatment parameters on microstructure−mechanical properties and corrosion behavior of maraging steel produced by direct metal laser sintering, Steel Research International 91 (10) (2020) 2000195.
[83] H.-C. Tran, Y.-L. Lo, M.-H. Huang, Analysis of scattering and absorption characteristics of metal powder layer for selective laser sintering, IEEE/ASME Transactions On Mechatronics 22 (4) (2017) 1807−1817.
[84] M.H. Rahman, P.R. Bhoi, An overview of non-biodegradable bioplastics, Journal of Cleaner Production (2021) 126218.
[85] systems, D, DuraForm® Flex Plastic, 3D Systems, Rock Hill, South Carolina, 2021. Available from: https://www.prototech.com/wp-content/uploads/2012/06/DS_DuraForm_Flex_Plastic.pdf.
[86] Stratasys, Selective Laser Sintering (SLS) Parts on Demand, Stratasys, Rehovot, Israel, 2021. Available from: https://www.stratasysdirect.com/technologies/selective-laser-sintering.
[87] Stratasys Nylon 12 AF, Laser Sintering Material Specifications, Stratasys, Rehovot, Israel, 2021. Available from: https://info.stratasysdirect.com/rs/626-SBR-192/images/LS_Nylon_12_AF_Material_Datasheet_201610.pdf.
[88] C.-Z. Yan, et al., Preparation and selective laser sintering of nylon-12-coated aluminum powders, Journal of Composite Materials 43 (17) (2009) 1835−1851.
[89] Nylon 12 AF. [cited 2021 1 October] Dinsmore Inc, Irvine, California, 2021. Available from: http://www.dinsmoreinc.com/wp-content/uploads/2017/01/Dinsmore_SpecSheet_Nylon12_AF.pdf.
[90] J.-P. Kruth, et al., Lasers and materials in selective laser sintering, Assembly Automation 23 (2003).
[91] I. Shishkovsky, K. Nagulin, V. Sherbakov, Laser sinterability and characterization of oxide nano ceramics reinforced to biopolymer matrix, The International Journal of Advanced Manufacturing Technology 78 (1−4) (2015) 449−455.
[92] T. Puttonen, M. Salmi, J. Partanen, Mechanical properties and fracture characterization of additive manufacturing polyamide 12 after accelerated weathering, Polymer Testing 104 (2021) 107376.
[93] S.L. Sing, et al., Direct selective laser sintering and melting of ceramics: a review, Rapid Prototyping Journal 23 (2017).
[94] D.F. Williams, On the mechanisms of biocompatibility, Biomaterials 29 (20) (2008) 2941−2953.
[95] T. Yamamuro, Bioceramics, in: D.G. Poitout (Ed.), Biomechanics and Biomaterials in Orthopedics, Springer London, London, 2004, pp. 22−33.
[96] Additive-X, What Is Selective Heat Sintering (SHS), and How Does it Work? 2021, 2016. Available from: https://www.additive-x.com/blog/selective-heat-sintering-shs-work/.
[97] C.Y. Yap, et al., Review of selective laser melting: materials and applications, Applied Physics Reviews 2 (4) (2015) 041101.

[98] J.-P. Kruth, et al., Benchmarking of different SLS/SLM processes as rapid manufacturing techniques, in: Proceedings of the International Conference Polymers & Moulds Innovations PMI 2005, 2005.
[99] A.K. Kushwaha, et al., Nanocrystalline materials: synthesis, characterization, properties, and applications, Crystals 11 (11) (2021) 1317.
[100] D. Goll, et al., Additive manufacturing of bulk nanocrystalline FeNdB based permanent magnets, Micromachines 12 (5) (2021) 538.
[101] A.M. Khorasani, et al., Analysis of thermal stresses in solidification of spherical SLM components, arXiv (2017) preprint arXiv:1706.08212.
[102] C. Yan, et al., Microstructure and mechanical properties of aluminium alloy cellular lattice structures manufactured by direct metal laser sintering, Materials Science and Engineering: A 628 (2015) 238–246.
[103] B.A. Raj, et al., Studies on heat treatment and electrochemical behaviour of 3D printed DMLS processed nickel-based superalloy, Applied Physics A 125 (10) (2019) 1–8.
[104] N. El-Bagoury, M.A. Amin, Q. Mohsen, Effect of various heat treatment conditions on microstructure, mechanical properties and corrosion behavior of Ni base superalloys, International Journal of Electrochemical Science 6 (12) (2011) 6718–6732.
[105] T. Hatch, What Is DMLS, and Why Is it Taking Over Aerospace? Fictiv, San Francisco, California, 2016. Available from: https://www.fictiv.com/articles/what-is-dmls-and-why-is-it-taking-over-aerospace.
[106] Understanding Additive Metals. [cited 2021 1 October], Stratasys Direct Manufacturing, Los Angeles, California, 2021 Available from: https://www.stratasysdirect.com/technologies/direct-metal-laser-sintering/dmls-understanding-additive-metal-manufacturing.
[107] M. Cerea, G.A. Dolcini, Custom-made direct metal laser sintering titanium subperiosteal implants: a retrospective clinical study on 70 patients, BioMed Research International 2018 (2018).
[108] R. Krishnan, et al., Experimental investigation on wear behavior of additively manufactured components of IN718 by DMLS process, Journal of Failure Analysis and Prevention 20 (5) (2020) 1697–1703.
[109] T. Zhang, et al., A finite element methodology for wear–fatigue analysis for modular hip implants, Tribology International 65 (2013) 113–127.
[110] C.M. Cheah, et al., Rapid prototyping and tooling techniques: a review of applications for rapid investment casting, The International Journal of Advanced Manufacturing Technology 25 (3) (2005) 308–320.
[111] A. Gratton, Comparison of Mechanical, Metallurgical Properties of 17-4PH Stainless Steel Between Direct Metal Laser Sintering (DMLS) and Traditional Manufacturing Methods, NCUR, 2012, 2012.
[112] B. Konieczny, et al., Challenges of Co–Cr alloy additive manufacturing methods in dentistry—the current state of knowledge (systematic review), Materials 13 (16) (2020) 3524.
[113] S. Sanchez, et al., Powder bed fusion of nickel-based superalloys: a review, International Journal of Machine Tools and Manufacture (2021) 103729.
[114] L. Gregurić, How Much Does a Metal 3D Printer Cost? All3DP GmbH, Munich, Germany, 2020 [cited 2021 1 October]; Available from: https://all3dp.com/2/how-much-does-a-metal-3d-printer-cost/.
[115] F. Calignano, et al., Influence of process parameters on surface roughness of aluminum parts produced by DMLS, The International Journal of Advanced Manufacturing Technology 67 (9–12) (2013) 2743–2751.

[116] V. Carlota, The Best Large FDM 3D Printers Available in 2021, 3Dnatives, Paris, France, 2021 [cited 2021 1 October]; Available from: https://www.3dnatives.com/en/best-large-fdm-3dprinters-2019-100120194/.
[117] 3DSourced, Direct Metal Laser Sintering: Everything To Know About DMLS 3D Printing. [cited 2021 1 October], 3DSourced, 2021. Available from: https://www.3dsourced.com/guides/direct-metal-laser-sintering-dmls/.
[118] A. Awad, et al., 3D printing: principles and pharmaceutical applications of selective laser sintering, International Journal of Pharmaceutics 586 (2020) 119594.
[119] BMP Medical, Thermoplastics used in medical device injection molding. [cited 2021 1 October], BMP Medical, Sterling, Massachusetts, 2021. Available from: https://www.bmpmedical.com/thermoplastics-used-for-medical-device-injection-molding/.
[120] B. Kianian, S. Tavassoli, T.C. Larsson, The role of additive manufacturing technology in job creation: an exploratory case study of suppliers of additive manufacturing in Sweden, Procedia CIRP 26 (2015) 93–98.
[121] B. Sedacca, Hand built by lasers [additive layer manufacturing], Engineering & Technology 6 (1) (2011) 58–60.
[122] M.H. Farshidianfar, A. Khajepour, A.P. Gerlich, Effect of real-time cooling rate on microstructure in laser additive manufacturing, Journal of Materials Processing Technology 231 (2016) 468–478.
[123] D. Mourtzis, et al., Knowledge-based estimation of manufacturing lead time for complex engineered-to-order products, Procedia CIRP 17 (2014) 499–504.
[124] P. Chen, et al., Investigation into the processability, recyclability and crystalline structure of selective laser sintered Polyamide 6 in comparison with Polyamide 12, Polymer Testing 69 (2018) 366–374.
[125] S.C. Joshi, A.A. Sheikh, 3D printing in aerospace and its long-term sustainability, Virtual and Physical Prototyping 10 (4) (2015) 175–185.
[126] L. Ciocca, et al., Direct metal laser sintering (DMLS) of a customized titanium mesh for prosthetically guided bone regeneration of atrophic maxillary arches, Medical & Biological Engineering & Computing 49 (11) (2011) 1347–1352.
[127] A. Jardini, et al., Improvement in cranioplasty: advanced prosthesis biomanufacturing, Procedia Cirp 49 (2016) 203–208.
[128] N. Donos, L. Kostopoulos, T. Karring, Alveolar ridge augmentation using a resorbable copolymer membrane and autogenous bone grafts, Clinical Oral Implants Research 13 (2) (2002) 203–213.
[129] A. Keshavarzkermani, M. Sadowski, L. Ladani, Direct metal laser melting of Inconel 718: process impact on grain formation and orientation, Journal of Alloys and Compounds 736 (2018) 297–305.
[130] B. Anush Raj, et al., Direct metal laser sintered (DMLS) process to develop Inconel 718 alloy for turbine engine components, Optik 202 (2020) 163735.
[131] G.H. Cao, et al., Investigations of γ', γ'' and δ precipitates in heat-treated Inconel 718 alloy fabricated by selective laser melting, Materials Characterization 136 (2018) 398–406.
[132] G.S. Sharma, et al., Influence of γ-alumina coating on surface properties of direct metal laser sintered 316L stainless steel, Ceramics International 45 (10) (2019) 13456–13463.
[133] P. Zhang, et al., Extraordinary plastic behaviour of the γ' precipitate in a directionally solidified nickel-based superalloy, Philosophical Magazine Letters 96 (1) (2016) 19–26.
[134] K.G. Prashanth, et al., Tribological and corrosion properties of Al–12Si produced by selective laser melting, Journal of Materials Research 29 (17) (2014) 2044–2054.

[135] N.G. Dudkina, Corrosion resistance of steel 45 subjected to electromechanical treatment and Surface plastic deformation, Metal Science and Heat Treatment 59 (9) (2018) 584−587.
[136] M.A. Khan, Electrochemical polarisation studies on plasma-sprayed nickel-based superalloy, Applied Physics A 120 (2) (2015) 801−808.
[137] C.K. Chua, K.F. Leong, C.S. Lim, Rapid Prototyping: Principles and Applications (With Companion CD-ROM), World Scientific Publishing Company, 2010.
[138] J. Delgado, J. Ciurana, C.A. Rodríguez, Influence of process parameters on part quality and mechanical properties for DMLS and SLM with iron-based materials, The International Journal of Advanced Manufacturing Technology 60 (5−8) (2012) 601−610.

Fundamentals of additive manufacturing of metallic components by cold spray technology

2

Mohammadreza Daroonparvar[1,2], *Charles M. Kay*[2], *M.A. Mat Yajid*[3], *H.R. Bakhsheshi-Rad*[3,4] *and M. Razzaghi*[4]
[1]Department of Mechanical Engineering, University of Nevada, Reno, NV, United States; [2]Research and Development Department, ASB Industries Inc., Barberton, OH, United States; [3]School of Mechanical Engineering, Faculty of Engineering, Universiti Teknologi Malaysia, Johor Bahru, Johor, Malaysia; [4]Advanced Materials Research Center, Department of Materials Engineering, Najafabad Branch, Azad University, Najafabad, Iran

2.1 Introduction

AM is a technology for layer-by-layer production of the parts that can be complicated geometrically [1,2], using a computer-aided designed 3D model format files like stereolithography. Unlike conventional manufacturing processes like milling, the AM process can efficiently prevent undesirable wastes in time and costs. Metal AM can be divided into two broad categories: (1) beamless or nonbeam technologies (or indirect methods) in which a binder is employed to bond the metal particles and postprocessing is required to eliminate the binder and finally consolidate the component and (2) beam-based technologies (or direct methods) in which ultimate component is right created without using a binder [3]. Likewise, the metal AM is currently dominated by beam-based technologies (e.g., SLM, EBM, *etc.*) compared to nonbeam metal AM methods [4]. Typically, the beam-based metal AM process has been utilized to produce complicated Ti, Fe, or Ni-based alloy components [5].

In a beam-based metal AM process, sources such as electron beam or laser with high energy are utilized to selectively melting down the powder bed, in which the resulted solidified metal can form the component. The most known metal AM methods include SLS, SLM, and EBM. However, as high energy sources are used in these methods, the disadvantages like undesirable phase transformations and residual stress because of high processing temperatures are unavoidable [1,2].

Compared to nonbeam metal AM techniques (e.g., binder jet 3D printing method, material jetting processes [4,6], *etc.*), the beam-based metal AM technique may not be a proper process for production of some nonferrous materials like Mg, Al, and Cu-based alloys due to its nature which utilizes the electron beam, laser, or arc for melting the materials [5]. Moreover, the processing of nonweldable metals through beam-based metal

Tribology of Additively Manufactured Materials. https://doi.org/10.1016/B978-0-12-821328-5.00002-0

AM systems was reported to be completely challenging [7,8]. Cold spray (as one of the beamless metal AM techniques [4]) is a newly developed technology that is able to produce the component via the solid-state deposition of powder particles [9–11].

In the CS process, which does not include the stages of powder melting and solidifying the melted metal powders (observed in beam-based metal AM methods), the feedstock powder is accelerated to elevated velocities by means of the supersonic flow effect. Following intensive plastic deformation of the particles in solid state due to high-velocity collide at temperatures lower than the powder melting point, adequate bonding could happen. Because of the unique "cold" characteristic, the CS process can prevent or minimize the severe thermal stresses (tensile stresses), oxidation, and phase transformations. The CS technique can be extensively used for the construction of functional coatings, MMCs, AM, and even for repairing the dimensional damage of the parts [1,2,5]. Some academic publications, including review articles and books, discuss different features and applications of CS technique [5,9].

Unlike traditional deposition processes which utilize high temperatures (e.g. prevalent beam-based metal AM methods and traditional thermal spraying processes [4,5,10]). In the CS technique, the deposition process typically uses the kinetic energy of the particle, before impinge, in place of thermal energy. The feedstock powder utilized in the CS process keeps its solid state in the whole deposition process. The deposition is mostly attained via mechanical interlocking and locally metallurgical bonding, which are formed by local plastic deformations at particle–substrate interfaces and at the interparticle boundaries. As mentioned, the relatively low temperature in the CS process prevents the common defects occurred in deposition processes that include high temperatures, like phase transformation, oxidation, residual thermal stress, grain growth, *etc* [5,10–12].

Mentioned remarkable advantages make CS a capable method for producing the coatings with a wide range of the materials (from low melting point to high melting point) consisting of most metals and their alloys (ferrous and nonferrous), nanostructured metals, and MMCs [13–15]. Also, there is no limitation in the growth of thickness in the CS coating process. So, besides the application of solid-state CS process for coating of surfaces and repairing structures, it can be employed as an AM process as well [11,16]. However, the significant geometrical accuracy is the major advantage of laser-based AM methods compared to CS technology [17]. In fact, both laser beam diameter and the powder particle size can define the completely fine spot size in SLM method. This can provide the probability of achieving highly complex customized shapes. But in the CS technology, the nozzle shape and geometry are the main parameters which can define the spray stream spot size. These parameters provide less precision on the spray spot (feature) size with the current technology [18] and cold spray nozzles which need to be further developed for attaining high precisions on the spray spot size. Nevertheless, in the CS process, target area for freestanding parts is mostly large-scale components with near net shape accuracy [4,19] that are not doable using prevalent powder bed AM methods. On the other hand, adequate machinability by standard milling and turning techniques was reported for CS near net-shape components (deposits) [18]. Moreover, considerably high deposition rates (300–400 cm^3/h and in some cases up to 1500 cm^3/h) were reported for CS technology. This can makes CS process a fairly scalable AM technology [20].

In this chapter, cold spray additive manufacturing (CSAM) is discussed and summarized in materials point of view; likewise, existing issues, problems, and prospects in the CSAM are explored. The information of this chapter can help researchers to increase their knowledge of CSAM and have an improved understanding of the microstructure and mechanical characteristics of CSAM components that can facilitate and expand the usage of the CS technology in industrial applications.

2.2 Cold spray versus other thermal spray processes

In the CS method, the velocity of the particle, spray plume temperature, and substrate temperature are different from traditional thermal spray methods, as shown in Fig. 2.1. The advantages of the CS process compared with traditional thermal spray routes could be summarized as below [9,21−24]:

1. Due to low deposition temperature, CS process is suitable for deposition of the materials that are sensitive to temperature like nanocrystalline (NC) and noncrystalline materials, the materials that are sensitive to oxygen like Al, Cu, and Ti, and the materials that are sensitive to phase transformation like carbide composites [25,26].
2. As a result of the effect of microshot-peening that makes compressive residual stress in the resulting deposited layers, the CS process of metals generally improves the fatigue resistance.
3. The CS deposition of metals, because of having inherent low-temperature and high-kinetic energy features, has high degrees of consolidation in their microstructures that may be comparable to wrought alloys [27].

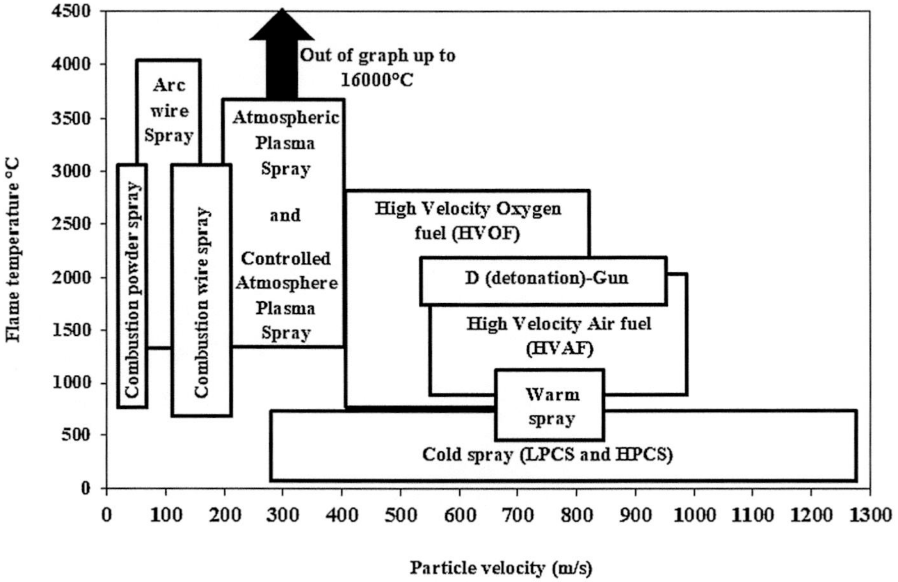

Figure 2.1 Flame temperature versus particle velocity in different types of spray process technologies.

4. Because of having a denser structure and lower content of oxides, the CSed deposits have higher electrical and thermal conductivities [9,28,29]:

   From oxygen content (%) point of view: flame spraying > wire arc spraying > plasma spraying > HVOF > CS. From porosity (%) point of view: flame spraying > wire arc spraying > plasma spraying > HVOF > CS.
5. The CS process features remarkably higher deposition efficiency (DE), but equal or inferior deposition rate compared to the other thermal spray processes [9,28].
6. Due to having the smaller nozzle (convergent section), spray spot size (nozzle exit), and standoff distances (normally with the diameter of 10–12, 5–7, and 25–30 mm, respectively), the CS process has higher accuracy in the control of the deposition area on the substrate and lower requirement of covering (masking) of the as-sprayed part.
7. Owing to the lower heat transfer into the substrate, which decreases the importance of the material type of the substrate, applying the CS process causes increase in the possibility of joining of different material types [28,30,31], Fig. 2.2.
8. Due to the higher compressively stresses, the layers with higher thicknesses (up to 13 mm or even more) could be deposited without debonding in the CSAM process [28].
9. CS process could impede the microstructural damage of the heat-vulnerable substrates such as Mg alloys which is often observed in the alternative thermal spray methods [32–36].
10. The ability to spray an extensive range of soft and hard metals or alloys, MMCs, and some cermets using the HPCS process which will be discussed in this section [9,28].

However, the CS process has some restrictions, which are as follows [9,24,28]:

1. The feedstock powder materials that have low capacity to plastically deform in low temperatures are not suitable to be applied by CS process. As the CS is a process that occurs completely in the solid state, just the materials that have adequate formability in the

Figure 2.2 Cold sprayed Ti coating on Mg alloy.

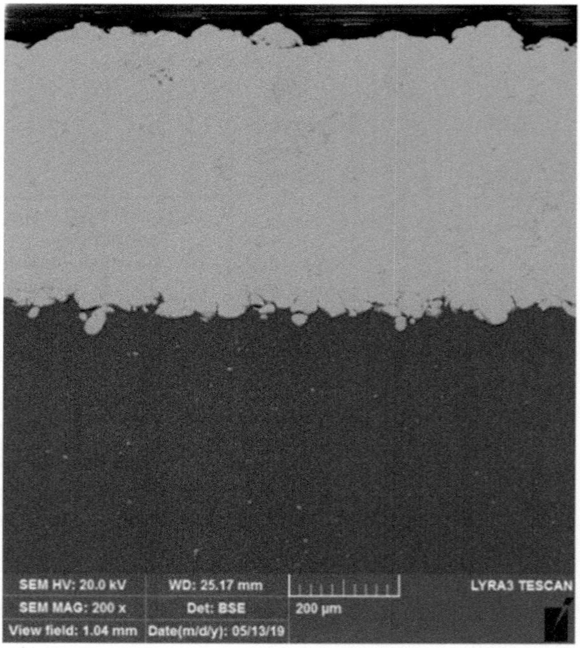

processing temperature window could be applied in the process. Nevertheless, recently, research attempts have tried to expand the range of the materials that can be used in the CS process, and the use of HPCS systems as an alternative of LPCS systems has allowed researchers to partly overcome the above-mentioned limitation.

2. During the CS process, feedstock powder particles typically experience high deformation, resulting in a reduction of formability of the CS-deposited material. Nevertheless, the mentioned issue could be partially moderated by utilizing appropriate size and temper of feedstock powder particles or performing post-CS thermal treatments.
3. The CS process, similar to the other spraying deposition methods, is a line-of-sight process that has a limitation to form deposition onto internal surfaces (internal diameters). Nevertheless, with an approximately 25 mm standoff distance commonly used in the CS process, appropriate nozzles to fit inside cavities can be designed that otherwise would be impossible for spraying. Currently, some commercial manufacturers of CS equipment can supply the nozzles which can spray powder materials inside the hollow parts with the diameter of as small as 90 mm.
4. In order to induce considerable plastic deformation of the impacted powder particles and attain acceptable attachment, the hardness of the substrate should be adequate compared to that of the powder particles feedstock.
5. Typical nozzle clogging always occurs in the CS process, when low melting point materials like aluminum, zinc, *etc.*, are sprayed by CS system. This can cause remarkable reduction in the working efficiency. Recently, water-cooled nozzles along with specialized hardware could significantly reduce the temperature of the nozzle wall and can considerably avoid the nozzle clogging during the spraying process.

2.3 Cold spray systems

Two main categories of the CS processes include LPCS and HPCS. The differences between the mentioned categories are briefly described in this chapter, and LPCS is individually discussed in this chapter as well. As shown in Fig. 2.3, in an HPCS process, high pressure small solid particles are typically injected into or before the convergent section of the cold spray nozzle and accelerated in a supersonic jet in the range of heated gas to elevated velocities, typically on the order of 300−1200 m/s, and then sprayed onto a substrate; that could be a metal, ceramic, or even a glass, polymer.

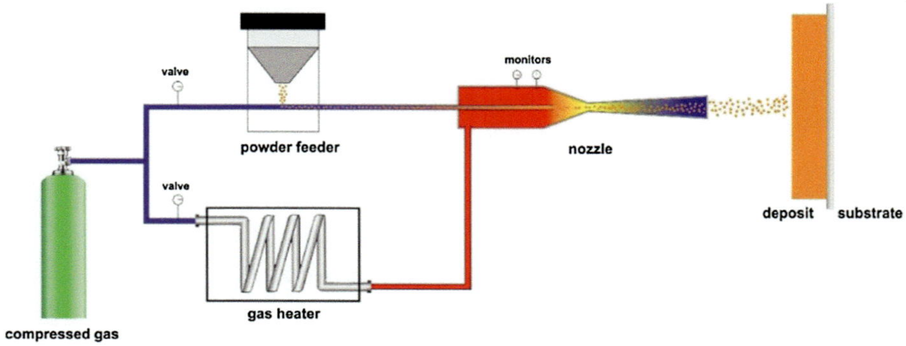

Figure 2.3 Schematic illustration of high-pressure cold spray (HPCS) system [10].

The powder particles are mostly spherical gas atomized, and in some rare cases, irregular-shaped metal powders [10], in the range of 5−50 μm in diameter. The powder particles undergo adiabatic heating and plastically deform at very high shear rates, causing them to flatten out bonds to the beneath surface in an adequately high velocity, and the appropriate material combination for the cold spray process, upon impinge [37,38]. The mentioned mechanism is very similar to the one which occurred in ASI found in the explosive welding but in the microscale (in case of CS process).

In the HPCS process, the process gas is heated. The process (propellant) gas is primarily heated for increasing the flow velocity of the gas in the cold spray nozzle and therefore increasing the velocity of feedstock powder particles. The process gas may have a mildly high temperature at start out; anyway, it cools off very quickly as it expands in the long diverging zone of the cold spray gun nozzle, and the temperature of the sprayed particles remains lower than the powder melting point. So, compared to traditional thermal spray, where the spraying particles are melted or nearly melted during the deposition process [39−45], CS has comparatively low process temperature, so it is called cold spray.

In an HPCS process, passing high pressure gas, normally in the range of 1−6 MPa, accelerates the jet of gas through a CD nozzle (de Laval nozzle) to attain supersonic flow. The diagram in Fig. 2.4 demonstrates the variations of gas pressure, flow

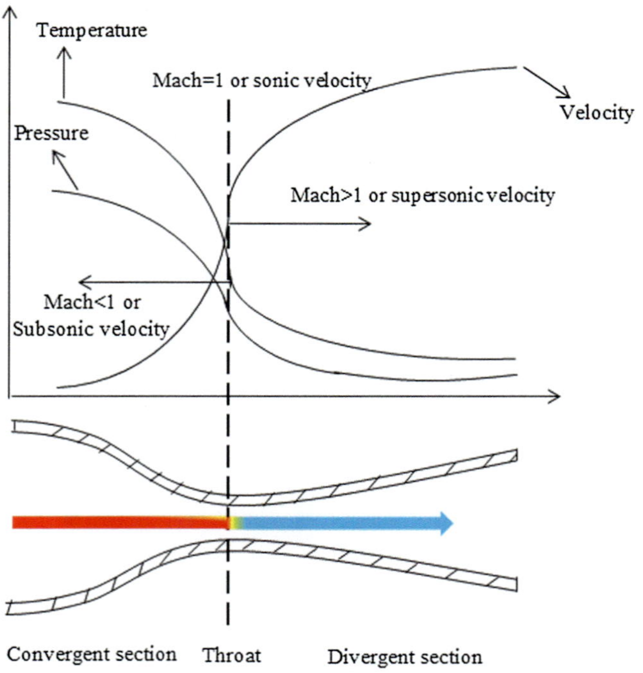

Figure 2.4 Variations in gas velocity, temperature, and pressure flowing through a converging−diverging nozzle (de Laval) nozzle (National diagram).
Modified from C.M. Kay, J. Karthikeyan, High Pressure Cold Spray: Principles and Applications, ASM International, 2016 ISBN: 978-1-62708-096-5.

velocity, and the temperature in a CD nozzle. The region that connects converging and diverging sections of a nozzle is commonly called nozzle throat, which is a critical element in the geometry of the nozzle. The gas will flow in the nozzle throat at exactly the local speed of the sound, which is sonic or Mach 1 flow, if the pressure drops at or above a critical minimum, from the upstream end of that region (nozzle throat) to its downstream end. Then the gas density decreases, and it cools off and accelerates to supersonic velocities as a result of the expanding the gas in the subsequent diverging section of the nozzle. In order to attain the highest velocities of in-flight powder particles, the solid particles are injected into the upstream of the nozzle throat, passing through both the sonic and supersonic regions of the de Laval nozzle.

The spray powder particle velocities, resulted from sonic velocity in the nozzle throat, are strongly related to the temperature and composition of the process gas supplied to the accelerating de Laval nozzle [46]. Most of the early CS systems operated primarily with pure helium or sometimes mixture of helium and nitrogen or helium and air, due to generally generating the higher velocities of the process gas and thus powder particles by the propellant gases with low molecular mass like helium. Nevertheless, limited availability of helium in some regions of the world and its high price were a remarkable limitation for many potential CS applications.

Higher velocities of the powder particles can be obtained by preheating the high-pressure process gas that is fed into the cold spray system, which increases the sonic velocity in the nozzle throat. So, the cold spray systems which utilized nitrogen process gas (readily available gas) with comparatively lower price, preheated to raise the powder particle velocities, were soon developed. Today, gas temperatures in such a system can be reached up to 1100°C. Despite the relatively high supply gas temperatures available in such systems, the gas temperature remarkably declines with expanding in the diverging section of the de Laval nozzle.

Although the sprayed particles experience high temperatures, the exposure time is relatively short, and they remain well below their melting point. Nevertheless, at relatively high process gas temperatures, heating of some powder particles can occur, which may have benefit as it tends to depress the critical velocity of powder particle resulting in the DE enhancement. However, with too high heating of the particles, there would be a risk of causing some oxidation of the metal surface if the spraying process happens in an ambient air environment [28].

Despite many similar features of LPCS and HPCS processes, they have a few slight but very important differences. In an LPCS process (Fig. 2.5), the compressed air or

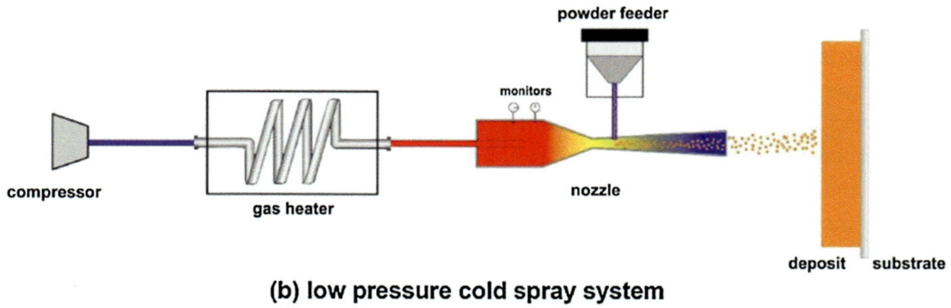

Figure 2.5 Schematic illustration of low-pressure cold spray (LPCS) system [10].

nitrogen with relatively lower pressure, commonly in the range of 0.5−1 MPa, is preheated to temperatures up to 550°C, then passed through a CD nozzle, and accelerated to velocities up to approximately 600 m/s. In LPCS, the electric gas heaters with comparatively lower weight and size systems are frequently incorporated into a portable spray gun (manual spray gun). In the LPCS process, unlike the HPCS, the feedstock powder is introduced into the downstream of the nozzle throat directly into the nozzle's diverging section. Complexity and cost of the powder-feeding equipment can be greatly reduced by injecting the powder into the downstream of the nozzle throat because the gas pressure in this region is much lower. However, the maximum attainable particle velocities in such a system are low. This could be related to a natural outcome of injecting the powder particles into the downstream of the nozzle throat and also utilizing moderately heated nitrogen or compressed air as propellant gas. So, the material types which can be used for the deposition in the LPCS system are comparatively more restricted [9,28].

2.4 Cold spray concepts

The implementation of the CS is comparatively simpler, compared to the traditional thermal spray technology. As previously mentioned, nitrogen or helium with high pressure is preheated and reaches supersonic velocities after passing through a CD nozzle. A fluidized feedstock powder, through a different gas line, is fed into the process gas stream. The powder is accelerated by the expanding gas and collides the substrate surface at velocities of up to 1200 m/s [37,47]. The velocity of the gas is mainly related to the type and temperature of the process gas, as well as the expansion rate of the cold spray gun nozzle.

The velocity of the gas is generated by the temperature of the propellant gas. This could be calculated using the following formula (2.1) [48]:

$$v = (\gamma RT/M_w)^{1/2} \tag{2.1}$$

where γ = fraction of the constant pressure to the constant volume-specific heats, which is considered 1.66 for monoatomic gases such as helium and 1.4 for diatomic gases such as oxygen and nitrogen, respectively. R = gas constant (8314 J/kmol·K), T = temperature of the gas, and M_w = molecular weight of the gas, which is the minimum for the gases like helium. Once the high-velocity powder particles collide on the substrate surface, the kinetic energy of the particle is changed into the thermal energy and then mechanical deformation [49,50]. It is obviously seen that helium gas with lower M_w and higher coefficient (γ) can easily reach higher velocities compared to N_2 process gas under the same gas temperatures. However, due to the high cost of helium and also its limited availability, this gas was gradually replaced with N_2 gas. So, the latest developments in the HPCS systems (which work with N_2 process gas) have led to the process optimization using standard powder particles (cold sprayable powder particles) with high DE.

The impingement of powder particles on a suitable substrate must happen at very high velocities to have an appropriate bonding. The critical velocity or the minimum velocity, which is needed for such a bonding, is related to the powder particle material [9,28].

The impacts of powder particles, in relatively lower velocities, cause abrasion of the substrates as a result of fatigue and fracture. With increasing impact velocity, as it exceeds the critical velocity (V_{crit}), the thermomechanical requirements for deposition of the material can be provided by the high kinetic energy of impacts. Anyway, in the velocities higher than erosion velocity ($V_{erosion}$), the material is removed from the substrate due to the hydrodynamic effects (Fig. 2.6).

The minimum velocity that can induce ASI (adiabatic shear instability) is defined as critical velocity [37]. Assadi et al. [37] suggested the following equation for calculating the critical velocity by using the material properties (formula 2.2):

$$V_{cr} = 667 - 14\rho_p + 0.08T_m + 0.1\sigma_u - 0.4T_i \quad (2.2)$$

where ρ = density in g/cm^3, T_m = melting point in °C, σ_u = UTS in MPa, and T_i = initial temperature of the particle in °C. It is clearly observed that the critical velocity rises with increasing the melting point and UTS and declines with raising the density and particle temperature. This formula is suitable for calculating the critical velocity of the materials like copper [38]. Another equation for calculating the critical velocity that covers a wide range of materials was later suggested by Schmidt et al (formula 2.3) [38].

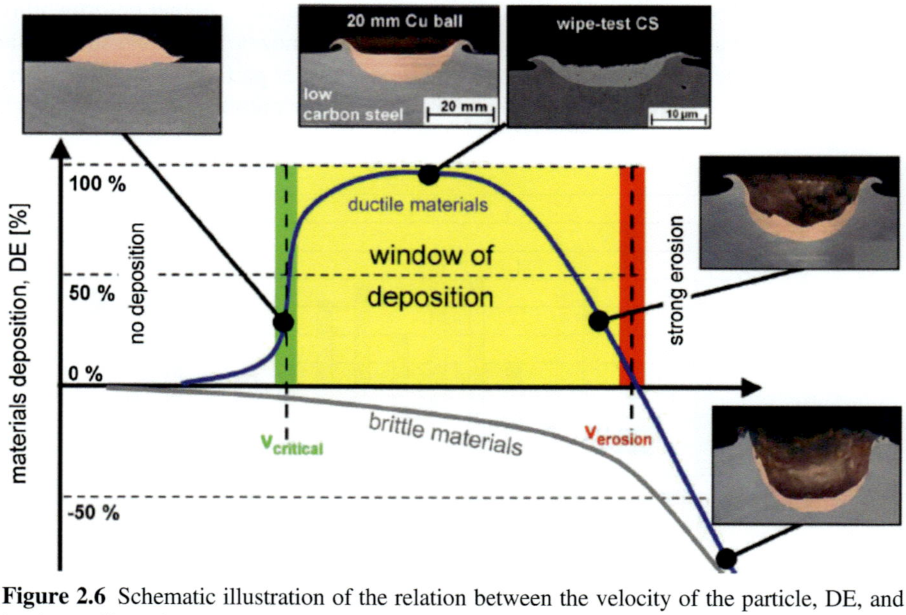

Figure 2.6 Schematic illustration of the relation between the velocity of the particle, DE, and impinge effects (at a constant temperature) [51].

$$v_{\text{crit}}^{\text{th,mech}} = \sqrt{\frac{F_1 * 4 * \sigma_{TS} * \left(1 - \frac{T_i - T_R}{T_m - T_R}\right)}{\rho} + F_2 * c_p * (T_m - T_i)} \tag{2.3}$$

ρ Density
T_m Melting point
σ_{TS} Tensile strength
T_i Impact temperature
c_p Specific heat
T_R Reference temperature (293 K)
F_1 Mechanical calibration factor (for cold spray 1.2)
F_2 Thermal calibration factor (for cold spray 0.3)

The results of calculating the critical velocity by Schmidt's formula have better agreement with the experimental results. Also, this formula predicts a more precise value of critical velocity for materials such as tin and tantalum compared to Assadi's equation [38]. Fig. 2.7 depicts the calculated results of the critical velocities for 25 μm particles (for different metallic materials) using Schmidt's formula (2.3).

The impact velocity of the powder particle should be upper than critical velocity v_{crit}. However, it should be lower than the erosion velocity $v_{erosion}$. Both v_{crit} and $v_{erosion}$ decrease with increasing the particle impact temperature [38] because of the thermal softening phenomenon [52], Fig. 2.8. The window of deposition lies between critical and erosion velocity in which the efficient deposition can occur.

As reported, increasing particle velocity enhances the quality of the coating [38]. Besides, the properties of the coating are enhanced by applying higher preheating temperatures [54], resulting in higher particle temperature and, thus, higher deformability and more robust adhesion [55]. Increasing velocity in the nozzle throat and supersonic velocity is attained in the nozzle's divergent zone. Powder particles reach the supersonic velocity in the pressurized gas flow. Because of the presence of the bow shock

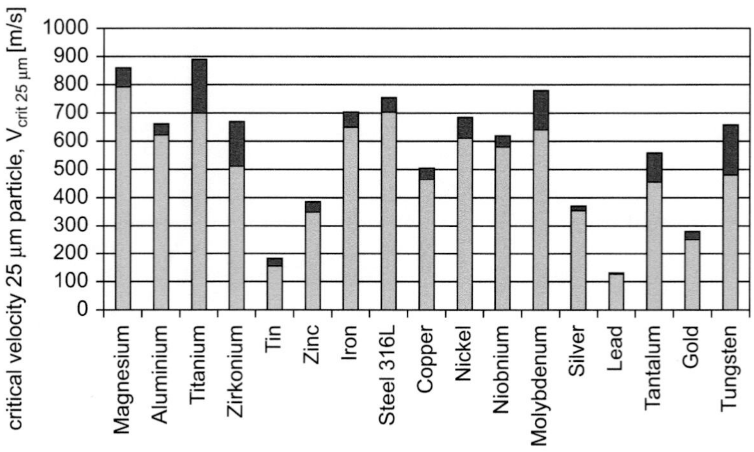

Figure 2.7 Calculated critical velocities for various powder material types [38].

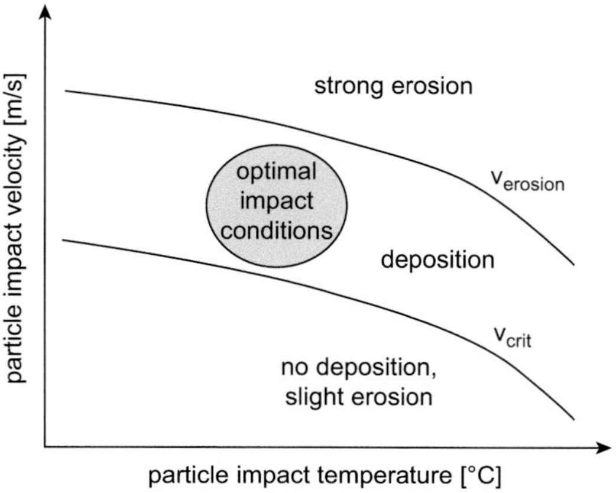

Figure 2.8 Changes of V_{crit} and $V_{erosion}$ as a function of particle impact temperature [38,53].

near the substrate surface, powder particle velocity declines in this area, and therefore, particles should have sufficient kinetic energy and velocity to pass the bow shock layer and to impact on the substrate surface.

The velocity of the gas rises at the throat and diverging zone of the nozzle. In fact, the gas attains higher velocities but lower temperatures compared to the powder particles [56]. It is interesting to note that the particle temperature considerably increases in the convergent section of the nozzle. However, its temperature quickly drops to the lower temperatures when powder particle enters the throat of the nozzle and even much lower at the divergent section of the nozzle. Nevertheless, the temperature of powder particles again rises close to the sprayed surface because of the bow shock presence (formed close to the surface). Gas and powder particle temperatures were noticed to have a robust influence on the bonding between the deformed particles and particle-substrate as well. Also, preheating the powder particles can raise the particle impact temperature [51].

The material cooling rate is influenced by powder particle size in which it declines with raising the particle size due to increasing the thermal inertia. It should be sufficiently low to boost the shear instability and, at the same time, should be sufficiently high to let the interface to cool down rapidly and finalize the bonding process. Shear instability could be hindered in small particles (having low thermal inertia and moments of inertia) because of high thermal slopes with their surroundings. Besides, smaller particles have higher quench rates during deposition, which raises the strength in these small particles. Also, smaller particles will generally show more significant levels of impurities after production because of higher ratios of surface-to-volume. Due to these parameters, there should be an optimum particle diameter above which the powder particles experience adequate rates of plastic deformation and quench. Schmidt et al. [38] suggested a formula (2.4) for the calculation of the critical diameter of the powder particles, above which particles will adhere to deposit as follows [24,38]:

$$d_{\text{crit}} = 36\frac{\lambda_p}{c_p \cdot \rho_p \cdot V_p} \tag{2.4}$$

where λ_p = thermal conductivity, c_p = specific heat of the particle, ρ_p = density of the particle, and V_p = velocity of the particle. The critical dimension (diameter) of different materials could be calculated based on formula 2.4. The calculated critical diameter (using formula 2.4) for various materials is shown in Fig. 2.9. These values show that thermal diffusion restricts attachment of small particles of tin, copper, silver, and gold, while spraying of steel 316L and titanium is less limited by the thermal diffusion phenomenon [38].

Two mechanisms that commonly are considered for the metallic bonding in the CS process are mechanical interlocking and metallurgical bonding as well [5]. As stated before, when the conditions are such that ASI is initiated, the bonding of sprayed particles could occur [37]. Fig. 2.10 depicts the experimental outcomes and simulation of the interaction between the substrate and the powder particles [37,38]. As can be seen, based on the simulations, the formation of "jet" is because of the ASI made by the high temperatures of adiabatic heating at the interfaces of particles and substrate, at which points bonding will happen. It means that if the thermal softening can excessively compensate the strain hardening and strain rate hardening of the material (resulting in ASI), the critical condition for bonding could be achieved. As shown in Fig. 2.10C, the calculated critical velocity of Cu is 400–500 m/s, which matches the experimental results [38]. ASI indicates that the material loses its shear strength and severely deforms, which results in changing the deformation mechanism from plastic to viscous flow. The resulted mechanism has been proposed by Assadi et al. [37] and Grujicic et al. [57,58]. The viscous flow because of shear localization at impact interfaces can cause some type of nano/microlength-scale mechanical material mixing/interlocking structures (roll-ups and vortices) [58]. This could be as a type of fluid flow concept, which leads to those interfacial configurations (illustrated in Fig. 2.11). Similar structures have been detected in explosive welding and cladding [59] processes, as depicted in Fig. 2.11 [59] but have not been found in the CS process.

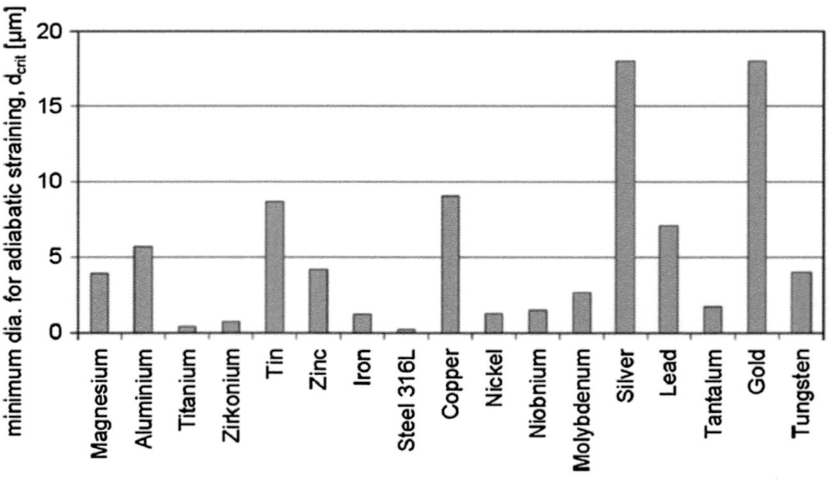

Figure 2.9 Critical powder particle diameter for ASI of various metallic materials [38].

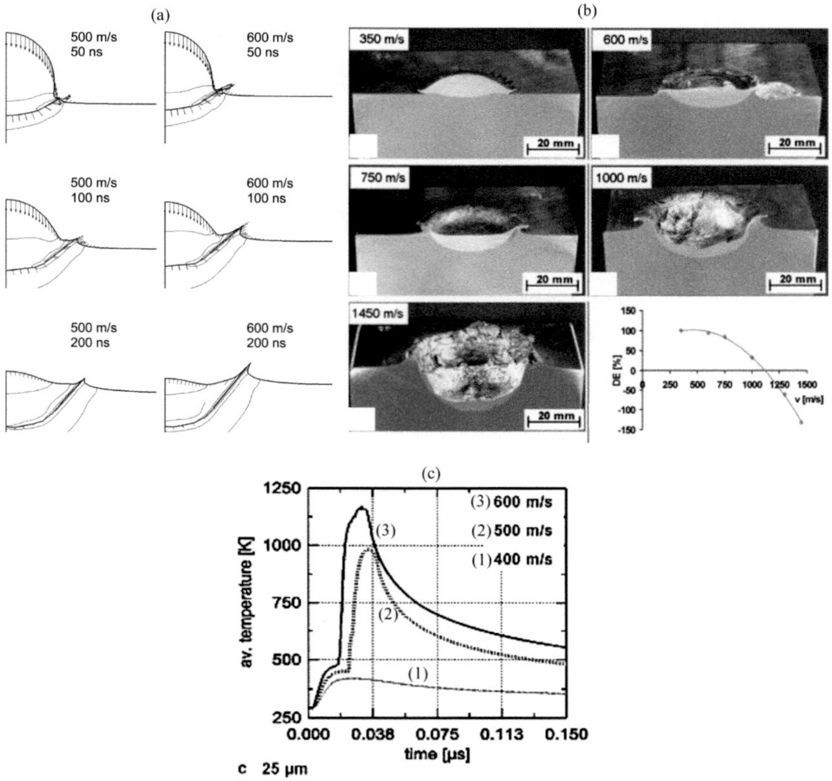

Figure 2.10 Formation of the material jet at impacting interface: (A) a Cu particle on a Cu substrate (a simulated impact) for 500 and 600 m/s (the initial impact velocities). The velocities of nodes at the respective surfaces of particle and Cu substrate are represented by the arrows, while the temperature distribution are shown by contours; (B) cross-sections of copper ball impacts on the low carbon steel plates (macro pictures) with different impact velocities, also diagram indicates the determined mass balance (DE) versus impact velocity; (C) changes of interface temperature, based on the time, for various impinge velocities (for Cu particle having 25 µm particle size) [37,38].

Besides, the intensive, and localized, plastic deformation because of the CS process causes thin oxide surface film disruption and facilitates to intimate conformal contact between the substrate and the powder particles. This, coupled with high contact pressures, is supposed to be required circumstances for bonding and also thick deposits formation (layer by layer) [60], Fig. 2.12. It is interesting to note that the material jet formation is well noticeable at spray angles about 90 degrees. Nevertheless, decreasing spray angle leads to the reduction of the DE and strength of the bulk deposited material, Fig. 2.13 [61].

It is interesting to note that ultrafine grains (even nano-scale) were seen at the particle-to-particle bonding region in the Ni (Figs. 2.14 and 2.15) and Al CS coatings (Fig. 2.16). Forming the mentioned grains with the nano-scale size is explained in

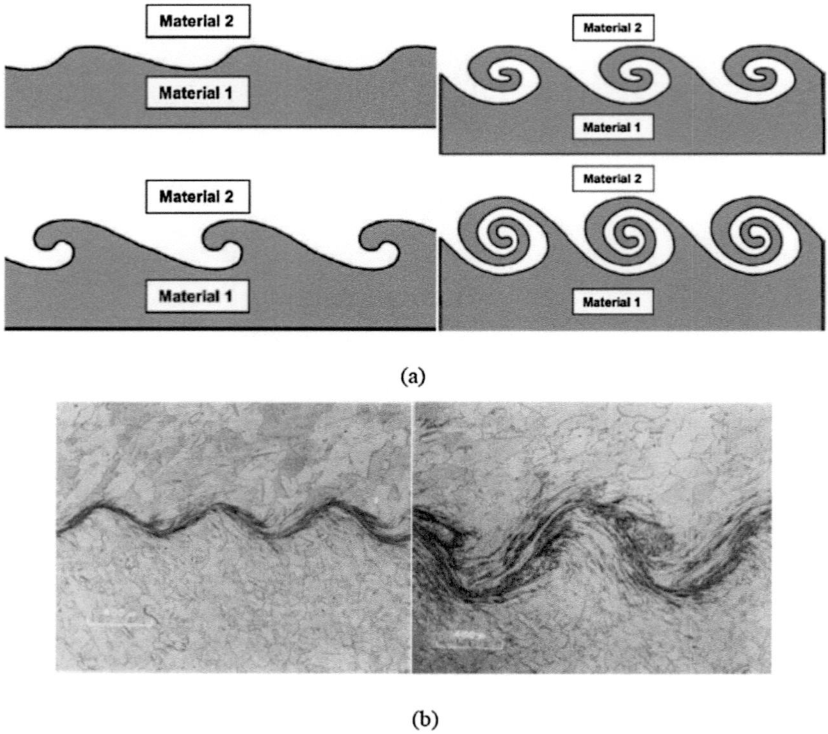

Figure 2.11 (A) A schematic illustration showing the evolution of the substrate and particle interface (designated with materials 1 and 2) based on instability and the accompanying formation of interfacial roll-ups and vortices. (B) Ni clad on Ni bonding areas made at two different velocities of impact (left: Vc = 1790 m/s; right: Vc = 2800 m/s) [58,59].

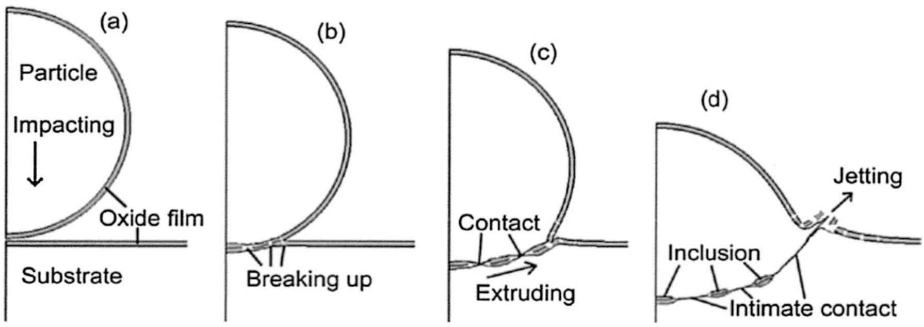

Figure 2.12 Illustration of the bonding development steps of CS-processed particle, including the breaking up and extruding the oxide film of the surface and formation of the material jetting [60].

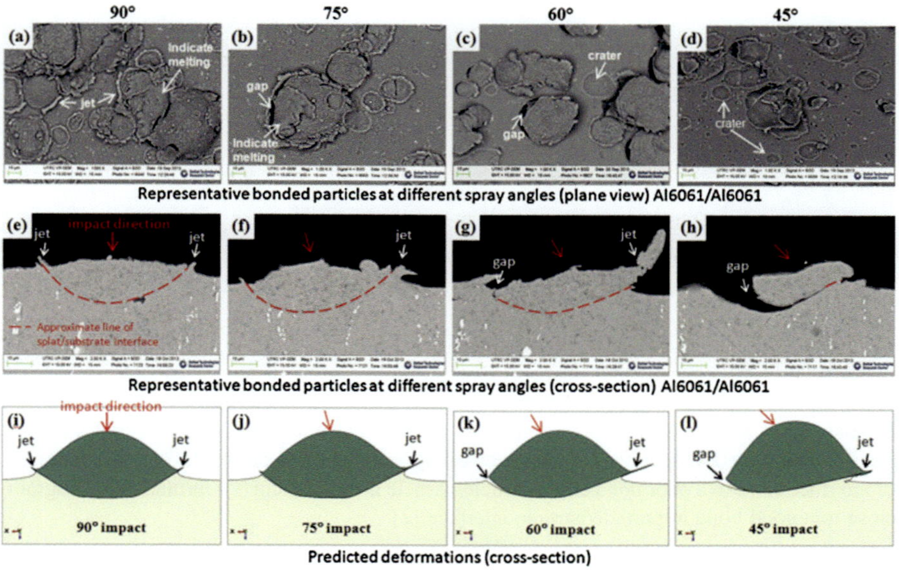

Figure 2.13 Spray angle effect on the bonding and corresponding deformation of impacted powder particles [61].

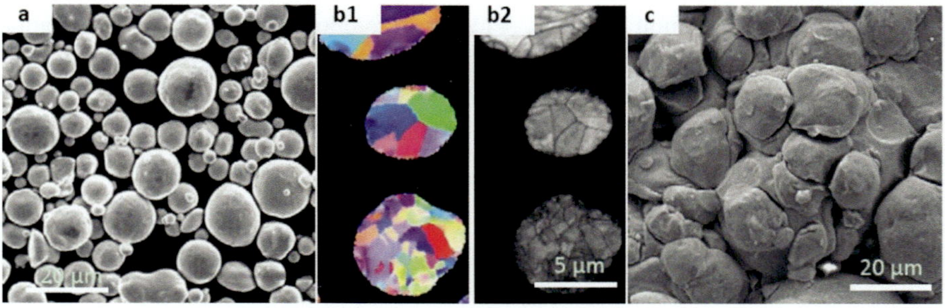

Figure 2.14 (A) SEM image, (B1) Euler angle map and (B2) pattern quality map of the Ni powder particles before spraying, (C) SEM image of the top surface of the CS coating [67].

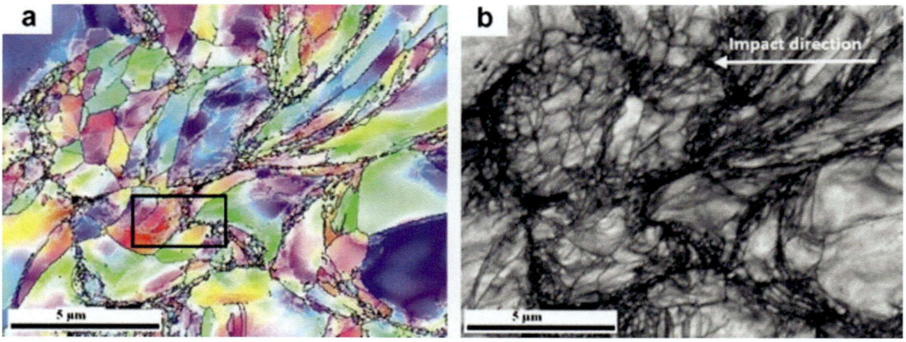

Figure 2.15 EBSD analysis of the cross-section of the CS-deposited Ni coating: (A) Euler angle map and (B) pattern quality map of the same area as 2(a), with low quality at boundaries of interparticle [67].

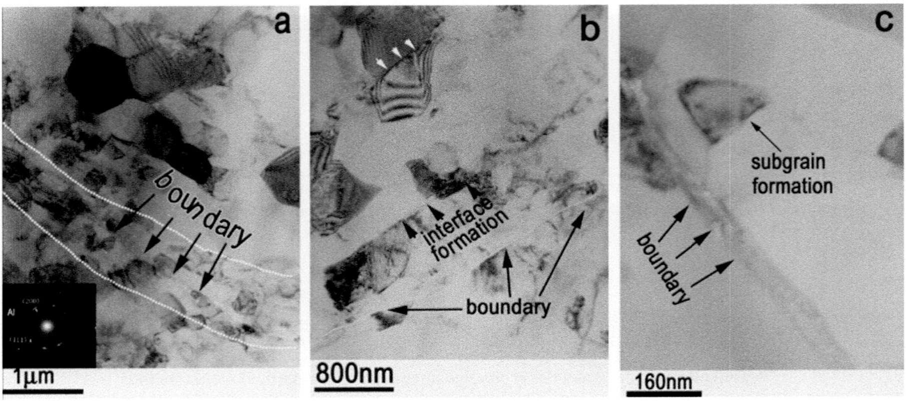

Figure 2.16 CS Al coating (TEM images) (A) grain refinement close to a good bond at particle/particle interface, (B) separation of a grain by a dense sliding dislocation (DDW) and formation of interface close to a poor bond at the particle/particle interface, and (C) formation of subgrain close to a good bond at particle/particle interface [2].

terms of dynamic recrystallization by lattice and subgrain rotation. The ultrafine grains and elongated subgrains which can be observed in the CS-processed Ni and Al deposits are similar to the ones in the adiabatic shear bands of the bulk materials under high strain rate deformation [62–64]. A model of rotational dynamic recrystallization for explaining the microstructural evolution in the process of the high strain rate deformation of bulk stainless steel and copper was proposed by Meyers et al. [65]. According to the EBSD results, it can be concluded that using this model to explain the formation of ultrafine grains in the Ni and Al CS coatings is reasonable. An illustration of the mentioned mechanism is presented in Fig. 2.17.

Before CS (impact), the density of the dislocations is comparatively low, and the microstructure is uniform (Fig. 2.17A). As can be seen in Fig. 2.17B, just after the collide happened, dislocations propagate and the lattice starts rotating progressively in the compression or shear direction (Fig. 2.17B). Elongated subgrains can be formed (Fig. 2.17C), when the high density of dislocations is accumulated and aligned in a short time. The mentioned subgrains are divided into the subgrains with equal dimensions in all directions, because of increasing the density of the dislocations (Fig. 2.17D). The misorientations between adjacent subgrains increase incrementally, and ultimately, the ultrafine grains with high-angle boundaries form to accommodate the further strain induced by the severe plastic deformation during CS process (Fig. 2.17E).

In situ grain refinement of impinged powder particles (during cold spray process) mainly due to the rotational dynamic recrystallization has been frequently reported for the cold sprayed metallic coatings [66].

Contrarily, the lower activation energy for recrystallization of the CS-processed Cu (Fig. 2.18), especially at higher temperatures of the process gas, shows the microstructure of static recrystallization and more uniform distribution of the hardness across the

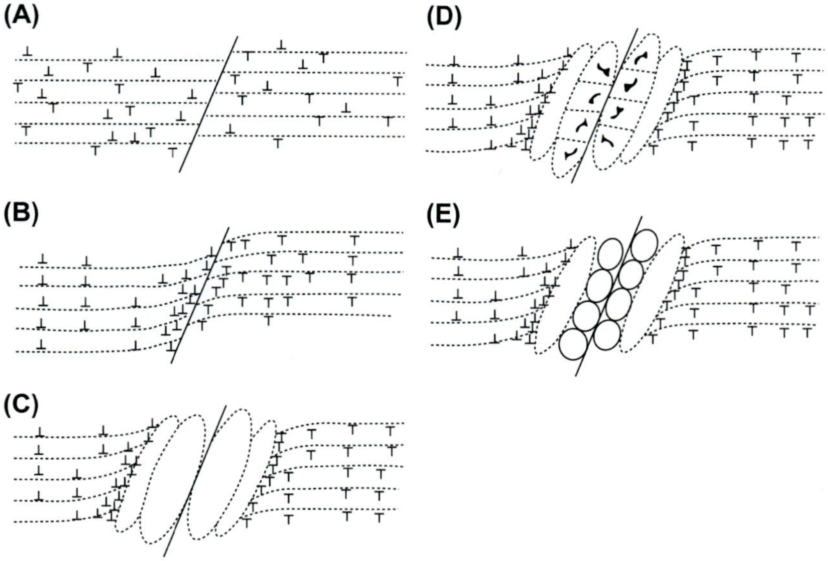

Figure 2.17 Illustration presenting the suggested mechanism of rotational dynamic recrystallization for Ni particles in CS process: (A) uniform microstructure with minimum dislocations before CS; (B) progressive lattice rotation and increasing dislocations just after collide; (C) accumulation of dislocations and forming of elongated subgrains to accommodate deformation; (D) subdividing of elongated subgrains into equiaxed subgrains and rotating to accommodate further deformation; (E) highly misoriented grains and equiaxed grains formation at interparticle boundaries [67].

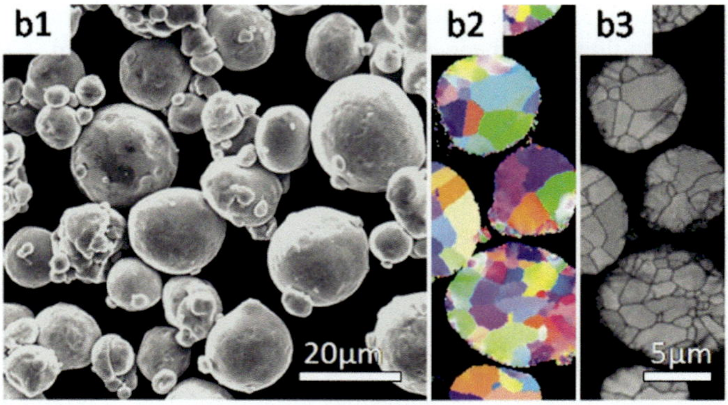

Figure 2.18 SEM images, Euler angle maps, and IQ maps of powder particles of Cu before CS (B1, B2, B3) [68].

coating structure. It is interesting to note that the high-quality EBSD patterns of as-sprayed Cu coating are mainly related to the typical feature of the formed grains (almost free of strain/defect) caused by the static recrystallization during CS process [68] (Fig. 2.19).

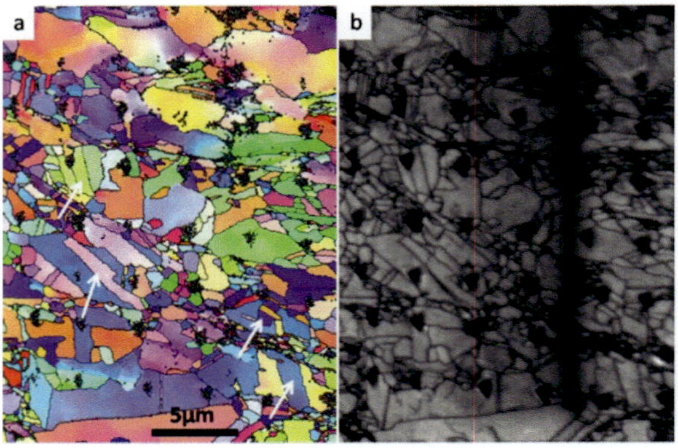

Figure 2.19 EBSD map of the cross-section of CS Cu coating after nanoindentation test: (A) Euler angle map, (B) IQ map with high quality pattern (sign of static recrystallization during CS process) [68].

So, it can be said that the powder particles material and also the cold spray parameters (especially process gas pressure and temperature) could affect the microstructure (type of recrystallization) and thus the mechanical properties of the cold sprayed coatings/deposits.

2.5 Cold spray additive manufacturing

As discussed in the previous sections, although CS originally was used as a coating process, it is able to produce the coating layers with no limit in thickness. Compared to the other AM methods, CS does not include either high-temperature processes, like DMD [69] and SLM [70,71], or ecologically troublesome chemical processes, like electroplating. Hence, 3D shapes with different geometries could be made by CS technology. This technology has been regarded as one of the standard AM processes (*ASTM F2793-12A*) that are represented in Table 2.1 [72,73].

According to the present literature, the applications of the CSAM process are discussed and summarized in a material point of view as follows.

2.5.1 Production of cold sprayed commercially pure titanium and titanium alloy components (by CSAM)

The high price of Ti metal and its alloys, as well as the difficulty of producing Ti-based components through traditional processes, has caused great attention on the AM process [73–75]. CSAM of Ti and its alloys has great potentials in producing the parts like prostheses for implants, pipes for desalination plants, and aerospace vehicles [76]. In general, the cold sprayed deposits of Ti and its alloys usually display high level of porosity (mainly due to relatively high hardness and microstructure of Ti and its alloys and the presence of lower slip systems in H.C.P crystal structures as well), poor strength, and toughness which can be rectified. The in situ shot peening during the CS process can generate a strong tamping effect, which results in the severe plastic deformation of the powder particles and thus considerable enhancement of the density and

Table 2.1 The list of AM processes defined in the ASTM F2793-12A standard [72].

Additive manufacturing methods	Explanations in ASTM F2793-12A standard
Directed energy deposition	The fused materials are deposited. In general, the materials are melted using focused thermal energy.
Powder bed fusion	The regions of a powder bed are fused by selectively thermal energy.
Vat photopolymerisation	Light-activated polymerization is used to selectively cure liquid photopolymer in a vat.
Sheet lamination	An object is formed by bonding the material sheets.
Binder jetting	Powder materials are selectively joined by depositing a liquid bonding agent (selectively).
Material jetting	A Part is manufactured by depositing the droplets of build material (selectively).
Material extrusion	Selective dispensation of material by using an orifice or nozzle
Cold spray	Powder particles with high velocities are propelled toward substrate surface. This can lead to the severe plastic deformation of powder and then material buildup (at window of deposition (WD) of materials).

microhardness. Moreover, internal residual stresses relieving and also the extensive interface bonding of the powder particles can be fulfilled by the commonly used post-cold spray thermal treatments. In order to develop CSAM Ti-based parts and also explore their applications, some studies have been practically performed in the past decade. Jahedi et al. developed Ti blocks using the CS process and the following annealing treatment [77]. According to the mechanical properties examinations (Fig. 2.20), tensile strength was reported to be 600 MPa, and elongation reached 8%

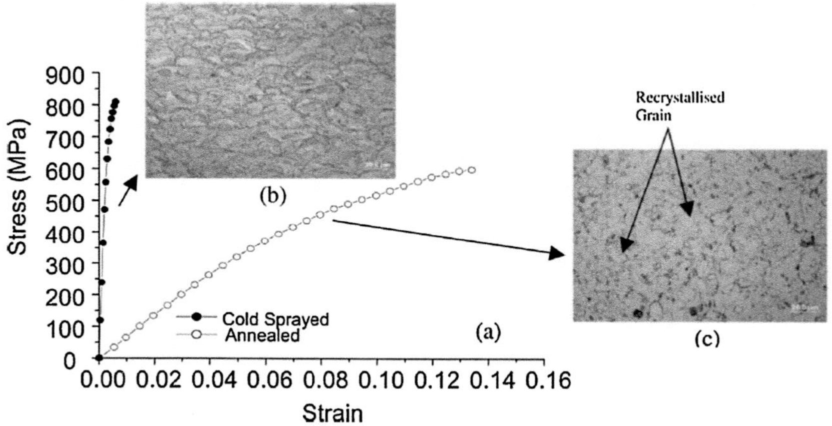

Figure 2.20 Tensile stress—strain curves of CS (as-sprayed) and annealed titanium deposits and relevant microstructures: (A) Tensile stress—strain curves (B) cross-section of as-sprayed and (C) cross-section of annealed Ti deposits [76].

after annealing at 550°C for 2 h [77]. The tensile strength of 380–550 MPa and elongation of 18% are required for the grade 3 Ti seamless pipes (according to the ASTM B861 standard). The requirements of the mentioned standard for the production of seamless pipes were fulfilled by the postcold spray heat treatment of CS-processed Ti at 850°C for 1 h. The tensile strength of this heat-treated part was reported to be 530 MPa [77].

Zahiri et al. [75,78] reported a similar result that the mechanical properties of CSAM deposited Ti coatings could be improved by the postannealing treatment. Apart from the simple block structure of the cold sprayed Ti which was mentioned before [73,76], CSAM produced Ti-based components like Ti billets and Ti seamless pipes and tubes [73,76].

It is interesting to note that the HIP treatment is accepted for reducing the internal defects, adjusting the microstructure, and improving the mechanical characteristics of Ti6Al4V deposits (produced by HPCS system utilizing high process gas pressure and temperature). Based on the XCT 3D reconstructions, the fully dense Ti6Al4V alloy could be attained via the high-pressure compacting and high-temperature diffusion of the HIP specimen (Figs. 2.21 and 2.22). After the HIP treatment, the severely deformed grains experienced significant growth with the homogenously distributed β precipitates (around equiaxed α grains). The outcomes of the tensile test revealed that the strength of CSed Ti6Al4V alloys could be considerably raised by the improved

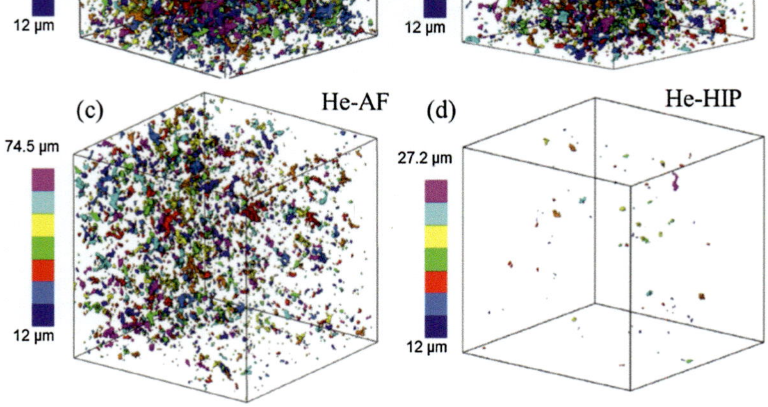

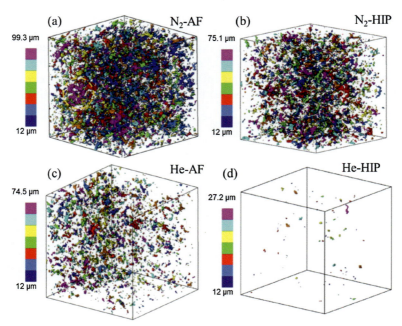

Figure 2.21 3D models representing of the pores in the CSed Ti6Al4V specimens in different conditions: (A) N_2-AF (as-fabricated), (B) N_2–HIP, (C) He-AF (as-fabricated), and (D) He–HIP, while the color scale bar shows the equivalent diameter of the represented pores [79].

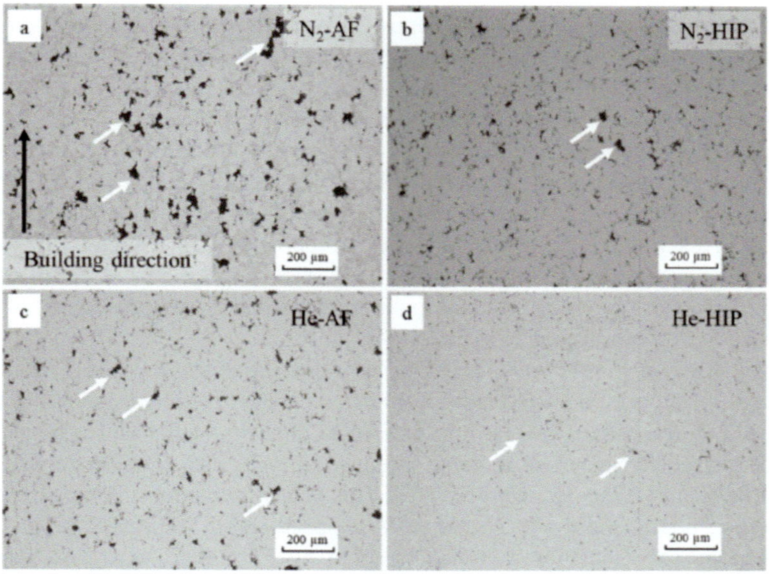

Figure 2.22 OM micrographs of the cross-section showing the morphology of CS Ti6Al4V specimens in different conditions), (A) N_2-AF (as-fabricated), (B) N_2—HIP, (C) He-AF (as-fabricated), and (D) He—HIP [79].

diffusion and resulting metallurgical bonding. HIP treatment extremely densified the morphology and adjusted microstructure of the CS-processed specimens that can enhance the mechanical characteristics [79] (Fig. 2.23).

Advanced materials allow the further use of FSP as a post-CS treatment, which will be available to process large Ti and Ti alloy deposits before and even after postcold spray heat treatments [80]. The issue is the control of the mentioned techniques to modify the properties of CS-processed Ti or Ti alloy deposits. As stated before, the microstructure of Ti alloy powders is one of the reasons that limits the deposition of them.

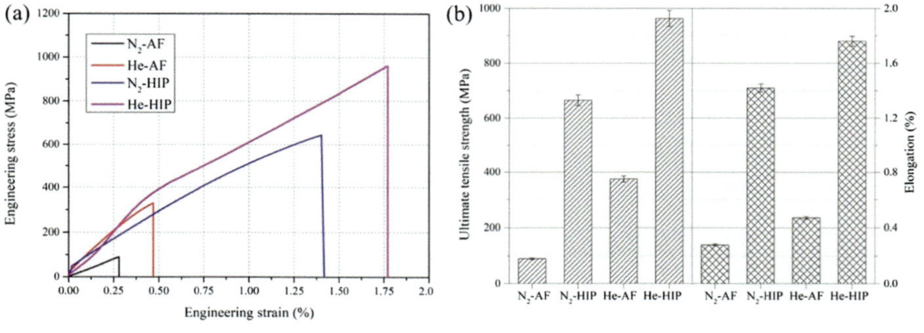

Figure 2.23 Changes of UTS and elongation of CS-processed Ti6Al4V specimens in different conditions [79].

The properties of Ti and Ti alloy deposits may be enhanced by using various pre-cold spray treatment routes like annealing and hydrogenation through modifying the microstructure of the powder particles. Another research direction can be a powder microstructure tailor to reduce its hardness and ease its deposition.

Ti and Ti alloy powder particles which are typically used for the CS process have spherical and irregular morphologies [81,82]. It was noticed that spherical Ti powder particles can reach lower velocities than irregular Ti particles under the same CS parameters. This was related to the higher drag coefficients for irregular Ti particles. Higher powder particle velocity simplifies the particle deposition and leads to the better DE. However, more porosity and lower microhardness were observed for the deposits produced with irregular particles [76].

Nozzle traverse speed is another parameter which influences the particle deposition. The porosity, deposit thickness, and substrate temperature are major properties that are influenced by the nozzle traverse speed. Two Ti deposits with a same thickness produced at two different nozzle traverse speeds: 150 mm/s and 5 mm/s, showed porosities of 0.9% and 0.1%, respectively [76]. This was attributed to the higher substrate temperature due to the lower nozzle traverse speed which improved the Ti particle deformation. This behavior was also reported by Choudhuri et al. [83], and Tan et al. [84].

Some examples of the CSAM Ti components are shown in Fig. 2.24 [73,76]. Pattison et al. [73] reported that the more complex Ti structure with internal channels could also be fabricated using the CSAM process. Fig. 2.25 shows the manufacturing

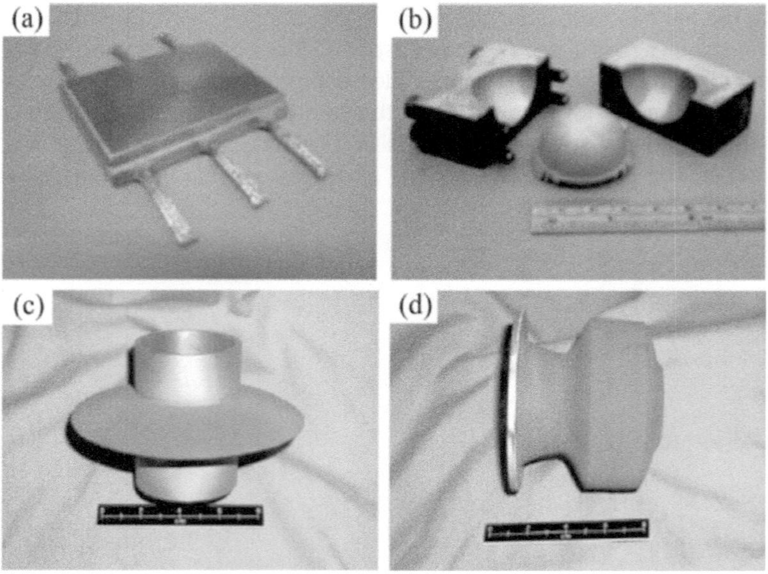

Figure 2.24 Some examples of CS fabricated components: (A) three covered thermocouples placed in a Ti part, (B) a mold-shaped Ti hemisphere, (C) pipes, and (D) one axisymmetric component [76].

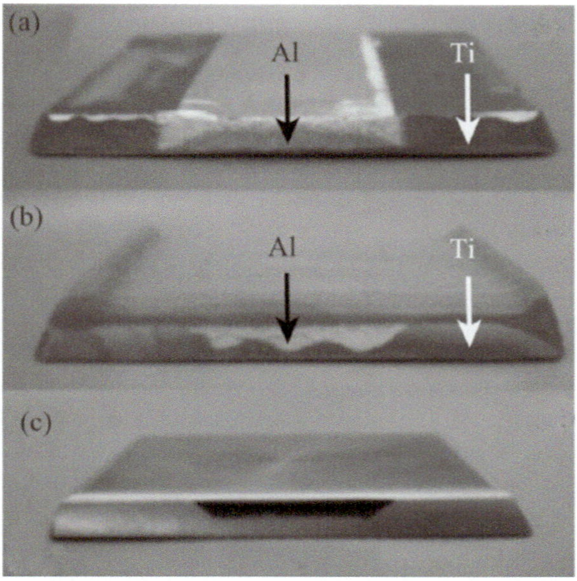

Figure 2.25 CS-processed Ti part fabricated using an internal channel [76].

process of the special structure [73]. At the first step, a Ti plate with a thickness of 5 mm was produced using the CS process, and then a groove was created in the central region of the plate by machining. Then, Al material was deposited on the machined Ti plate using CS, and the top surface was machined for having a flat external surface. Thereafter, a thin layer of Ti was deposited by CS on the machined surface. Lastly, to achieve the internal structure, the inserted aluminum was removed by alkaline solution [73]. Moreover, large or complex Ti-based components can be produced by CSAM. A CS-processed Ti cylindrical spattering target produced by Plasma Giken Co. Ltd. is shown in Fig. 2.26 [76]. Fig. 2.26 [76] also shows a Ti ring with a diameter of 140 mm fabricated with CSAM as another example.

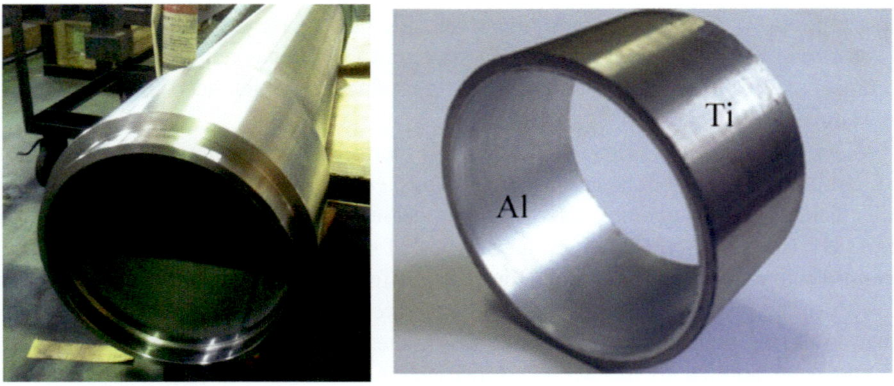

Figure 2.26 CSAM fabricated components (A) Ti cylindrical sputtering target [159] (B) 140 mm diameter Ti ring, having aluminum support to be removed after fabrication [76].

2.5.2 Production of cold sprayed aluminum alloy components (by CSAM)

Because of high specific strength and relatively high corrosion resistance (depends on the type of secondary phases in the Al matrix), aluminum alloys are extensively used as high-performance engineering alloys in automobile and aerospace industries. Mechanical characteristics (especially tensile strength) of AM parts are of significant concern for industrial applications. Mechanical properties of the CS-processed metallic components are often pertained to the interparticle bonding and levels of defects [85].

The detailed spray parameters and relevant porosity and mechanical properties including YS, UTS, elongation, and E of the CS-processed AA6061 and other Al−alloy deposits from literature are summarized in Table 2.2. As can be seen, the porosity commonly exists in the CS-processed Al alloy depositions (especially in 6061 Al alloy) when N_2 propellant gas (CS−N_2) is utilized, due to insufficient particle plastic deformation during deposition [85] and comparatively lower density of Al than that of Cu which has lower critical velocity than Al.

As an example, Aldwell et al. [86] showed that at process gas temperature and pressure of 400°C and 2 MPa, respectively, the porosity of CS-N_2 AA6061 deposits would be as high as about 8%. Two main strategies can be usually employed to solve these issues: i) thermally softening of the feedstock powder to lower its resistance to plastically deform and ii) increasing the velocity of the particle for improving the "driving force" for more plastic deformation. Normally, using higher gas pressure and temperature results in more compact deposits due to increased particle temperatures and velocities [85]. For instance, high-density CS-N_2 Al alloy C355 (Al−5Si−1Cu−Mg), with the porosity of as low as 1%, using N_2 gas at high gas temperature of 500°C and pressure of 6 MPa was prepared by Murray et al. [87]. However, for Al alloys with a comparatively lower melting point, the probability of the nozzle clogging and thus interrupting the continuous spraying process and also reduction of the deposition quality can be increased when higher process gas temperatures are used. A more straightforward approach to increase the particle velocity is the usage of the He gas (CS−He) in place of N_2 as processing gas. This would result in the denser deposits due to the lower gas density and higher specific heat of He gas compared to N_2 gas [10]. However, using the expensive and nonrenewable He gas requires the installation of a gas recycling facility for industrial-scale production, which needs massive capital investments [9].

Newly, CSAM deposition of AA6061 high-performance aluminum alloy was made utilizing comparatively cheaper N_2 process gas, enabled by an in situ MF effect (MF-CS). Based on the outcomes, the MF-CS processed AA6061 showed high UTS and E and very low porosity. Moreover, it consisted of superior equiaxed submicron fine Al grains (with random orientations). However, the formability of the deposition was very low, resulted in the severe work hardening induced by intensive particle plastic deformation during deposition. For solving this issue, three different types of heat treatments were applied to the MF-CS processed AA6061. These postcold spray heat treatments consisted of stress relieving, recrystallization annealing, and T6. The heat

Table 2.2 The detailed cold spray parameters and resultant porosity in the sprayed coatings and mechanical properties, including YS, UTS, elongation, and E of cold sprayed aluminum alloy deposits sprayed by different propellant gases [85].

Methods	Materials	Gas temp. °C	Gas pres. (MPa)	Porosity (%)	YS (MPa)	UTS (MPa)	Elongation (%)	E (GPa)
CS-N_2	AA6061	350	3	14.4	-	-	-	-
		400	2	8.27	-	-	-	-
	AA2024	500	3.5	0.41	-	-	-	68.6
	C355	500	6	1.07	-	~200	~0.65	-
	A357	450	5.7	~0	-	183–217	~0	62–63
CS-He	AA6061	400	2	~0	~290	~340	~3	-
		400	2.8		-	~440	~3	-
		400	2		262	286.8	2	67.5

treatment generated complex effects on both interparticle bonding and inner-particle microstructure (grain size, dislocation density, and precipitation), and different heat treatments resulted in different mechanical properties of AA6061 depositions. T6 thermal treatment resulted in the best improvement on the overall mechanical properties among three different strategies in which the UTS and E were comparable, and only the ductility was slightly decreased in comparison with the pertinent T6 bulk aluminum [85], Figs. 2.27 and 2.28.

Particle velocities, their positions in the spray stream, and particle size measurements are valuable information about spraying conditions which could be monitored by the HWCS2 online spray monitoring system (Fig. 2.29). This system also improves repeatability and reliability of the coating production. Recently, H. Koivuluoto et al., studied on the cold spraying Al6061 powder and the correlations between particle inflight properties and coating characteristics such as structures and mechanical properties. Higher particle deformation and hence, higher coating quality, denser structures, as well as enhanced adhesions (bond strengths), were archived when powder particle velocities increased. Increasing velocities of powder particles also considerably raised the Al 6061 powder DE. So, it is anticipated that this accurate quality control by online monitoring with the aid of optimized cold spray process conditions could be used for the CSAM [88].

Despite the previous broadly studying of the solution and aging treatment of Al alloys bulks, the research studies on CSAM Al alloys are still very disorganized, and just limited range of Al alloys in CSAM have been studied and investigates on other type of Al alloys such as 2000 and 5000 Al alloys series are needed to be performed. CSAM can produce large Al-based components. As shown in Fig. 2.30, Al alloy tube and flange were produced by the HPCS process [10].

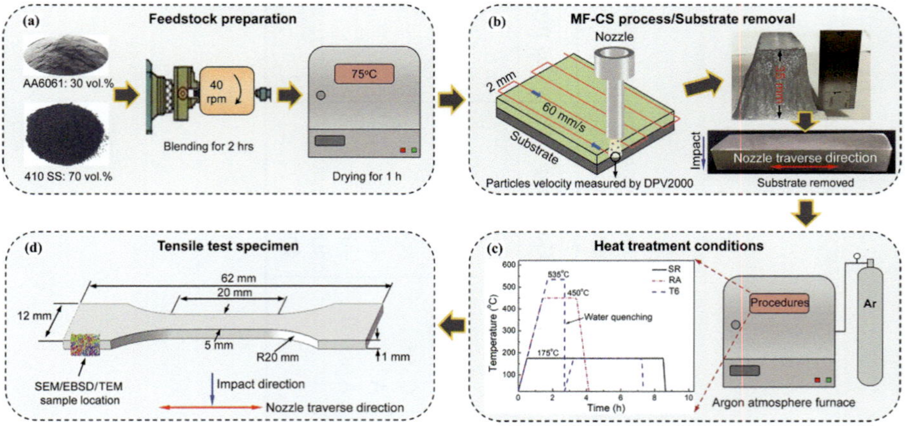

Figure 2.27 Schematically illustrations of (A) preparation of feedstock powder, (B) process of MF-CS and images of the deposit AA6061 before and after of removal of the substrate, (C) conditions of the postcold spray heat treatments, and (D) geometry of the tensile test specimen (dog bone) [85].

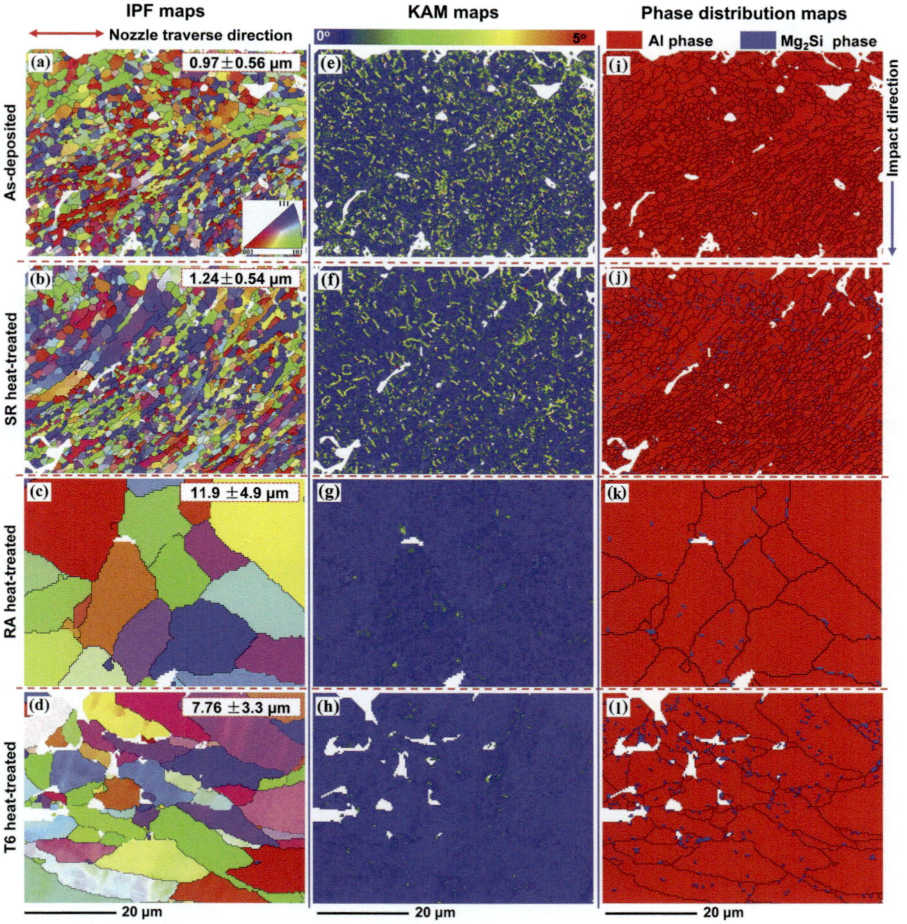

Figure 2.28 IPF, KAM, and phase distribution maps of (A, E, and I) as-deposited (B, F, and J) stress relieved (SR), (C, G, and K) recrystallization annealed (RA), and (D, H, and L) T6 heat-treated AA6061 deposits, respectively. The white spots are the regions that could not be identified [85].

2.5.3 Production of cold sprayed steel components (by CSAM)

Normally, fabrication of dense steel or stainless steel is difficult due to their comparatively high strength. However, it is easily possible to deposit dense steel at high particle velocities using high pressure and temperature N_2 gas (Fig. 2.31), or more expensive He process gas. Many studies have been performed for investigating the effects of powder particle size, annealing postcold spray heat treatment, and CS parameters (like process gas temperature and powder particle velocity) on the deposition behavior and mechanical performance of the alloy.

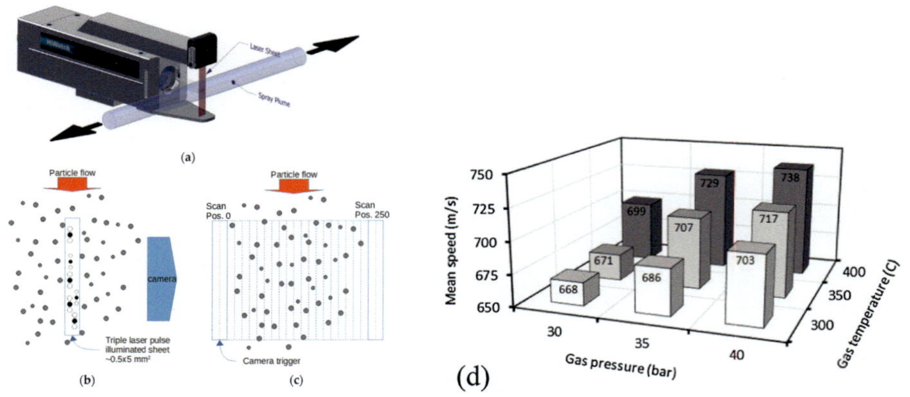

Figure 2.29 (A) HW CS2 system for online spray monitoring, (B) geometry of particle imaging, (C) 2D mapping (by scanning plume), and (D) Particle velocities (average) sprayed with diverse cold spray parameters (gas and temperature of process gas) [88].

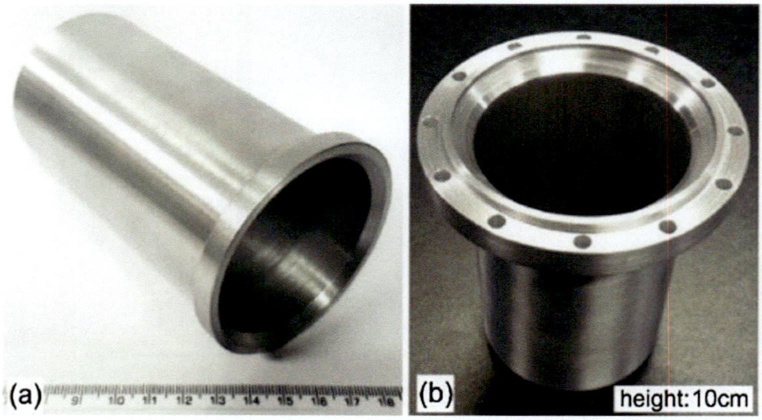

Figure 2.30 The machined finished part of CSAM processed Al alloy, using HPCS system: (A) a pipe (B) a flange [10].

For example, Chaoyue Chen et al. [89], recently fabricated the maraging steel 300 (MS300) by CSAM, followed by solution-aging heat treatment, Figs. 2.32 and 2.33. For the specimen sprayed at 5 MPa and 1000°C (for process gas in CS process), the XCT (as none destructive evaluation test) result showed the porosity of 0.168%. After solution heat treatment of this specimen, the porosity was decreased to 0.139% from 0.168%, Fig. 2.30. Reduction of the porosity was attributed to the grain growth and coalescence of small pores. XRD and EBSD phase characterizations revealed that the phase composition of the feedstock powder particles has been well retained in the fabricated samples. As highly martensitized microstructure (due to the solution-aging treatment) was observed in the MS300. The EBSD characterization showed

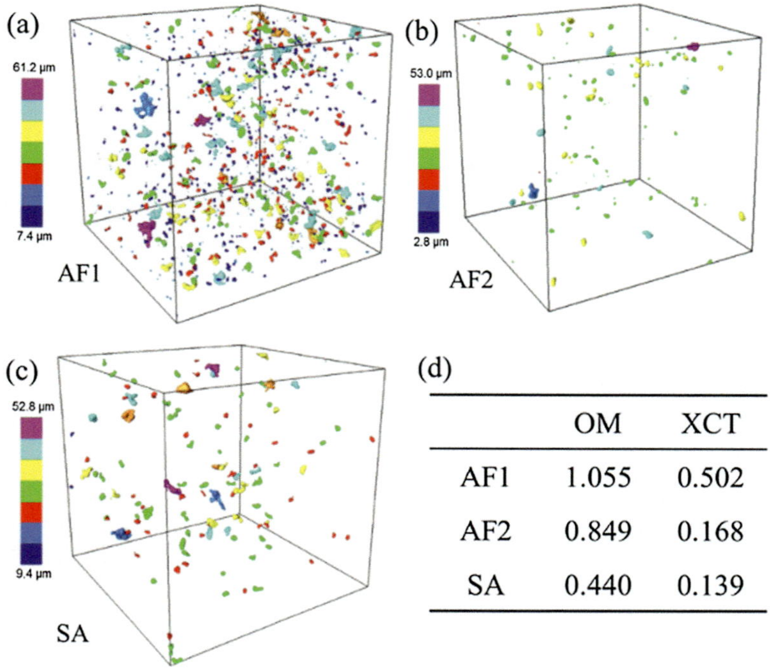

	OM	XCT
AF1	1.055	0.502
AF2	0.849	0.168
SA	0.440	0.139

Figure 2.31 XCT results of the CS MS300 deposits in different CS process parameters and postcold spray treatment conditions with a color map indicating the equivalent diameter of the pores: (A) AF1 (5 MPa and 900°C for N2 as process gas), (B) AF2 (5 MPa and 1000°C for N2 as process gas), (C) SA (solution aged samples), and (D) the percentage of pore fractions measured by OM and XCT (as NDE test) [89].

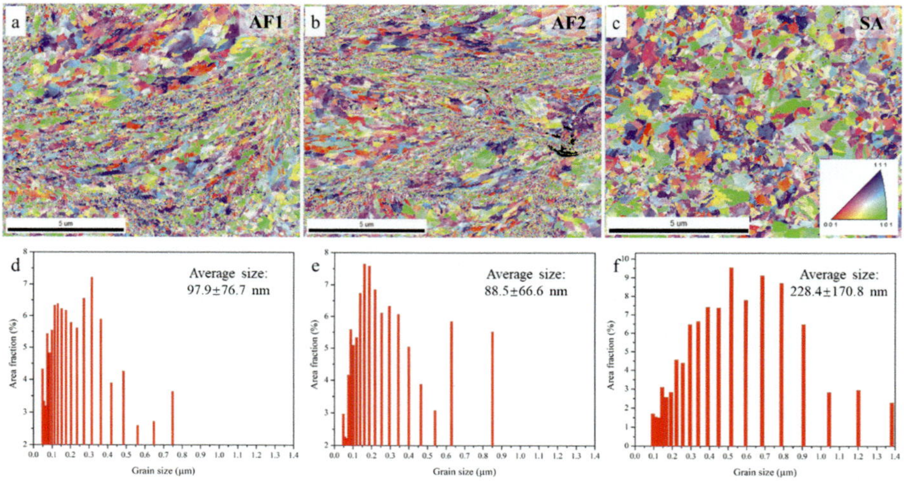

Figure 2.32 EBSD characterization: (A–C) IPF maps and (D–F) grain size distribution of the CS MS300 samples: first column is AF1 (50bar 900C), second one is AF2 (50bar 1000C), last one is SA (solution-age hardened) [89].

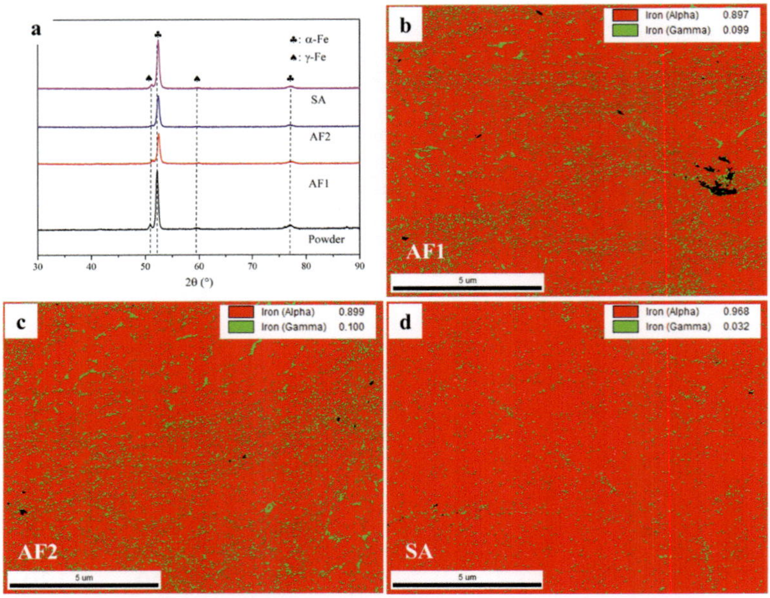

Figure 2.33 (A) XRD patterns and (B−D) EBSP phase maps of the MS300 specimens produced by the CS process at different conditions: (B) AF1, (C) AF2, and (D) SA [89].

the ultrafine grain structure of the fabricated samples that can be attributed to dynamic recrystallization phenomenon during the CS process. Tensile strength test showed remarkably improved tensile properties, including UTS and elongation resulted in the cohesive diffusion and precipitation strengthening after solution-aging heat treatment. Furthermore, the fracture morphology of the specimen contained widespread dimples as a sign of metallurgical bonding at interparticle boundaries. Likewise, it was reported that the precipitation hardening despite the coarsened microstructure can considerably improve the wear resistance of the CS MS300 deposit [89].

S. Yin et al. [90], for improving the microstructure and mechanical characteristics of CS-processed 316L stainless steel deposits, analytically investigated three different annealing strategies: air annealing, vacuum annealing, and HIP treatments, Figs. 2.34

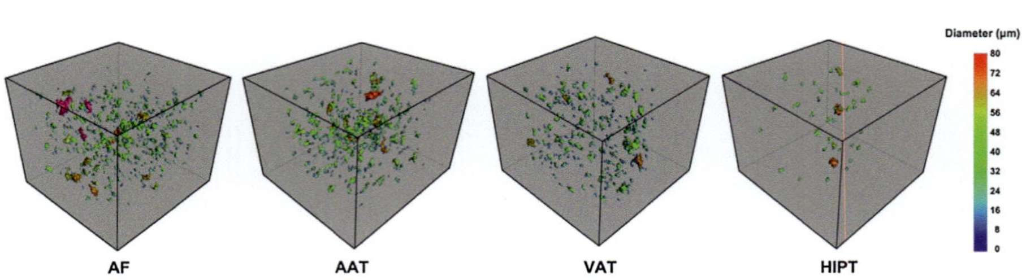

Figure 2.34 3D µCT construction of the pores inside $0.5 \times 0.5 \times 0.5$ mm^3 spaces in the 316L deposits before and after different annealing treatments (*AAT*, air annealing treatment; *AF*, as-fabricated state; *HIPT*, HIP treatment; *VAT*: vacuum annealing treatment) [90].

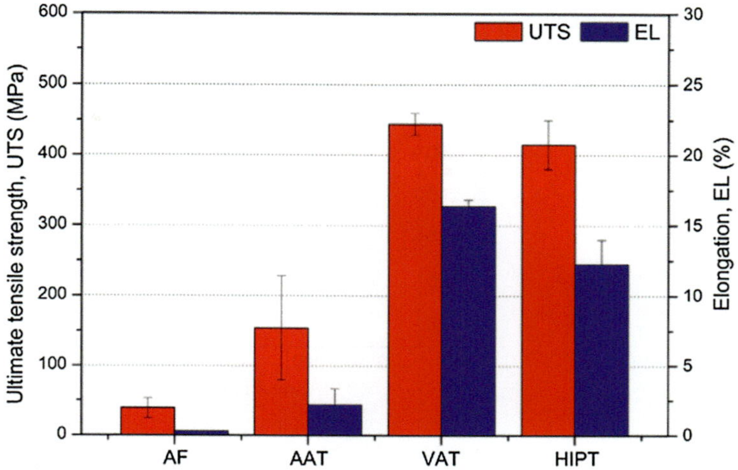

Figure 2.35 UTS and elongation of the 316L deposits before and after different annealing treatments [90].

and 2.35. The HIP heat treatment is a process that is commonly used in fusion-based metals AM (like SLM) for the densification of the AM fabricated parts. The results indicated that the mechanical characteristics of the CS-processed 316L stainless steel deposits can be enhanced by all of the mentioned annealing treatments. This was attributed to the grain recrystallization and diffusion at oxide-free interparticle boundaries. This improvement is less pronounced for the air annealing treatment due to the formation of oxide inclusions which prevented the full interparticle metallurgical bonding during annealing. The mentioned negative effect is suppressed in the vacuum annealing treatment. This resulted in a considerable enhancement in the tensile strength and elongation as well. Although HIP treatment remarkably enhanced the density of the deposit, it could not enhance the mechanical characteristics greater than those after vacuum annealing treatment. Based on the results, the researchers noticed that the role of improved particle grain structure and interparticle bonding on the strengthening of CS-processed 316L SS deposits is more compared to the reduction of porosities in the deposit [90].

Furthermore, the researches presented that high-temperature vacuum annealing treatment might also break the oxide films from the particle surfaces and may cause additional enhancement of the interparticle metallurgical bonding [91,92]. These studies describe the reason of the higher strength and ductility of VAT and HIPT specimens than those of as-fabricated (AF) and air-annealed (AAT) specimens [90].

Meng et al. [93,94] investigated the effect of thermal treatment on CS-processed 304L stainless steel (Fig. 2.36). The outcomes showed that raising the annealing temperature improves the ultimate strength of the deposits.

In summary, thick and comparatively dense stainless-steel deposits could be made by CSAM (Fig. 2.37). Despite the fact that the mechanical characteristics are not as good as to those of typical wrought materials, they could be efficiently enhanced via proper post-CS thermal treatments. Hence, the applications of cold sprayed stainless steels are strongly encouraged in future work.

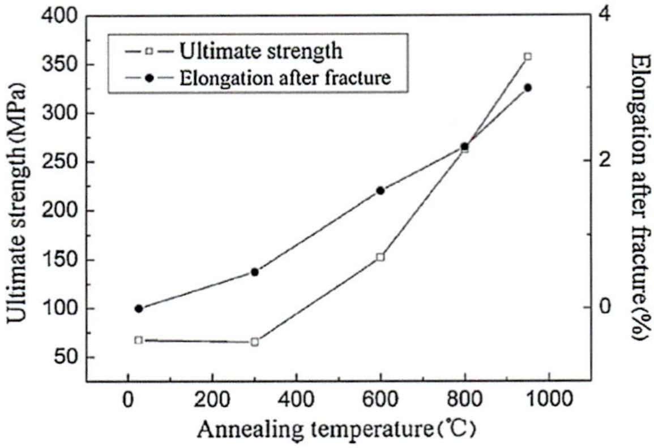

Figure 2.36 Ultimate strength and elongation after fracture of CS-processed and annealed 304L at different temperatures [94].

Figure 2.37 Freestanding stainless-steel component made by high-pressure CS system (top image indicate the size of the component, while the bottom image shows the cross-sectional thickness). It demonstrates that the high-pressure cold-spray (HPCS) can be used to manufacture freestanding large-scale steel components without visible defects.
Courtesy: UNR-ASB Industries, Inc collaborative work.

2.5.4 Production of cold sprayed Ni-based superalloy components (by CSAM)

Superalloys are generally employed at high temperatures and pressures, e.g., in gas turbines and jet engines. Difficulties in machining and shaping of the components with Inconel material type, using the traditional fabrication processes, result in the relatively high cost of the production of them; therefore, finding a capable AM method (e.g., CSAM) for the fabrication of high-quality Inconel-based materials would be motivating [95]. Nevertheless, in the CS process, the superalloy particles hardly deform after impact, even at elevated process gas temperatures. This can prevent the formation of a dense deposits using CS process. Therefore, these deposited superalloys have inferior mechanical strength and limited formability (elongation). Hence, the probability of getting superior mechanical and physical characteristics of CSAMed superalloys via possible post-CS treatments should be studied and evaluated [96]. The influence of heat treatment (in 990°C, for 4h) on CS-processed Inconel 718 coatings was studied by Ma et al. [97]. The results revealed the enhancement of both tensile and adhesive strengths after the heat treatment. Likewise, the influence of process gas (including nitrogen and helium) on the mechanical properties of the cold sprayed IN718 deposits was investigated by Huang and Fukanum [98]. Cohesion and adhesion strength test results depicted that the fabricated deposits using the He process gas had higher quality compared to the deposited material using N_2 process gas; however, improved mechanical properties after postcold spray thermal treatment were also obtained for the deposited material using N_2 (as process gas). S. Bagherifard et al. used the HPCS system (Impact Spray System 5/11; upgraded up to 60 bar; Impact Innovations GmbH, Haun, DE, N_2 process gas, at 1000°C and 55bar) to produce IN718 deposits and compare them with similar specimens manufactured by SLM technique. Various thermal treatments were utilized on the CS-processed specimens (HTA: 1050C, 3h, and HTB: 1200C, 1h both under 100%pure Ar gas) for improving their mechanical properties [20,95].

The outcomes confirmed that the CS process could be potentially used as a complementary AM technique (compared to SLM) for the construction of freestanding components with high quality where lower process temperatures, scaled-up final size, and higher deposition rates are needed [20,95].

In another research, Wong et al. [99] investigated the tensile characteristic of CS-processed IN718. Primary outcomes obviously showed that the tensile strength of CSAM IN718 using He propellant gas is noticeably superior to that utilizing N_2 process gas. Besides, regardless of the propellant gases utilized, the ductility (engineering strain) of the deposited IN718 increased with raising the heat treatment temperature. This could be because of improved interparticle bonding as a result of the sintering at high temperatures. With regard to the high costs related to He as propellant gas in the cold spray process, Xia-Tao Luo *et* al. [100] produced cold sprayed IN718 superalloy deposits by utilizing comparatively low-price nitrogen process gas and aid of in situ MF technique. For this purpose, micron-size 410 stainless steel powder particles (410SS, average size: 150 μm) were mixed mechanically into the IN718 powder (average size: 13.5 μm). So, the MF particles can hammer and considerably densify the deposited IN718 layer during the CS process.

The authors claimed that the possible contamination of 410 stainless steel particles was prevented during CS process, and the IN718 deposits were free of the mentioned (MF) particles. Moreover, enhancement of the DE due to the oxide scale removal was reported by Xia-Tao Luo et al. for the first time. Likewise, outstanding improvement in ultimate strength from 96 to 464 MPa was attained because of the lower porosity, and improved interparticle bonding resulted from the in situ MF effect. The specimens after a heat treatment at 1200°C for 6h had a ductile fracture manner and fractured at relatively high stress of about 1089 MPa [100], Figs. 2.38 and 2.39.

Further to the peening effect by blending large MF particles (different materials than feedstock powder) in the feedstock powders, the unbonded feedstock powder particles (mostly large particles) may also produce peening effects (tamping effects) on the already formed deposits (which is encouraged for the future researches). It was reported that [1] the strengthened peening effect of subsequent particles (on the underneath

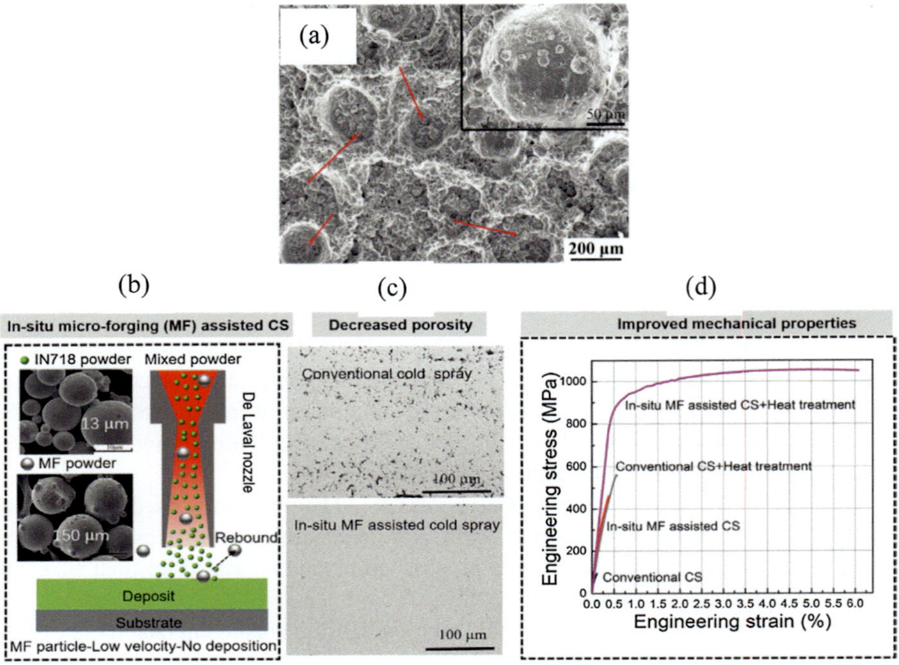

Figure 2.38 (A) Morphology of the surface of the 50 vol.% MF particles-assisted CS-processed IN718 deposit displaying the large-sized 410 stainless steel particle impact-induced craters; the inset is a closer view of an individual crater, (B) in situ microforging (MF)-assisted cold spray process, (C) porosity reduction in the in situ MF-assisted cold sprayed deposit, (D) improved mechanical properties of sprayed samples after postcold spray heat treatment (changing fracture behavior to ductile from brittle (with elongation about less than 0.5%)) [100].

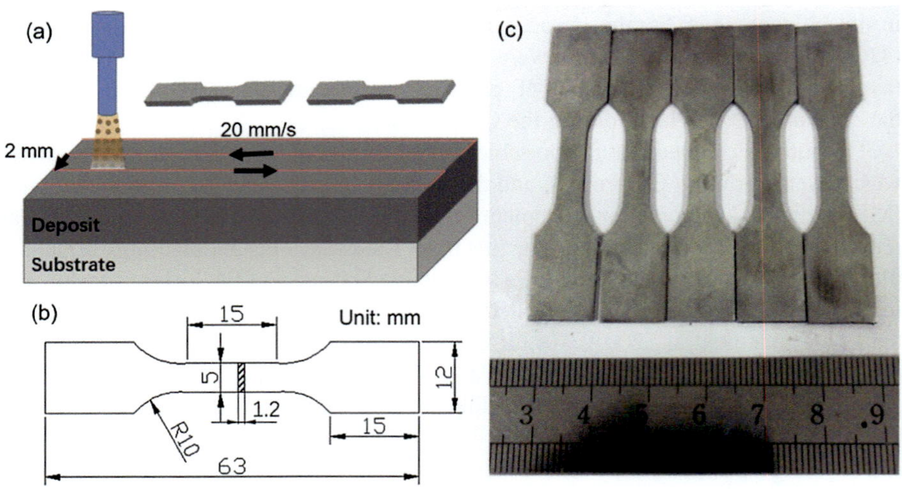

Figure 2.39 Explanation of the tensile test specimens; (A) illustration of the CS pathway and sampling direction of the tensile specimen (dog bone samples); (B) dimensions and geometry of the tensile specimen; and (C) actual image of the tensile test specimens [100].

deposited layer) with successive and strong collide energy may be the determining factor for forming of metallurgical bonding during the CS process. In fact, strengthened peening effect remarkably enhanced the quality of the CS coatings via improved metallurgical bonding, which was evidenced by declining porosity and increasing adhesion strength as well [1]. Likewise, apart from applying He as a costly process gas and high-temperature post-CS heat treatments (e.g., sintering), the other postcold spray heat treatments such as HIP should be performed and studied for the further improvement of the mechanical characteristics of CSAM Inconel-based deposits.

2.5.5 Production of cold sprayed metal matrix composite components by CSAM

For preventing the degradation of MMCs characteristics in high temperatures of processing (e.g., the powder bed-based AM techniques using electron beam or laser), the CS technology has attracted growing attention as a deposition technique in solid state [101]. The CS process has been effectively utilized in the construction of the materials that are sensitive to temperature such as Al [102,103], Cu [104], high-performance alloys such as Ti6Al4V [105], Inconel [100], and different MMCs with various reinforcements [106].

The effects of the hard particles morphology on the microstructure of CS-processed Al_2O_3/A380 composite deposits were investigated by X. Qiu et al. [107]. Apart from the microtamping effect of spherical Al_2O_3 particles, limited particles were retained in the cold sprayed deposits. Hence, low porosity, improved interface bonding and significant increase in strength (about 390 MPa) were observed for the resultant deposits in comparison with the pure A380 deposit. In contrast, irregular Al_2O_3 particles disclosed an embedding effect with repetition of fragmentation occurrences inside the deposit during CS process. Likewise, a slight increase in strength (about 330 MPa) and relatively lower porosity (compared to the pure A380 deposit) were also observed for the irregular Al_2O_3 particles containing deposits [108]. This was related to the Al_2O_3 particles fragmentation and weakly bonded interfaces between Al_2O_3 and A380 in the cold sprayed deposits [108].

Tariq et al. [109] employed the CSAM technique to deposit B_4C/Al composite coating with 6 mm thickness (by using Al+40%B_4C powder mixture, Figs. 2.40 and 2.41) on a 6061-T6 cylindrical substrate (Fig. 2.42). They showed that the as-deposited coating had

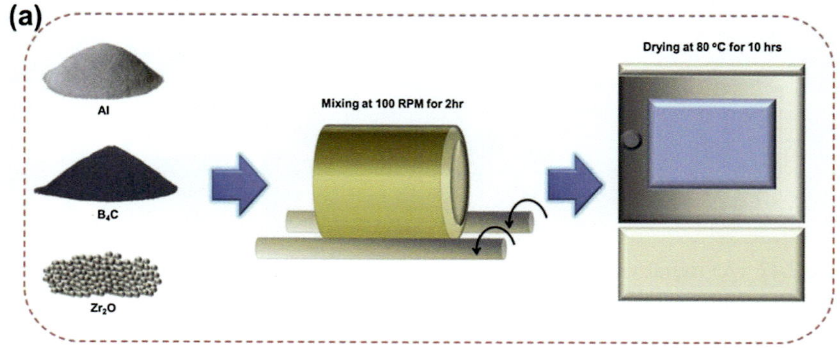

Figure 2.40 Schematic illustration of the preparation process of the feedstock MMC powder [109].

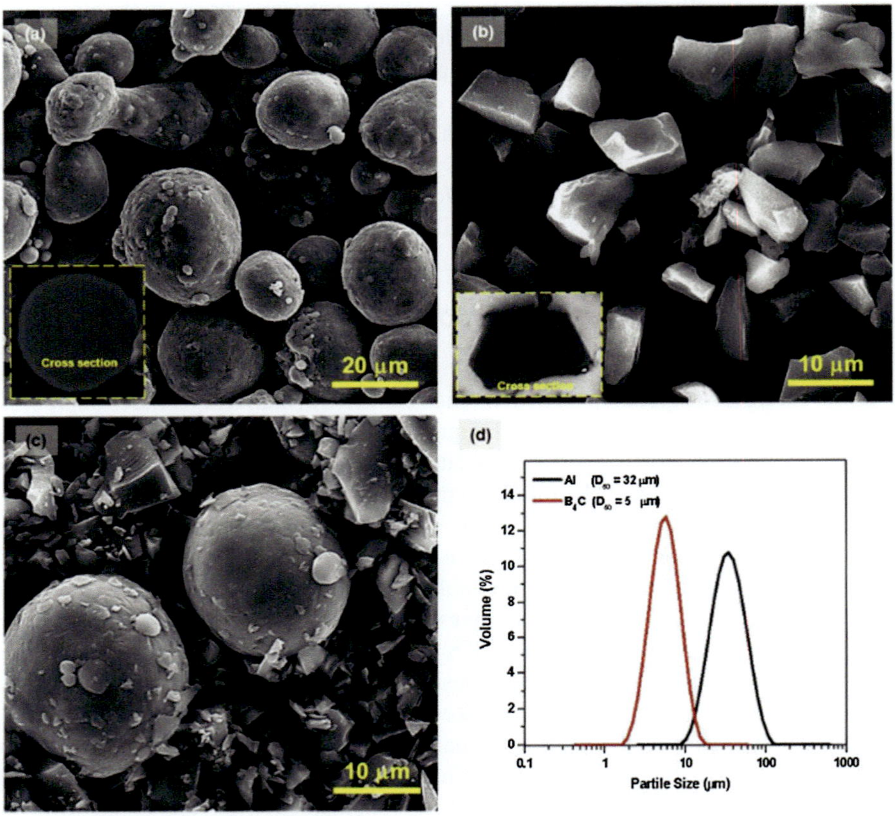

Figure 2.41 SEM micrographs displaying the morphology of powders before CS: (A) Al, (B) B4C, (C) Al+40%B4C blended powder, (D) the size distribution of Al and B4C powders [109].

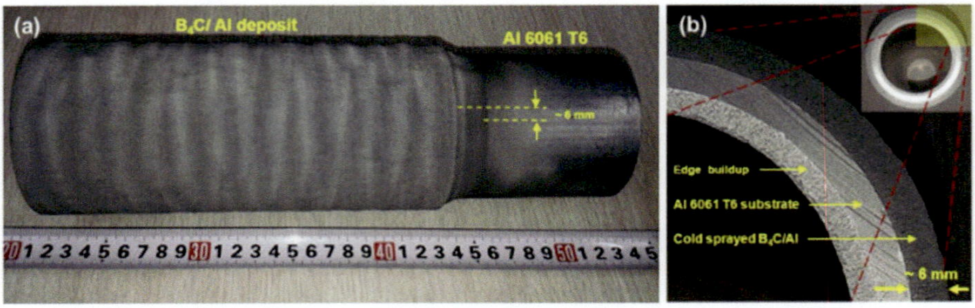

Figure 2.42 (A) lateral view and (B) cross-sectional view of thick CS-processed B4C/Al composite coating on Al6061 T6 substrate [109].

low strength and a brittle fracture manner because of the existence of a large number of defects in interparticle boundaries and severe cold working as well. Moreover, the increasing temperature of thermal treatment resulted in progressively recovering the ductility of the deposited coating. Likewise, progressive healing and bonding of intersplat boundaries (by recrystallization and recovery phenomena) considerably enhanced the strength. Superior ductility (1.4%) and strength (60 MPa) with a low porosity level of 1.9% were achieved for CS composite coating by a heat treatment at 500°C [109].

In addition to the above-mentioned work, the same research group has also employed hot rolling method (as postcold spray treatment, Fig. 2.43) on the cold sprayed Si/A380 and B4C/Al MMC deposits [110,111]. Healing the microdefects at the interparticle boundaries and considerable improvement of the mechanical properties of the MMCs were observed as a result of the usage of this method. After hot rolling, in addition to the thickness reduction of the MMC deposits, extensive refined recrystallized grains and substructured grains (as a hybrid microstructure) were seen in the composites. This observation was attributed to the continuous dynamic recrystallization (Fig. 2.43B–E). Furthermore, enhancement of diffusion activity resulted in the substantial improvement of bonding between Al/Al particles and B4C/Al interfaces during this method. These phenomena finally led to the simultaneous enhancement of the yield strength, ultimate tensile strength, and elongation of the MMC deposits. In general, hot rolled MMCs showed better tensile strength and ductility than as-sprayed MMCs [112].

A combination of hot rolling and hot compression methods (Fig. 2.44A) was also employed to improve the mechanical properties of cold sprayed B_4C/Al MMCs. Hot compression treatment at 500°C followed by a hot rolling treatment led to the further deformation of the Al matrix and closing up the interconnected pores (at the

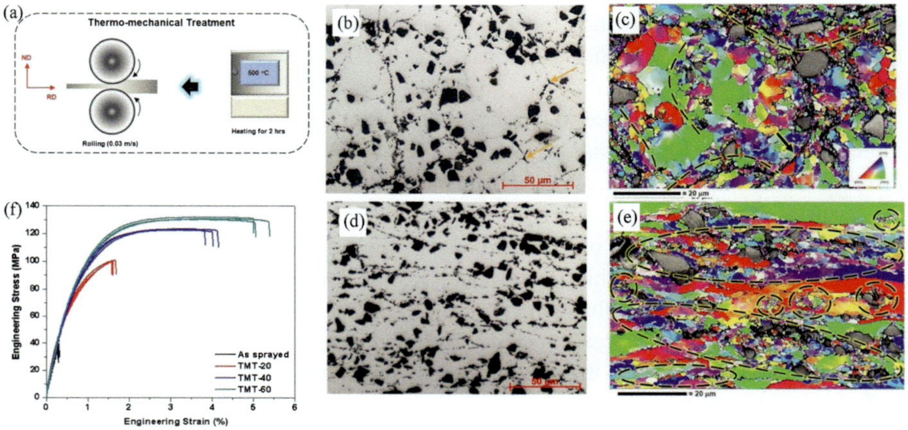

Figure 2.43 (A) Illustration of hot rolling process, (B–E) micrographs (by OM) and Euler angle maps (by EBSD) of cross-sections of (B and C) as-deposited samples and (D and E) cold sprayed samples after hot rolling, and (F) stress–strain curves of as-deposited samples and cold sprayed samples after hot rolling (adapted from Ref. [110]).

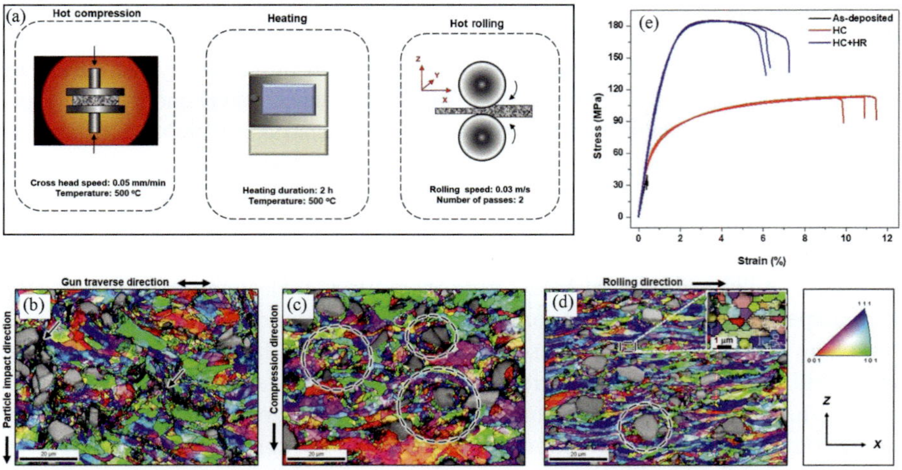

Figure 2.44 (A) Illustrations of a combination of hot rolling and hot compression methods (a hybrid method), (B−D) EBSD results of polished cross-section of (B) as-sprayed sample, (C) as-sprayed sample after hot compression and (D) as-sprayed sample after hot compression and hot rolling, and (E) stress−strain curves of Al/B 4C samples under various conditions [113].

interparticle boundaries within the MMC microstructure). Moreover, the simultaneous operation of continuous dynamic recrystallization and geometric dynamic recrystallization mechanisms (Fig. 2.44B−D) [113] caused the considerable enhancement of the bonding at the Al/Al and Al/BC interfaces, the Al grains refinement, and uniform dispersion of B_4C (as reinforcement) in the Al matrix. Thus, the ultimate tensile strength and elongation of the as-sprayed samples were concurrently enhanced to 185 MPa from 37 MPa to 6.2% from about 0.3%, respectively (Fig. 2.44E). The strength and elongation of hot rolled/hot compressed MMCs were three times and four times (respectively) higher than those of isothermally heat-treated MMC deposits [112,113].

Chen et al. [101] produced WC reinforced MS300 composite utilizing the HPCS route (Fig. 2.45). Using higher pressure of the propellant N_2 gas (5 MPa, at 900°C) resulted in the improvement of the tensile strength and increased the wear resistance of the as-fabricated specimens. This was related to the severe plastic deformation and substantial retainability of WC particles in the deposit structure (Fig. 2.46). The results disclosed that the higher temperature of the solution heat treatment (at 1000°C) could lead to the enhancement of the cohesive bonding in the deposited composite. A conspicuous wear resistance due the enhanced WC bonding and precipitation hardening was observed for the solution-aged composite [101].

In a concise summary, it was noticed that the morphology of hard powder particles (mostly as reinforcement particles in MMCs), postcold spray heat treatments, and cold spray parameters could noticeably affect the microstructure and mechanical properties of MMC deposits produced by CSAM process. However, the mechanical properties of MMC deposits have not been compared to those of produced MMCs by other available

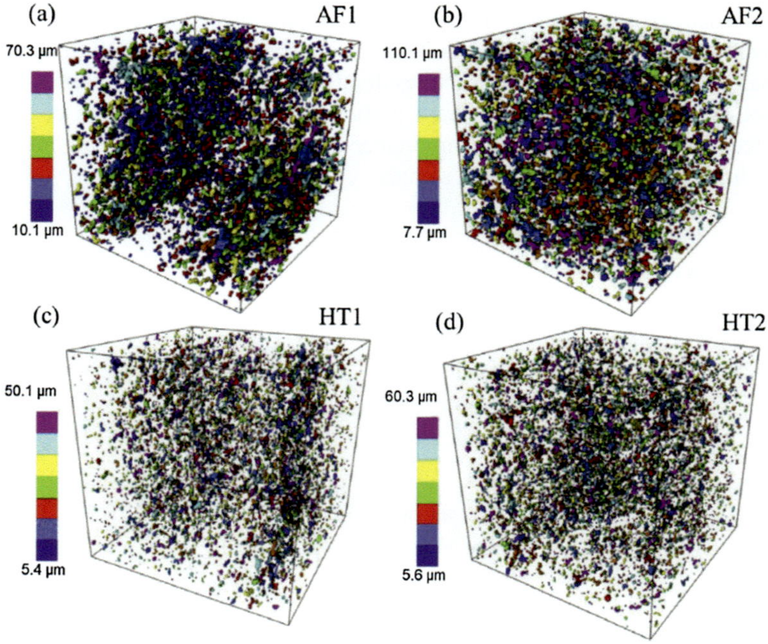

Figure 2.45 Dispersion of WC reinforcement particles in the WC reinforced MS300 composite specimens (size: 1 × 1 × 1 mm^3) manufactured using the CS process in different conditions with color showing the equivalent size. (For a detailed description of the color references in this figure legend, please refer to the web version of the research.) [101].

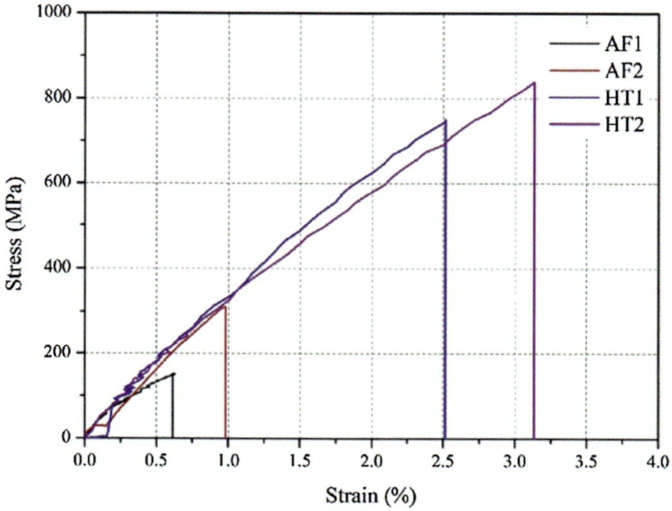

Figure 2.46 Stress−strain curves of WC reinforced MS300 composite in different AF (as-fabricated) and HT (heat-treated) conditions [101].

methods such as powder bed-based AM techniques using electron beam or laser. Moreover, it is well known that the soft matrix of MMCs has more DE than reinforcement hard particles of MMCs. So, the resultant cold sprayed deposit shows the lower retainability of the reinforcement hard particles in the microstructure compared to MMC feedstock powders. So, this importance is encouraged to be deeply investigated and modified in the future research studies.

Abbreviations

AM	Additive manufacturing
ASI	Adiabatic Shear Instability
CAD	Computer-aided design
CD	Converging-diverging
CP	Commercially pure
CS	Cold spraying
CSAM	Cold spraying additive manufacturing
CSed	Cold sprayed deposit/coating
DE	Deposition efficiency
DMD	Direct Metal Deposition
E	Elastic modulus
EBM	Electron beam melting
EBSD	Electron backscatter diffraction
EBSP	Electron backscatter patterns
FSP	Friction stir processing
HIP	Hot isostatic pressing
HIPT	Hot isostatic pressing treatment
HPCS	High-pressure cold spray
HRTEM	High-resolution transmission electron microscopy
HVOF	High velocity oxygen fuel
HWCS	Hiwatch cold spray
IN718	Inconel 718
IPF	Inverse pole figures
IQ	Image quality
KAM	Kernel average misorientation
LPCS	Low-pressure cold spray
MF	Microforging
MMC	Metal matrix composite
NC	Nanocrystalline
NDE	None destructive evaluation
OM	Optical microscopy
SAD	Selected area electron diffraction
SLM	Selective laser melting
SLS	Selective laser sintering
STL	Stereolithography
UTS	Ultimate tensile strength
WC	Tungsten carbide

WOD	Window of deposition
XCT	X-ray computed tomography
XRD	X-ray Diffraction
YS	Yield Strength

References

[1] Y. Xie, et al., Strengthened peening effect on metallurgical bonding formation in cold spray additive manufacturing, Journal of Thermal Spray Technology 28 (2019) 769–779.
[2] Y. Tao, et al., Microstructure and corrosion performance of a cold sprayed aluminum coating on AZ91D magnesium alloy, Corrosion Science 52 (2010) 3191–3197.
[3] D. MacDonald, R. Fernandez, F. Delloro, B. Jodoin, Cold spraying of armstrong process titanium powder for additive manufacturing, Journal of Thermal Spray Technology 26 (2017) 598–609.
[4] M. Vaezi, et al., Beamless metal additive manufacturing, Materials 13 (2020) 922.
[5] W. Li, et al., Solid-state additive manufacturing and repairing by cold spraying: a review, Journal of Materials Science & Technology 34 (2018) 440–457.
[6] R. Karunakaran, et al., Additive manufacturing of magnesium alloys, Bioactive Materials 5 (2020) 44–54.
[7] J.H. Martin, et al., 3D printing of high-strength aluminium alloys, Nature 549 (2017) 365.
[8] L.E. Murr, Metal fabrication by additive manufacturing using laser and electron Beam Melting, Journal of Materials Science and Technology 28 (2012) 1–14.
[9] Z. Monette, A.K. Kasar, M. Daroonparvar, P.L. Menezes, Supersonic particle deposition as an additive technology: methods, challenges, and applications, The International Journal of Advanced Manufacturing Technology 106 (2020) 2079–2099.
[10] S. Lin, et al., Cold spray additive manufacturing and repair: fundamentals and Applications, Additive Manufacturing 21 (2018) 628–650.
[11] H. Assadi, et al., Cold spraying – a materials perspective, Acta Materialia 116 (2016) 382–407.
[12] Z. Arabgol, et al., Analysis of thermal history and residual stress in cold-sprayed coatings, Journal of Thermal Spray Technology 23 (2013) 84–90.
[13] K. Kang, Mechanical property enhancement of kinetic sprayed Al coatings reinforced by multi-walled carbon nanotubes, Acta Materialia 60 (2012) 5031–5039.
[14] S. Dosta, Cold spray deposition of a WC-25Co cermet onto Al7075-T6 and carbon steel substrates, Acta Materialia 2 (2013) 643–652.
[15] W.Y. Lie, Significant influences of metal reactivity and oxide films at particle surfaces on coating microstructure in cold spraying, Applied Surface Science 253 (2007) 3557–3562.
[16] V. Champagne, et al., The unique abilities of cold spray deposition, International Materials Reviews 61 (2016) 437–455.
[17] M.E. Lynch, et al., Design and topology/shape structural optimisation for additively manufactured cold sprayed components: this paper presents an additive manufactured cold spray component which is shape optimised to achieve 60% reduction in stress and 20% reduction in weight, Virtual Physical Prototyping 8 (2013) 213–231.
[18] A. Sova, S. Grigoriev, A. Okunkova, I. Smurov, Potential of cold gas dynamic spray as additive manufacturing technology, The International Journal of Advanced Manufacturing Technology 69 (2013) 2269–2278.

[19] C.A. Widener, et al., Structural repair using cold spray technology for enhanced sustainability of high value assets, Procedia Manufacturing 21 (2018) 361–368.
[20] S. Bagherifard, Cold spray deposition for additive manufacturing of freeform structural components compared to selective laser melting, Materials Science & Engineering A 721 (2018) 339–350.
[21] I.M. Zulkifli, M.M. Yajid, M. Idris, M.B. Uday, M. Daroonparvar, A. Emadzadeh, A. Arshad, Microstructural evaluation and thermal oxidation behaviors of YSZ/NiCoCrAlYTa coatings deposited by different thermal techniques, Ceramics International 46 (2020) 22438–22451.
[22] M. Daroonparvar, M.A.M. Yajid, N.M. Yusof, H.R. Bakhsheshi-Rad, M. Sakhawat Hussain, E. Hamzah, Evaluation of normal and nanolayer composite thermal barrier coatings in fused vanadate-sulfate salts at 1000 C, Advances in Materials Science and Engineering (2013), 790318.
[23] M. Daroonparvar, C.M. Kay, J. Karthikeyan, Modified bond coatings improve service life of plasma sprayed thermal barrier coatings and protective performance of overlay coatings, Advanced Materials & Processes 176 (2018) 40–43.
[24] M.R. Rokni, Review of relationship between particle deformation, coating microstructure, and properties in high-pressure cold spray, Journal of Thermal Spray Technology 26 (2017) 1308–1355.
[25] M.F. Smith, Comparing cold spray with thermal spray coating technologies, in: V.K. Champagne (Ed.), The Cold Spray Materials Deposition Process: Fundamentals and Applications, first ed., Woodhead, Cambridge, 2007.
[26] J. Villafuerte, Modern Cold Spray: Materials, Process, and Applications, Springer, Berlin, 2015.
[27] N.M. Chavan, B. Kiran, A. Jyothirmayi, P.S. Phani, G. Sundararajan, The corrosion behavior of cold sprayed zinc coatings on mild steel substrate, Journal of Thermal Spray Technology 22 (2013) 463–470.
[28] C.M. Kay, J. Karthikeyan, High Pressure Cold Spray: Principles and Applications, ASM International, 2016. ISBN: 978-1-62708-096-5.
[29] M. Daroonparvar, M.A. Mat Yajid, C.M. Kay, H. Bakhsheshi-Rad, R.K. Gupta, N.M. Yusof, H. Ghandvar, A. Arshad, I.S. Mohd Zulkifli, Effects of Al_2O_3 diffusion barrier layer (including Y-containing small oxide precipitates) and nanostructured YSZ top coat on the oxidation behavior of HVOF NiCoCrAlTaY/APS YSZ coatings at 1100°C, Corrosion Science 144 (2018) 13–34.
[30] M. Daroonparvar, M.U. Farooq Khan, Y. Saadeh, C.M. Kay, R.K. Gupta, A.K. Kasar, P. Kumar, M. Misra, P.L. Menezes, H.R. Bakhsheshi-Rad, Enhanced corrosion resistance and surface bioactivity of AZ31B Mg alloy by high pressure cold sprayed monolayer Ti and bilayer Ta/Ti coatings in simulated body fluid, Materials Chemistry and Physics 256 (2020) 123627.
[31] M. Daroonparvar, M.U. Farooq Khan, Y. Saadeh, C.M. Kay, A.K. Kasar, P. Kumar, M. Misra, P.L. Menezes, P.R. Kalvala, R.K. Gupta, H.R. Bakhsheshi-Rad, L. Esteves, Modification of surface hardness, wear resistance and corrosion resistance of cold spray Al coated AZ31B Mg alloy using cold spray double layered Ta/Ti coating in 3.5 wt%-NaCl solution, Corrosion Science (2020) 109029.
[32] M. Daroonparvar, M.A. Mat Yajid, N.M. Yusof, H.R. Bakhsheshi-Rad, E. Hamzah, Microstructural characterisation of air plasma sprayed nanostructure ceramic coatings on Mg−1% Ca alloys (bonded by NiCoCrAlYTa alloy), Ceramics International 42 (2016) 357–371.

[33] M. Daroonparvar, M.A.M. Yajid, N.M. Yusof, H.R. Bakhsheshi-Rad, Fabrication and properties of triplex NiCrAlY/nano $Al_2O_3 \cdot$ 13% TiO_2/nano TiO_2 coatings on a magnesium alloy by atmospheric plasma spraying method, Alloys and Compounds 645 (2015) 450–466.
[34] H.R. Bakhsheshi-Rad, E. Hamzah, A.F. Ismail, M. Daroonparvar, M.A.M. Yajid, M. Medraj, Preparation and characterization of NiCrAlY/nano-YSZ/PCL composite coatings obtained by combination of atmospheric plasma spraying and dip coating on Mg–Ca alloy, Journal of Alloys and Compounds 658 (2016) 440–452.
[35] M. Daroonparvar, M.A.M. Yajid, N.M. Yusof, H.R. Bakhsheshi-Rad, E. Hamzah, T. Mardanikivi, Deposition of duplex MAO layer/nanostructured titanium dioxide composite coatings on Mg–1% Ca alloy using a combined technique of air plasma spraying and micro arc oxidation, Journal of Alloys and Compounds 649 (2015) 591–605.
[36] M. Daroonparvar, M. Azizi Mat Yajid, R. Kumar Gupta, N. Mohd Yusof, H.R. Bakhsheshi-Rad, H. Ghandvar, E. Ghasemi, Antibacterial activities and corrosion behavior of novel PEO/nanostructured ZrO_2 coating on Mg alloy, Transactions of Nonferrous Metals Society of China 28 (2018) 1571–1581.
[37] H. Assadi, F. Gärtner, T. Stoltenhoff, H. Kreye, Bonding mechanism in cold gas spraying, Acta Materriala 51 (2003) 4379–4394.
[38] T. Schmidt, F. Gärtner, H. Assadi, H. Kreye, Development of a generalized parameter window for cold spray deposition, Acta Materriala 54 (2006) 729–742.
[39] S.M. Zulkifli, M.A.M. Yajid, M.H. Idris, M. Daroonparvar, H. Hamdan, TGO Formation with NiCoCrAlYTa bond coat deposition using APS and HVOF method, Advanced Materials Research 1125 (2015) 18–22.
[40] M. Daroonparvar, M. Sakhawat Hussain, M.A. Mat Yajid, The role of formation of continues thermally grown oxide layer on the nanostructured NiCrAlY bond coat during thermal exposure in air, Applied Surface Science 261 (2012) 287–297.
[41] M. Daroonparvar, M.A.M. Yajid, N.M. Yusof, M.S. Hussain, H.R. Bakhsheshi-Rad, Formation of a dense and continuous Al_2O_3 layer in nano thermal barrier coating systems for the suppression of spinel growth on the Al_2O_3 oxide scale during oxidation, Journal of Alloys and Compounds 571 (2013) 205–220.
[42] M. Daroonparvar, M.A.M. Yajid, N.M. Yusof, S. Farahany, M.S. Hussain, H.R. Bakhsheshi-Rad, Z. Valefi, A. Abdolahi, Improvement of thermally grown oxide layer in thermal barrier coating systems with nano alumina as third layer, Transactions of Nonferrous Metals Society of China 23 (2013) 1322–1333.
[43] M. Daroonparvar, M.A.M. Yajid, N.M. Yusof, H.R. Bakhsheshi-Rad, E. Hamzah, M. Nazoktabar, Investigation of three steps of hot corrosion process in Y_2O_3 stabilized ZrO_2 coatings including nano zones, Journal of Rare Earths 32 (2014) 989–1002.
[44] M. Daroonparvar, M.A.M. Yajid, N.M. Yusof, H.R. Bakhsheshi-Rad, Z. Valefi, E. Hamzah, Effect of Y_2O_3 stabilized ZrO_2 coating with tri-model structure on bi-layered thermally grown oxide evolution in nano thermal barrier coating systems at elevated temperatures, Journal of Rare Earths 32 (2014) 57–77.
[45] M. Daroonparvar, Effects of bond coat and top coat (including nano zones) structures on morphology and type of formed transient stage oxides at pre-heat treated nano NiCrAlY/nano ZrO_2-8% Y_2O_3 …, Journal of Rare Earths 33 (2015) 983–994.
[46] R.C. Dykhuizen, et al., Gas dynamic principles of cold spray, Journal of Thermal Spray Technology 7 (1998) 205–212.
[47] D.L. Gilmore, et al., Particle velocity and degradation efficiency in the cold spray process, Journal of Thermal Spray Technology 8 (1999) 576–582.

[48] M. Grujicic, C.L. Zhao, C. Tong, W.S. Derosset, D. Helfritch, Materials Science and Engineering, A-Structural Materials Properties Microstructure and Processing 368 (2004) 222–230.
[49] V.K. Champagne (Ed.), The Cold Spray Materials Deposition Process, Woodhead Publishing Ltd., CRC, 2007.
[50] J.G. Legoux, E. Irissou, C. Moreau, Effect of substrate temperature on the formation mechanism of cold-sprayed aluminum, zinc and tin coatings, Journal of Thermal Spray Technology 16 (2007) 619–626.
[51] T. Schmidt, et al., From particle acceleration to impact and bonding in cold spraying, Journal of Thermal Spray Technology 18 (2009) 794–808.
[52] T. Stoltenhoff, et al., An analysis of the cold sprayed process its coatings, Journal of Thermal Spray Technology 11 (2002) 542–550.
[53] T. Schmidt, et al., New developments in cold spray based on higher gas and particle temperatures, Journal of Thermal Spray Technology 15 (2006) 488–494.
[54] S. Shin, S. Yoon, Y. Kim, C. Lee, Effect of particle parameters on the deposition characteristics of a hard/soft-particles composite in kinetic spraying, Surface and Coatings Technology 201 (2006) 3457–3461.
[55] P. Sudharshan Phani, D. Srinivasa Rao, S. Joshl, G. Sundararajan, Effect of process parameters and heat treatments on properties of cold sprayed copper coatings, Journal of Thermal Spray Technology 16 (2007) 425–434.
[56] T. Stoltenhoff, H. Kreye, H. Richter, An analysis of the cold spray process and its coatings, Journal of Thermal Spray Technology 11 (4) (2001) 542–550.
[57] M. Grujicic, C.L. Zhao, W.S. DeRosset, D. Helfritch, Adiabatic shear instability based mechanism for particle/substrate bonding in the cold-gas dynamic-spray process, Materials & Design 25 (2004) 681–688.
[58] M. Grujicic, J.R. Saylor, D.E. Beasley, W.S. DeRosset, D. Helfritch, Computational analysis of the interfacial bonding between feed-powder particles and the substrate in the cold-gas dynamic-spray process, Applied Surface Science 219 (2003) 211–227.
[59] G. R Cowan, O.R. Bergmann, A.H. Holtzman, Mechanism of bond zone wave formation in explosion –clad metals, Metallurgical Transactions 2 (1971) 3145–3155.
[60] W.Y. Li, C.J. Li, H. Liao, Significant influence of particle surface oxidation on deposition efficiency, interface microstructure and adhesive strength of cold-sprayed copper coatings, Applied Surface Science 256 (2010) 4953–4958.
[61] X. Wang, F. Feng, M.A. Klecka, M.D. Mordasky, J.K. Garofano, T. El-Wardany, A. Nardi, V.K. Champagne, Characterization and modeling of the bonding process in cold spray additive manufacturing, Additive Manufacturing 8 (2015) 149–162.
[62] J.A. Hines, K.S. Vecchio, S. Ahzi, Metallurgical and Materials Transactions A-Physical Metallurgy and Materials Science 29 (1998) 191–203.
[63] A. Mishra, et al., High-strain-rate response of ultra-fine-grained copper, Acta Materialia 56 (2008) 2770–2783.
[64] J.F.C. Lins, et al., Materials Science and Engineering, A-Structural Materials Properties Microstructure and Processing 457 (2007) 205–218.
[65] M.A. Meyers, et al., Microstructural evolution in adiabatic shear localization in stainless steel, Acta Materialia 51 (2003) 1307–1325.
[66] Y. Kang Wei, Y. Juan Li, Y. Zhang, et al., Corrosion resistant nickel coating with strong adhesion on AZ31B magnesium alloy prepared by an in-situ shot-peening-assisted cold spray, Corrosion Science 138 (2018) 105–115.

[67] Y. Zou, et al., Dynamic recrystallization in the particle/particle interfacial region of cold-sprayed nickel coating: electron backscatter diffraction characterization, Scripta Materialia 61 (2009) 899–902.
[68] Y. Zou, D. Goldbaum, J.A. Szpunar, S. Yue, Microstructure and nanohardness of cold-sprayed coatings: electron backscattered diffraction and nanoindentation studies, Scripta Materialia 62 (2010) 395–398.
[69] J. Mazumder, D. Dutta, N. Kikuchi, A. Ghosh, Closed loop direct metal deposition: art to part, Optic Lasers Engineering 34 (2000) 397–414.
[70] H. Attar, et al., Effect of powder particle shape on the properties of in situ Ti–TiB composite materials produced by selective laser melting, Journal of Materials Science and Technology 31 (2015) 1001–1005.
[71] R. Casati, et al., Microstructure and fracture behavior of 316L austenitic stainless Steel produced by selective laser melting, Journal of Materials Science and Technology 32 (2016) 738–744.
[72] ASTM F2792-12a, 2015, pp. 15–17.
[73] J. Pattison, et al., Cold gas dynamic manufacturing: a non-thermal approach to freeform fabrication, International Journal of Machine Tools and Manufacture 47 (2007) 627–634.
[74] F.H.(Sam) Froes, Titanium powder metallurgy: a review – part 2, Advanced Materials & Processes 170 (2012) 26–29.
[75] S.H. Zahiri, et al., Elimination of porosity in directly fabricated titanium via cold gas dynamic spraying, Journal of Materials Processing Technology 209 (2009) 922–929.
[76] W. Li, C. Cao, S. Yin, Solid-state cold spraying of Ti and its alloys: a literature review, Progress in Materials Science 110 (2020) 100633.
[77] M.Z. Jahedi, S.H. Zahiri, S. Gulizia, B. Tiganis, C. Tang, D. Fraser, Direct manufacturing of titanium parts by cold spray, Materials Science Forum 618 (2009) 505–508.
[78] S.H. Zahiri, et al., Recrystallization of cold spray-fabricated CP titanium structures, Journal of Thermal Spray Technology 18 (2009) 16–22.
[79] C. Chena, et al., Effect of hot isostatic pressing (HIP) on microstructure and mechanical properties of Ti_6Al_4V alloy fabricated by cold spray additive manufacturing, Additive Manufacturing 27 (2019) 595–605.
[80] F. Khodabakhshi, et al., Surface modification of a cold gas dynamic spray-deposited titanium coating on aluminum alloy by using friction-stir processing, Journal of Thermal Spray Technology 28 (2019) 1185–1195.
[81] T. Hussain, Cold spraying of titanium: a review of bonding mechanisms, microstructure and properties, Key Engineering Materials 533 (2013) 53–90.
[82] W. Wong, et al., Effect of particle morphology and size distribution on cold-sprayed pure titanium coatings, Journal of Thermal Spray Technology 22 (7) (2013) 1140–1153.
[83] A. Choudhuri, et al., Bio-ceramic composite coatings by cold spray technology, International Thermal Spray Conference (2009) 391–396.
[84] T. WY, et al., Effects of traverse scanning speed of spray nozzle on the microstructure and mechanical properties of cold sprayed Ti6Al4V coatings, Journal of Thermal Spray Technology 26 (4) (2017) 1–14.
[85] Y.K. Wei, Solid-state additive manufacturing high performance aluminum alloy 6061 enabled by an in-situ micro-forging assisted cold spray, Materials Science & Engineering A 776 (2020) 139024.
[86] B. Aldwell, E. Kelly, R. Wall, A. Amaldi, G.E. O'Donnell, R. Lupoi, Machinability of Al 6061 deposited with cold spray additive manufacturing, Journal of Thermal Spray Technology 26 (7) (2017) 1573–1584.

[87] J. Murray, M. Zuccoli, T. Hussain, Heat treatment of cold-sprayed C355 Al for repair: microstructure and mechanical properties, Journal of Thermal Spray Technology 27 (1−2) (2018) 159−168.
[88] H. Koivuluoto, et al., Cold-sprayed Al6061 coatings: online spray monitoring and influence of process parameters on coating properties, Coatings 10 (2020) 1−16.
[89] C. Chen, et al., Microstructure evolution and mechanical properties of maraging steel 300 fabricated by cold spraying, Materials Science & Engineering A743 (2019) 482−493.
[90] S. Yina, Annealing strategies for enhancing mechanical properties of additively manufactured 316L stainless steel deposited by cold spray, Surface and Coatings Technology 370 (2019) 353−361.
[91] G. Meng, et al., Vacuum heat treatment mechanisms promoting the adhesion strength of thermally sprayed metallic coatings, Surface and Coatings Technology 344 (2018) 102−110.
[92] B. Zhang, et al., Dependence of scale thickness on the breaking behavior of the initial oxide on plasma spray bond coat surface during vacuum, Applied Surface Science 397 (2017) 125−132.
[93] X. Meng, et al., Influence of gas temperature on microstructure and properties of cold spray 304SS coating, Journal of Materials Processing Technology 27 (2011) 809−815.
[94] X. Meng, et al., Influence of annealing treatment on the microstructure and mechanical performance of cold sprayed 304 stainless steel coating, Applied Surface Science 258 (2011) 700−704.
[95] S. Bagherifard, Cold spray deposition of freestanding Inconel samples and comparative analysis with selective laser melting, Journal of Thermal Spray Technology 27 (1−2) (2018) 159−168.
[96] W. Sun, Improving microstructural and mechanical characteristics of cold sprayed Inconel 718 deposits via local induction heat treatment, Journal of Alloys and Compounds 797 (2019) 1268−1279.
[97] W. Ma, et al., Microstructural and mechanical properties of high-performance Inconel 718 alloy by cold spraying, Journal of Alloys and Compounds 792 (2019) 456−467.
[98] R. Huang, H. Fukanum, Study of the properties of cold sprayed Inconel718 deposits, in: International Thermal Spray Conference, 2016. Shanghai, 2016, pp. 299−304.
[99] W. Wong, et al., Cold spray forming Inconel 718, in: International Thermal Spray Conference (ITSC), 2012, pp. 243−248.
[100] X.T. Luo, Deposition behavior, microstructure and mechanical properties of an in-situ micro-forging assisted cold spray enabled additively manufactured Inconel 718 alloy, Materials and Design 155 (2018) 384−395.
[101] C. Chen, Cold sprayed WC reinforced maraging steel 300 composites: microstructure characterization and mechanical properties, Journal of Alloys and Compounds 785 (2019) 499−511.
[102] C. Chen, S. Gojon, Y. Xie, S. Yin, C. Verdy, Z. Ren, H. Liao, S. Deng, A novel spiral trajectory for damage component recovery with cold spray, Surface and Coatings Technology 309 (2017) 719−728.
[103] W.Y. Li, R.R. Jiang, C.J. Huang, Z.H. Zhang, Y. Feng, Effect of cold sprayed Al coating on mechanical property and corrosion behavior of friction stir welded AA2024-T351 joint, Materials and Design 65 (2015) 757−761.
[104] S. Yin, et al., Deposition behavior of thermally softened copper particles in cold spraying, Acta Materrialia 61 (2013) 5105−5118.
[105] P. Vo, et al., Mechanical and microstructural characterization of cold-sprayed Ti-6Al-4V after heat treatment, Journal of Thermal Spray Technology 22 (2013) 954−964.

[106] S. Yin, et al., Advanced diamond-reinforced metal matrix composites via cold spray: properties and deposition mechanism, Composites B 113 (2017) 44–54.
[107] X. Qiu, et al., Effects of dissimilar alumina particulates on microstructure and properties of cold-sprayed alumina/a380 composite coatings, Acta Metallurgica Sinica -English Letters 32 (2019) 1449–1458.
[108] X. Qiu, et al., A hybrid approach to improve microstructure and mechanical properties of cold spray additively manufactured A380 aluminum composites, Materials Science & Engineering A 772 (2020) 138828.
[109] N.H. Tariq, et al., Cold spray additive manufacturing: a viable strategy to fabricate thick B_4C/Al composite coatings for neutron shielding applications, Surface & Coatings Technology 339 (2018) 224–236.
[110] N.H. Tariq, et al., Thermo-mechanical post-treatment: a strategic approach to improve microstructure and mechanical properties of cold spray additively manufactured composites, Materials and Design 156 (2018) 287–299.89.
[111] X. Qiu, et al., In-situ Sip/A380 alloy nano/micro composite formation through cold spray additive manufacturing and subsequent hot rolling treatment: microstructure and mechanical properties, Journal of Alloys and Compounds 780 (2019) 597–606.
[112] W. Sun, et al., Post-process treatments on supersonic cold sprayed coatings: a review, Coatings 123 (2020) 1–35.
[113] N.U. Tariq, et al., Achieving strength-ductility synergy in cold spray additively manufactured Al/B_4C composites through a hybrid post-deposition treatment, Journal of Materials Science and Technology 35 (2019) 1053–1063.

Fundamentals of stereolithography: techniques, properties, and applications

3

Amanendra K. Kushwaha [1], *Md Hafizur Rahman* [1], *David Hart* [1], *Branden Hughes* [1], *Diego Armando Saldana* [1], *Carson Zollars* [1], *Dipen Kumar Rajak* [2] *and Pradeep L. Menezes* [1]
[1]Department of Mechanical Engineering, University of Nevada, Reno, NV, United States; [2]Department of Mechanical Engineering, Sandip Institute of Technology & Research Centre, Nashik, Maharashtra, India

3.1 Introduction

In recent years, additive manufacturing has rapidly evolved to establish its place in the manufacturing sector. It has significantly improved its capability to rapidly prototype functional machine components for planning, design, research, and testing purposes. 3D printing allows users to design and convert a computer-aided design (CAD) model into a usable prototype within a few hours, through layer by layer printing of the material on a built platform [1]. Stereolithography (SLA) is one of the most widely used 3D printing technologies in the industrial sector. This technique of printing employs a precise laser to cure a liquid polymer, layer by layer, onto an inverted built platform. This implies that the object is printed upside-down, and the platform is moved up by single-layer thickness after each subsequent curing of the plastic monomers by the laser. This process of curing polymer using a laser or ultraviolet (UV) light is known as photopolymerization [2]. After the completion of the printing process, it is removed from the resin vat and placed inside a UV oven to complete the curing process. This step is particularly important if the manufactured part needs to have high mechanical and thermal properties [3]. Depending on the intended application, the produced part may also be sanded down afterward if needed. While this process is relatively expensive for resin and the printer itself, the parts produced using an SLA process are much stronger, smoother, and more precise when compared to a typical filament 3D printer.

SLA printing is an innovative process, and since its creation in the mid-1980s, the uses and scientific applications have increased tremendously [4]. SLA printers are versatile at their capability to print with a wide variety of resins, including standard (detailed and smooth, but brittle), tough resin (much stronger and resistant to heavy loads), heat-resistant resin (temperature resistant but brittle), rubber-like resin (high flexibility but degrades over time), to name a few. This chapter discusses the tribological (friction and wear) and corrosion properties of SLA-manufactured products and how the different types of resins may change the effects of these properties. This is

Tribology of Additively Manufactured Materials. https://doi.org/10.1016/B978-0-12-821328-5.00003-2

particularly important with industries that use SLA-manufactured products to replace traditional materials. The ease of manufacturing complex geometries with high surface precision and good mechanical properties gives SLA-manufactured parts an edge over other materials and production methods [5].

SLA printing is typically used for intricate prototyping and modeling in various industries such as medical, automotive, aerospace, and manufacturing. Prototyping is a very useful tool for any type of product design or development. This allows the opportunity to test a product or model at a lower cost or potentially on a lower scale before moving into the manufacturing or production process. The medical benefits of SLA printing include accurate diagnoses, preoperative planning, and determining the best path of treatment [6]. SLA-printed parts are used in other biological applications such as a print part having indirect or temporary contact with a living body to provide structure and support. Development of SLA to print using biodegradable materials for biomedical uses is currently under research [7]. Although still new, SLA 3D printing technique has a very promising future.

3.2 Stereolithography

3.2.1 Upside-down SLA

In general, the upside-down (inverted) SLA is the most common technique employed by SLA printers. Unlike in traditional 3D printing technique such as fused deposition modeling (FDM), which uses a roll of filament wire that is melted and dispensed through a nozzle onto a built surface, inverted SLA has a building platform placed upside-down into a vat of liquid resin. At the beginning of SLA, the printing process was originally always right-side up, and a scraper would dispense fresh resin onto the build surface [8]. Although minimal in numbers, this type of SLA is still in use at very large-scale manufacturing industries. The problem with this type of printing is that it requires a high amount of initial capital investment, high maintenance to keep running, and a high volume of resin in the tank at all times. This led to the creation of inverted SLA, which can be employed for economical printing of small parts leading to a more affordable and easier-to-use desktop printer approach.

The first step in creating a 3D-printed object is to create a 3D CAD model of the part and slice it into layers using a slicing software or printer's inbuilt slicing software. For the next step, the printing polymeric resin is added into a small, clear tank inside the printing area. This resin acts as the "ink" for the printer that will be hardened layer by layer to take the form of the object being printed. Fig. 3.1 shows the basic schematic diagram of different parts of an inverted SLA printer and the printed part in the resin tank. During the printing process, the build platform moves down into the vat of resin leaving just enough space between the platform and the bottom of the resin tank to form one layer. Two mirror galvanometers then direct a powerful UV laser to a precise coordinate on the build surface as per the CAD model. The laser cures the resin and bonds it to either the build platform or the previous layer of cured resin. The build platform then moves up by one layer thickness to allow for the creation of another layer.

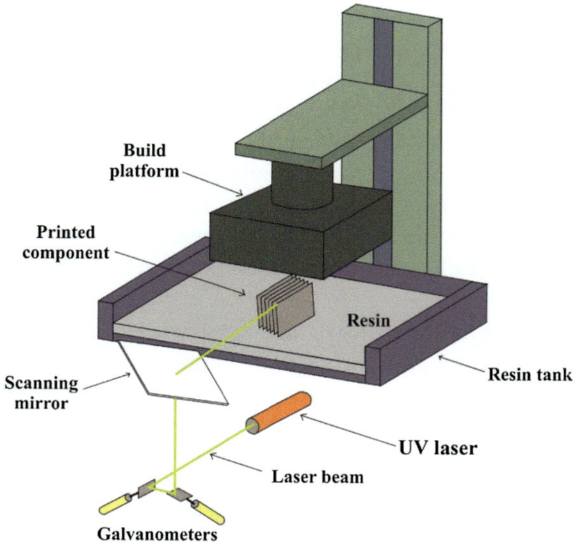

Figure 3.1 Schematic diagram showing different parts of an inverted SLA printer.

This process is repeated until all layers of the part are cured in the desired shape. Since the entire structure sits in a liquid resin, an intricate combination of solid base and supports is also printed alongside the part (determined by the printing software used) to provide adequate structural support and avoid any deformities or damage to the printed part. After the completion of the printing process, a scraper is used to pop out the completed part off the build platform. The support structures are generally meshed and can be easily broken to set the part free. Further, small pieces of support structure still attached to the part can be removed using a small pair of snips, and eventually, the surface can be smoothened using sandpaper for the final surface finish [9].

Inverted SLA has several benefits over the conventional FDM process that includes precise, smooth, and stronger 3D-printed components. These surface features provide enhanced tribological properties to SLA-printed objects [10]. Inverted SLA also offers many benefits over regular SLA in terms of compact size and cost. This type of printing process requires a smaller vat of resin at all times in the tank, such that the part to be printed can significantly exceed the volume of the vat. Due to their small size and compact design, they are much affordable to buy and maintain in the workplace. These features enable the usage of these machines as desktop printers to print components in bulk [11].

The potential downsides of inverted SLA compared to regular SLA are often attributed to the peel forces due to the gravity, which pulls the part away from the build platform, causing intersurface deformities. To avoid this failure, a carefully planned solid base and supports made of resin are used to support the printed structure [12]. This gravitation force also limits the size of the parts that can be printed because larger/heavier parts will start to peel away from the build plate. The prints are also not quite as precise as using a right-side-up plate.

SLA printers can cater to several types of resins that allow for a much larger range of material properties such as high melting point, flexibility, etc., that can be obtained from the printed parts depending on their intended application [13]. There are primarily two types of resins used in SLA: (1) The standard SLA resins and (2) the engineering SLA resins. The standard SLA resins are the most commonly used resin types in SLA printers. These resins come in many colors (including clear) and can print intricate designs with a smooth surface texture. These resins are relatively cheaper and are perfect for small prototype parts. However, the potential downsides of these resins are that they are very brittle, melt at relatively low heats, and have low impact strength [14].

The engineering SLA resins are often more expensive than the standard SLA resins; thus, the printed end products are generally used for specialized purposes. The parts printed with these resins are often used in production areas. The most common resin in this category is "tough resin" which is much stronger than the standard resin. It can withstand much higher stress and strain (tensile strength of about 55.7 MPa) and is more suitable for cyclic loading. However, tough resins are not usually recommended to print parts with small, thin walls (>1 mm). "Durable resin" is yet another type of resin that is flexible and not as strong as the tough resin but is perfect for wear-resistant, low-friction applications. The "heat-resistant resin" has similar properties to other resins; however, it is mostly used in high-temperature applications. The heat-resistant resins can withstand temperatures up to 200°C−300°C without undergoing any plastic deformation [15]. The "rubber-like resin" is highly flexible and is well suited for parts that need to be compressed or bent. The drawback of this type of resin is that it does not have all the natural properties of rubber and will degrade over time when exposed to UV light (sunlight). The "ceramic-filled resins" have relatively high stiffness, moderate heat resistance, and are great for very fine-featured parts. However, they still have a low impact strength and are brittle materials [16]. With this wide range of resins, there are limitless possibilities to the types and the applications of parts that can be printed using an SLA 3D printer.

3.2.2 SLA workflows

The typical SLA workflow is an intricate process that starts with a design, which is then interpreted by the printing software; this interpretation is sent to the machine to be printed and finally undergoes postprocess treatment. The first step of SLA printing is the same for every type of 3D printing, that is, to create a design. The design can be created practically using any 3D CAD software and exported in the Standard Tessellation Language (STL) format [17]. Conversion to this format essentially means that the CAD design is converted into a collection of triangular-shaped meshes, which represent the surface geometry of part [18]. The number and size of triangles can be adjusted to change the resolution of the print. More triangles result in a larger file but a more precise print; however, a finer mesh is not always beneficial. Generally, the printers have a limited level of precision that they can print with. The STL file is then opened into a slicing software to prepare the model for printing. Typically, this interpretation involves breaking the design horizontally into thin layers, checking for obstructions that would impede the printing process, and creating a print path. This information is then uploaded to the printer either wirelessly or hard connected through a cable. The design, print path, and inclusions of proper support structures are extremely important for a successful print [8,19]. For example, a design requiring a high level of precision usually requires supports to be printed first, which then supports the structure while it is being printed. These structures can later be removed after the print part has completely cured. SLA-modeled designs are intentionally kept hollow to save material and expedite the printing time. This is common since SLA printing produces such a sturdy print, that making the whole design solid does not increase the strength substantially enough to make it worth the time or extra material.

The second stage of an SLA process is the printing itself. As mentioned in the prior section, an SLA apparatus prints the imported design using a laser to solidify liquid resin contained in a vat below. This form of additive manufacturing builds upon the previously solidified layer. This is why the preprint software divides the entire part into thin layers, so the laser can trace the current layer's shape on the liquid surface, while the prior layers reside submerged below. After the UV laser completes each layer, a recoating mechanism swipes across the print surface, leaving a fresh layer of resin on the recently solidified layer. Also, the elevator, which the print is built upon and is submerged in the resin vat, slightly lowers the portion of the print already completed to make room for additional layers. This printing method allows a wide range of products to be made since it can utilize resins with different properties (heat resistant, flexible, transparent, etc.) and deliver a precise print regardless. After confirming that the first layer of your print is accurate, the printer can be left unattended in a well-ventilated area to complete the print [20,21].

The last stage of the SLA workflow process is posttreatment. After the printing has concluded, the elevator raises the submerged print from the resin vat. Shortly after that, the part must be rinsed using isopropyl alcohol to remove any excess uncured resin still left on the part. The rinse is a time-sensitive step and must be performed shortly after the completion of the print. This must be carried out to prevent the leftover resin from drying up on the surface or crevices of the printed part. If this leftover resin is allowed to dry up on the surface or crevices of the printed part, it can ruin the desired finish of the print or potentially compromise the precision needed for a printed component to fit into a larger design. Some designs may require additional time to cure after the isopropyl rinse because SLA printing yields an isotropic product, which may take a slightly longer time to set.

Unlike traditional 3D printing, the laser in an SLA compacts the resin particles so tightly that the result forms covalent bonds, which create isotropy (uniform material in all directions) and an incredibly strong print [22]. The isotropic nature of an SLA print also means that there are no tiny gaps between print layers; this is important for creating watertight parts and is crucial in a variety of medical field applications [23]. Many manufacturers can expedite the curing process by using SLA printers for mass production by utilizing a UV curing oven. This allows faster production and ensures that the covalent bonds are fully developed before the supports are removed [24]. The burrs left behind after the curing process are smoothened out with sandpaper for a uniform surface finish.

Planning and execution of each of the steps of an SLA printing process are equally crucial in obtaining good print results. For example, a print may not cure correctly or could potentially break during printing without a proper design due to a lack of structural integrity. Similarly, the printing speed is crucial in determining the printed layer's surface precision and structural integrity. The curing process is perhaps the most essential step, which provides structural strength to the part. If the isopropyl rinse is skipped or the part was not allowed to cure for an appropriate amount of time before applying forces or putting into use, the part could deform or collapse. All of these factors combine to create one of the most streamlined forms of the additive manufacturing process. The current progress of SLA printing and its applications have grown exponentially in the past decades and will only continue to push the boundaries in the field of 3D printing.

3.2.3 SLA: recent advancement

In the past few years, the use of SLA has grown with the introduction of different resins, making it useable for varied applications. In the medical field, the polymers made with SLA are being combined with aliphatic polyesters for their biodegradable nature [25]. Aliphatic polyesters are commonly used in medicine to create capsules that contain different drugs inside. SLA printers are also used to print with multiple resins by exchanging them during the printing process [26]. The SLA printers now can store x and y print coordinates while the resin being used is exchanged [26,27]. This includes removing resin vats, cleaning up, and then introducing the remaining resin vats as needed. Another recent development in this area is to introduce a two-resin mixture vat into an SLA printer with two corresponding light sources. The individual monomers within each resin will react to either of the light sources. Alternating between the two light sources allows for the creation of two different materials without the need to swap resin vats.

SLA is often used over casting due to its faster production time and to create prototypes. However, one of the biggest problems with SLA is the decreased mechanical and thermal properties of the produced parts when compared to other processes such as casting [28]. In recent years, there have been several experiments conducted to improve the properties and curing time of SLA-produced parts. One of these experiments was conducted using a prepolymer composed of 2-hydroxyethyl methacrylate, isophorone diisocyanate, and polyethylene glycol. Three different diluents were added to the prepolymer to see how each affected the properties of the produced material. The three diluents used in the experiment were ethylene glycol monophenyl ether, 2-hydroxyethyl acylate, and ethylene glycol. The addition of ethylene glycol to the prepolymer produced a material that is much denser than PVC, which is the main factor for its appeal in applications. The addition of 2-hydroxyethyl acylate to the prepolymer produced materials with a higher level of accuracy. The prepolymer with 2-hydroxyethyl acylate also produced parts with shrinkage levels smaller than 1.3%. The glycol monophenyl diluent did not show any noticeable characteristics, so it was studied further under different laser powers [29]. Fig. 3.2 shows the effect of increasing laser power on the microstructural features of the ethylene glycol

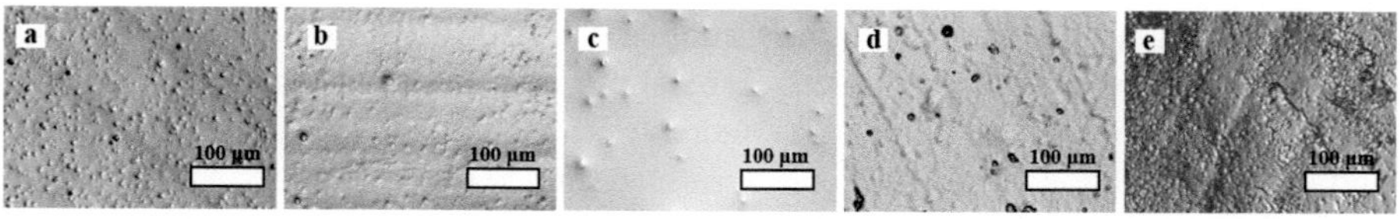

Figure 3.2 Microstructure of ethylene glycol monophenyl ether diluent due to increasing laser power from (A)–(E) [29].

monophenyl ether diluent. The laser powers used during the experiment are (a) 138 mW, (b) 196 mW, (c) 350 mW, (d) 580 mW, and (e) 1000 mW, respectively. The microstructure of **2(c)** showed similar characteristics to that of most SLA-printed parts. As the laser power was increased to 580 mW, many bubbles were seen due to the gasification of the resin as shown in **2(d)**. Further, at the maximum laser power of 1000 mW, the printed part microstructure showed a much more densified microstructure **2(e)**. When comparing the results in terms of laser power, it is clear that the material becomes much denser at higher laser power leading to an increase in hardness. It is also important to note that a variable that was not tested but could potentially alter the results in this experiment is the speed of the laser during the curing process [29].

Another notable result from this experiment was the effect of viscosity on the quality of the final product. In basic SLA printing, resin with a viscosity below 5 Pa·s is preferred to properly produce parts. All of the three diluents used in the above experiment were under 2 Pa·s and did not cause any unexpected results during normal SLA printing conditions. However, the prepolymer mixed with other diluents, which had a viscosity well under 5 Pa·s, showed defects in the final product due to shrinkage. The cause of the shrinkage was not discovered, but it was useful in knowing that viscosity alone cannot be used to guarantee functional parts.

This study showed that cross-linking using prepolymers with different diluents was capable of producing finished products with better properties than other commonly used materials. It also showed that diluents that did not show a noticeable property initially could be further tested with other variables, such as laser power, to produce a noticeable change in the material properties. From a functional testing and prototyping aspect, the creation of fully functional ready-to-use parts would make it a viable option when compared to other manufacturing methods such as casting.

3.3 Properties: manufactured products

3.3.1 Mechanical properties

The mechanical properties of a material determine the usability and probable applications of a 3D-printed part. Determining the mechanical properties of a printed part includes examining the yield strength, ultimate tensile strength, and impact strength of a specimen. The process parameters during the printing process such as layer thickness, hatch spacing, laser power, laser speed, and the orientation at which the specimen was constructed can greatly impact the mechanical properties of the printed specimen [30]. They also depend on the print material being used, the process to create the specimen, and any modifications to the original print material [31].

To further distinguish how each of these factors alters the mechanical properties, experiments were conducted using the SL5530 epoxy resin. Multiple tests were conducted on three specimens with layer thickness values of 100, 125, and 150 μm. All the specimens were constructed in the same built orientation with a similar hatch spacing value. The tensile tests were conducted on 10 samples of each specimen type (100, 125, and 150 μm) at a load of 200 kN. Impact tests were then conducted using the same number of test pieces with the following parameters: 360 J of impact energy, 120 degrees drop angle, 4.67 kg in effective pendulum weight, and a striking velocity of 2.45 m/s. The results of these tests are shown in Table 3.1. In general, the mechanical properties increase with the decrease in the layer thickness. This experiment provided a baseline for the epoxy resin behavior while only changing one of its process parameters [32].

While lowering the layer thickness increased the strength of the material, more significant deviations were found in the mechanical properties when more than one parameter was altered. Three levels of specimens were observed with varying layer thickness, orientation, and hatch spacing. These three sets of parameters were chosen so that the Taguchi method could be used to find the optimal parameters [33]. This method was used to determine the set of parameters with the most optimal results from the L_9 orthogonal array. It was necessary to also include the percent contribution of each parameter toward density and tensile strength to further distinguish the stimulating parameters [34].

After narrowing the selection, the gray relational analysis was used to observe possible combinations of parameters to further optimize its properties. The optimal set of parameters contains a value from each of the three variations. This proves that the methodology used was efficient in finding the optimal combination while only requiring a fraction of direct experimentation [35]. Furthermore, it was found that the layer thickness typically contributes the most to the density and tensile strength of the specimen. However, in terms of tensile strength, the orientation is more significant to the test piece.

Table 3.1 Results from the tensile and impact tests showing how the layer thickness affects an SLA specimen's mechanical properties [32].

	Average values of 10 specimens				
Test number	**Lt (μm)**	**YS (N/mm^2)**	**UTS (N/mm^2)**	**E (J)**	**IS (J/m)**
1	100	68.24	70.05	0.2926	29.25
2	125	60.76	68.50	0.2539	25.39
3	150	49.92	56.00	0.2125	21.25

3.3.2 Tribological properties

Tribology is the study of surface interactions for two bodies in contact with one another. This includes properties such as friction, wear, and lubrication. The need for a durable printing material is highly desired among the manufacturing, prototyping, and medical industries. Since many printed parts are used as a component in a larger system or machine, the surface interactions between the parts are generally high. Thus, the primary requirement to be able to print a wear-resistant industrial part is to use a wear-resistant printing material that can hold up against the constant friction forces with time. A more durable part means a longer service life, which means a more reliable system that requires less frequent maintenance and lasts much longer.

There are several advantages of using an SLA printer in terms of durability. The SLA printing process compactly packs the resin particles to form covalent bonds, creating isotropy (uniform material in all directions) [13]. This means that as the printer puts down layers, the laser melds the particles together to make an incredibly strong print surface that can hold up against wear and friction. SLA-printed parts have a very uniform and low friction smooth surface finish due to the high precision of the printer and the postprint isopropyl alcohol bath that the finished part receives. This helps reduce friction in a scenario where two parts are rubbing against each other in a moving system, especially if both of those parts have been printed using an SLA.

3.3.3 Corrosion properties

The 3D-printed parts used in industries are often exposed to corrosive chemicals and other environmental factors that might corrode the printed part and degrade its quality. To prevent exposure to the corrosive environment and achieve corrosion resistance, the SLA-printed part surface can be coated through a process known as plating [36]. Plating is accomplished by depositing metal onto the plastic surface (relatively high-temperature resistance material) in a plating bath. There are two common methods of plating: electroless plating and electroplating. Electroless plating is generally used in most cases since it does not require an electric current and is more cost-effective. Electroplating, on the other hand, requires a current, which means that the print's surface must be conductive. This is an added condition to achieve corrosion resistance, which makes the process longer and more expensive. However, electroplating yields a thicker and more durable plating. The process of plating is mostly used in an industrial setting since it is crucial in increasing overall durability and providing higher resistance to corrosion and abrasion. Without plating, most resins are either too brittle or have too low of a melting point to hold up against high speeds or applied forces over time [37].

3.3.4 Tribo-corrosion properties

SLA is convenient because it greatly decreases production time, but it usually does not produce parts with similar or better properties than other production methods. However, using different types of resins or adding different substances to the resin can improve the properties of printed parts, making SLA a viable option. This section primarily focuses on different methods used to improve the corrosive and tribo-corrosive properties of SLA-manufactured products [38,39]. The tribo-corrosion is the degradation of material as a resultant effect of wear and corrosion that degrade the material over time.

One method used to improve SLA's corrosive properties is the addition of nanofillers such as graphene [40]. These nanofillers can be prepared using various nano particle manufacturing techniques such as chemical precipitation and chemical vapor deposition (CVD) [40a]. The advantages of using graphene are increased electron conductivity, excellent optical properties, high surface area, high mechanical properties, and superior thermal conductivity [41]. The efficiency of using graphene depends on how well the nanomaterial is dispersed on the applied surface area. The best and most efficient overall results come from the use of a single layer of graphene. However, this is difficult because graphene can wrinkle if not properly applied to a surface. Thus, multiple layers of graphene can be applied to the surface. This is a convenient procedure for materials that require increased stiffness, strength, and hardness, but it is generally expensive. On the contrary, using a single layer would yield better results at a low cost compared to using multiple layers.

Another way to improve the corrosive properties of SLA-manufactured parts is to choose a corrosion-resistant resin. Acrylic resins are known to provide improved corrosion resistance properties [42]. The SLA-manufactured parts can also be plated with metals and alloys to improve their surface properties. The benefits of plating SLA-manufactured parts include improved heat deflection, higher strength, and higher chemical resistance. SLA-manufactured products can be plated with electroplating or electroless plating. Electroplating is a procedure used specifically for corrosion through the process of hydrolysis [39,43,44]. The common metals used for plating processes are nickel, copper, and gold. However, only nickel and gold are used specifically for corrosion resistance. Nickel alloys such as tin and tungsten are also commonly used for electroplating to provide corrosion resistance. An important factor to consider when using electroplating is the variation in surface finish and dimensions of the final product due to the buildup of plating metal on the edges of the part, resulting in the nonuniform thickness.

In the medical field, SLA is being used to produce retainers, aligners, prosthetics, and orthodontic models using different dental resins [45]. The choice of resin determines the flexibility, transparency, and strength of the final part. These resins are used in dental work due to their resistance to wear and fracture. A related study was conducted in an attempt to toughen alumina (Al_2O_3) with zirconia (ZrO_2), which is commonly used for dental work. The study results showed that strong bonding between the zirconia and alumina particles resulted in improved mechanical and chemical properties [46]. The material produced with the zirconia compound showed improved corrosion resistance when in water or body fluids. The toughness of the material is also improved with the zirconia compound. Another study was conducted that involved the use of a quasicrystal-resin composite to manufacture SLA parts [47]. A quasicrystal is a material that has a structured order when observed under a microscope. However, the structure will not be periodic, which means that the structures observed will not repeat themselves at certain intervals. This means that there is no symmetrical geometry within the structure. This is different from crystals because they have a structure that is both ordered and periodic [48]. Quasicrystals were discovered in 1983 by the Nobel Prize winner Daniel Shechtman, and more than 100 quasicrystals have been found ever since [49]. The quasicrystal—resin composite analyzed in the study was composed of quasicrystalline particles of the icosahedral Al-Cu-Fe phase. A small portion of the aluminum atoms was replaced with boron atoms to increase properties such as hardness, fracture strength, and friction. This is paired with the use of aluminum quasicrystals, which have been proven to provide improved corrosion resistance. Another notable result from the study was a decrease in the wear loss of material by 40% when the quasicrystal resin was used to print parts [50].

3.4 Applications of SLA

3.4.1 Prototyping

Prototyping is one of the first experimental steps of manufacturing in the design process that takes an idea and transforms it into a physical part that can be tested. Prototyping is essential to the design process since it allows for quick and inexpensive initial testing so that design flaws can be found and improved upon before the part makes it to full-scale production [51]. SLA can print intricate designs with precision and accuracy. This method of printing uses an UV laser that cures paper-thin layers of resin allowing for the most intricate and small details to be included with extremely narrow tolerances. SLA process can print parts with a high level of complexity using a simple and user-friendly process. Compared to alternative methods of additive manufacturing, the majority of SLA printers allow for bigger print sizes. Most standard FDM printers are small and rated for print size of $200 \times 200 \times 200$ mm [52]. For bigger print sizes, the printer options get increasingly expensive. SLA printers are similar in the aspect of small print capacities and are expensive too, so the bigger prints are generally outsourced. At an industrial scale, the SLA printers can print up to $736 \times 635 \times 533$ mm. The quality in which SLA produces these bigger prints is far superior to FDM printing something of a similar size [53].

There are several traditional methods of creating prototypes that include casting, molding, and machining [54—57]. However, the SLA 3D printing process is a good choice for initial prototyping due to the ease of design change and a variety of material options [51]. SLA-printed parts can also be postprocessed by methods such as metal plating to make them more resistant to friction and corrosion [58]. Even with these surface modifications, an SLA printed part will never be as strong as a machine-produced part. SLA printing is best suited for fast and low-cost prototyping applications that require high levels of accuracy but moderate levels of strength. Utilizing additive manufacturing when prototyping allows creating an initial design that can be quickly improved upon [59]. Other forms of prototyping such as casting, or molding, generally have a much longer production cycle to make any design changes and manufacture a new prototype. Thus, SLA saves time in the long term by easing the design change process [60]. In terms of cost-efficiency, an SLA printer is not the cheapest form of prototyping since the printers themselves cost anywhere between \$200 and \$10,000 depending on the quality and size [61,62]. Apart from the setup cost, the resin approximately costs around \$50 per liter depending on the desired quality and strength. However, it is important to note that a liter of resin can produce many prototypes. SLA prototyping saves money in the long term since the printer will begin to pay for itself by saving in comparison to other types of prototyping. For instance, the machinery or resources necessary to prototype using rapid CNC prototyping and casting could be much expensive for a company compared to SLA. The costs increase because most designs usually go through several improvisations until a final design is set and approved. Therefore, the ability to quickly make changes in CAD files to reprint to prototype saves time and cost [63,64].

Prototyping is one of the most important components of the product design process. The ability to produce an accurate and strong part in a timely and cost-efficient manner is essential in the prototyping process. The SLA creates parts with intricate designs and strong prints that can be used for prototyping at a much cheaper rate in the longer run. Prototyping is just one of the many applications of SLA printing, another industry that is beginning to push the envelope of what is possible with a 3D printer is the medical field utilizing the technology to create models.

3.4.2 Medical modeling

3D modeling is used in different procedures and for preoperative training purposes in the medical field. The most common applications include creating 3D-printed models of organs and parts of a body for surgeons for preoperative planning, 3D ultrasounds, implants, surgical tools, and biological organs [65–70]. This aids in fixing injuries, diagnosing diseases, and developing better medical processes inside and outside of hospitals. The use of SLA has grown tremendously in the recent years for medical applications [71]. Thus, a lot of research is being conducted recently to manufacture organs and tissues that can implanted into the human body [72–74].

Each surgery has its complications in the surgical field and can drastically vary from one to another. The use of X-rays, magnetic resonance imaging, computed tomography scans, ultrasounds, etc., can provide a 2D image of the body. However, there is not any easy way to visualize them in 3D. Now with SLA, those scans can be used to create a 3D model of the whole area that the surgeon will be working on. Surgeons can now look at the problem in 3D and perform preoperative planning [75]. With more mainstream surgeries, models can be produced in mass and can be distributed to universities, hospitals, and other professionals to allow them to more easily practice surgeries. Several researchers and companies are looking to create expandable models that can be used to practice surgeries such as brain surgery. A surgery practice that is being developed for a c-spine (top four vertebrae in the spine) surgery is facilitated by SLA [76]. Those were made with SLA using layers of fat, muscle, and skin-like materials, which were placed on top of the printed skeleton. These materials were printed out of Dragon Skin, a highly stretchable, strong high-performance silicon rubber that was developed specifically for 3D printers [77]. Another similar use of 3D printing is the creation of a

3D version of ultrasound images. This allows to visualize the image of an unborn baby in the womb and detect tumors as well [78–80]. Apart from these direct benefit, the 3D printing technology can assist people with disabilities in countless different ways and make their life easier. A few examples are custom 3D-printed hearing aids, exoskeleton used by patients, and 3D-printed model for a visually impaired expectant person [81,82].

SLA is also being used to 3D print body implants. Different types of printing materials allow for many different uses in the human body such as materials used by dentists, surgeons, etc. [69,83]. A dentist can take an X-ray of a patient's mouth and create parts that can simulate structures such as a jaw or support different areas in the mouth to help correct the alignment of teeth [84]. An emerging application for SLA is the use of re-creating full organs for the use as implants for patients in need [85]. As someone's organs begin to fail, they often need a lifesaving procedure that includes an organ donated from another person. The receiver sometimes needs to wait months until a perfect matching donor is found. So, for years, scientists have been looking for ways to recreate or grow organs outside of the human body; however, due to the incredible complexity of the human body, most attempts have completely failed; however, SLA has brought a new life to this idea and is known as bioprinting. The idea is to print structures using biocompatible resins that can house living cells that can grow into organs for the body [86]. This is carried out by having the printer pipette an exact amount of living cells in an area and printing the cells in layers until it forms a whole organ or a layer of the skin or fat. Researchers are trying to recreate some organs: kidneys, livers, bladders, heart valves, tracheal structures, and lungs [74]. Other than being used for transplants, one of the applications for these organs is for pharmaceutical companies to test their drugs on "real" organs before sending their drugs into clinical trials to see the effect on a specific organ. This could also help expedite the clinical trial process, so that lifesaving drugs could potentially hit the market much further and save even more lives and money in the process. This broadens the applications of SLA far beyond the manufacturing and modeling environment into other critical fields such as medical and biological 3D printing.

3.4.3 Biological applications

One of the constant demands in this industry is cost-efficient prototyping and fabrication of lab-on-a-chip (LOC) devices. LOC technology is typically a device that carries out data collection or experimental research on a very small scale. The use of additive manufacturing has gained popularity in the LOC field since it allows a way for cheap and easy prototyping. LOC technology is essentially microfluidic devices that are designed with the intent to be applied in human biology. There is still a lot of hesitation in the industry to use 3D printing as a solution since the production of these chips is a lengthy and complex process. The current LOC also lacks applications, which is an important aspect that would make the LOC superior to alternative internal data collection methods [87]. Within 3D printing, SLA is one of the preferred methods due to the intricacies and strong bonds the printing apparatus can achieve. Some SLA applications of LOC devices include integrating an LOC into molds used in dental implants and surgical planning [88]. These LOCs are typically used to analyze DNA or detect pathogens through processing, mixing, or separating the fluids that flow through it [89].

Using SLA-generated parts in direct contact with the living body has always been regarded as a difficult process due to the low biocompatibility of the photosensitive resin used in SLA printing. When placed into the human body, these resins are known to have a certain level of toxicity that could cause inflammation [90]. This was initially resolved by treating the printed part with a carbon dioxide solution to reduce resin toxicity, leading to a far more biocompatible implant. More recently, developments have created a biocompatible resin that is completely safe and nontoxic [45,91]. However, this still does not solve all the problems, when implementing SLA parts into human biology, many parts of the body have high biological activity, so the print needs to be in accordance to this activity and not impede it by disturbing the natural system of the body. This requires a delicate balance when designing a supportive print structure that must also be flexible enough to move in sync with the system it exists within and not interrupt natural body processes.

The applications of SLA printing in the biomedical field are ever-growing, and new research is constantly opening up new avenues for new applications. A few examples of current print technology that can be implemented into the human body are bionic eyes, heart pumps, skin replacements, elastic bones, to name a few [92–96]. SLA also has the bonus of using many different types of materials in the printing process to accommodate the desired material properties, and new developments have made these materials biocompatible and free of toxins [45]. One of the latest ideas being explored in SLA printing is the concept of bioprinting, where the biocompatible resin is combined with the cells and other biomaterials to create mock tissue. The advantage of using SLA in bioprinting is the use of digital light to cure bio-ink results in a high level of accuracy and efficiency. Unfortunately, the intensity of the UV light also creates disadvantages, such as the radiation given off from the UV laser that cures the resin and the requirement of postprint treatment. SLA printing has a fairly medium cost compared to other forms of additive manufacturing when applied to bioprinting. Bio ink has solved the main issue of compatible materials; the special composition includes cellular material, growth additives, and in some cases, a supportive scaffold [97]. This allows the bio ink to be a diverse and adaptable material that can be tailored to fit, regardless of varying circumstances. The main application of bio ink is printing micro-tissues that requires a fine balance between maintaining the structural integrity while also not deteriorating cell health [98].

The biological applications of SLA printing are spreading and growing every day. New scientific developments continue to push the boundaries of what can be achieved through photo-solidification manufacturing. The current scope of the 3D printing biological field includes LOC technology, which can be broken down into microfluidic devices that can detect pathogens and analyze DNA. Other applications include 3D printing of bionic eyes, heart pumps, replacement skin tissue, elastic bone, etc. Although the field seems limitless, there are still many obstacles left to navigate in this industry. The main problems are the cost and the biocompatibility of the resins being used. SLA printing has solved many problems in the biological field, but there is still lots of research to be performed to optimize these processes and realize their usage for new potential applications.

3.5 Conclusions

In conclusion, SLA 3D printing can be performed with right-side-up printers or upside-down (inverted) printers. Right-side-up printers have the base submerged in the resin vat for the entire time because it is gradually lowered to allow multiple layers to be formed on top. Upside-down printers will begin with the base submerged and gradually elevate it to form layers underneath. Upside-down SLA printing is more commonly used due to its precision, surface finish, and overall strength compared to right-side-up SLA printing. The main problem with inverted SLA printing is the peel forces that can be generated with large or heavy prints. These forces can cause peeling or tearing in the material that can lead to deformities in the overall product. Both of the SLA printing methods use the same printing process with the only difference being the direction that the part is produced. The basic SLA printing process begins with loading the printer with the desired type of resin that gives the desired end product material properties. UV lights are then used to cure the resin of either the top or bottom depending on the type of printer. The final step is postprocessing, which includes sanding, a UV oven, or a bath, depending on the required surface finish of the product. Both SLA printers make parts with the help of software such as CAD. SLA is an expensive 3D printing process, but it greatly shortens production time compared to other commonly used manufacturing methods. SLA is also commonly used to create prototypes.

SLA's properties produced parts with basic resin that might not be superior to those produced by traditional methods, such as casting. However, manipulating the resins used or using post production processes can help achieve the desired properties. For example, placing SLA-produced parts in isopropyl alcohol baths produces a smooth low friction surface that eventually reduces the part's wear. The use of nanofillers such as graphene can improve the corrosive, mechanical, electrical, and thermal properties of the SLA-printed parts. SLA parts can also be plated to increase heat deflection, strength, and chemical resistance. The two types of plating are electroplating and electroless plating. The drawback to this method is imperfections on the surface that could arise due to the process itself. If an even surface is desired, other methods such as resin made from acrylic-modified zircon with alumina powders, or quasicrystal-resin composites, can be used to improve the mechanical and tribo-corrosive properties of SLA-produced parts. SLA parts are mainly used to create prototypes due to high precision and low production time.

SLA process is also used in the medical field to create 3D model of organs. Therefore, doctors can create a specific part or group of organs for preoperative planning and research. With the use of SLA, the medical field has seen a lot of development from prototyping of model to producing finished functional. Further research and testing with the SLA process could potentially allow for 3D printing of intricate organs with photocurable biocompatible materials that could revolutionize the medical industry. Further research and development of the SLA process can yield materials and processing techniques that would help to replace traditional manufacturing techniques for other applications as well.

References

[1] N. Shahrubudin, et al., An overview on 3D printing technology: technological, materials, and applications, Procedia Manufacturing 35 (2019) 1286−1296.
[2] A. Bagheri, et al., Photopolymerization in 3D printing, ACS Applied Polymer Materials 1 (4) (2019) 593−611.
[3] B. Yu, et al., Enhanced thermal and mechanical properties of functionalized graphene/thiol-ene systems by photopolymerization technology, Chemical Engineering Journal 228 (2013) 318−326.
[4] A. Su, et al., History of 3D printing, in: 3D Printing Applications in Cardiovascular Medicine, Elsevier, 2018, pp. 1−10.
[5] J.A. Schönherr, et al., Stereolithographic additive manufacturing of high precision glass ceramic parts, Materials 13 (7) (2020) 1492.
[6] F. Zhang, et al., The recent development of vat photopolymerization: a review, Additive Manufacturing (2021) 102423.
[7] S.A. Skoog, et al., Stereolithography in tissue engineering, Journal of Materials Science: Materials in Medicine 25 (3) (2014) 845−856.
[8] P.-T. Lan, et al., Determining fabrication orientations for rapid prototyping with stereolithography apparatus, Computer-Aided Design 29 (1) (1997) 53−62.
[9] J. Son, et al., Preliminary study on polishing SLA 3D-printed ABS-like resins for surface roughness and glossiness reduction, Micromachines 11 (9) (2020) 843.
[10] M. Gonçalves, et al., Study of tribological properties of moulds obtained by stereolithography, Virtual and Physical Prototyping 2 (1) (2007) 29−36.
[11] Y. He, et al., Rapid fabrication of paper-based microfluidic analytical devices with desktop stereolithography 3D printer, RSC Advances 5 (4) (2015) 2694−2701.
[12] C. Kirschman, et al., Computer aided design of support structures for stereolithographic components, in: International Design Engineering Technical Conferences and Computers and Information in Engineering Conference, American Society of Mechanical Engineers, 1991.
[13] R. Hague, et al., Materials analysis of stereolithography resins for use in rapid manufacturing, Journal of Materials Science 39 (7) (2004) 2457−2464.
[14] S. Park, et al., Mechanical and thermal properties of 3d-printed thermosets by stereolithography, Journal of Photopolymer Science and Technology 32 (2) (2019) 227−232.
[15] R. Hague, et al., Material and design considerations for rapid manufacturing, International Journal of Production Research 42 (22) (2004) 4691−4708.
[16] C. Hinczewski, S. Corbel, T. Chartier, Ceramic suspensions suitable for stereolithography, Journal of the European Ceramic Society 18 (6) (1998) 583−590.
[17] P. Papaspyridakos, et al., Complete digital workflow in prosthesis prototype fabrication for complete-arch implant rehabilitation: a technique, The Journal of Prosthetic Dentistry 122 (3) (2019) 189−192.
[18] E. Béchet, et al., Generation of a finite element MESH from stereolithography (STL) files, Computer-Aided Design 34 (1) (2002) 1−17.
[19] V. Canellidis, et al., Pre-processing methodology for optimizing stereolithography apparatus build performance, Computers in Industry 57 (5) (2006) 424−436.
[20] M.N. Cooke, et al., Use of stereolithography to manufacture critical-sized 3D biodegradable scaffolds for bone ingrowth, Journal of Biomedical Materials Research Part B: Applied Biomaterials 64 (2) (2003) 65−69.

[21] J. Huang, et al., A review of stereolithography: processes and systems, Processes 8 (9) (2020) 1138.
[22] Y.Y.C. Choong, et al., Curing characteristics of shape memory polymers in 3D projection and laser stereolithography, Virtual and Physical Prototyping 12 (1) (2017) 77–84.
[23] B.T. Phillips, et al., Additive manufacturing aboard a moving vessel at sea using passively stabilized stereolithography (SLA) 3D printing, Additive Manufacturing 31 (2020) 100969.
[24] G. Salmoria, et al., Stereolithography somos 7110 resin: mechanical behavior and fractography of parts post-cured by different methods, Polymer Testing 24 (2) (2005) 157–162.
[25] T. Zhao, et al., Aliphatic silicone-epoxy based hybrid photopolymers applied in stereolithography 3D printing, Polymers for Advanced Technologies 32 (3) (2021) 980–987.
[26] J.-W. Choi, et al., Multi-material stereolithography, Journal of Materials Processing Technology 211 (3) (2011) 318–328.
[27] B. Grigoryan, et al., Development, characterization, and applications of multi-material stereolithography bioprinting, Scientific Reports 11 (1) (2021) 1–13.
[28] C. Mendes-Felipe, et al., Evaluation of postcuring process on the thermal and mechanical properties of the Clear02™ resin used in stereolithography, Polymer Testing 72 (2018) 115–121.
[29] R. Ni, et al., A cross-linking strategy with moderated pre-polymerization of resin for stereolithography, RSC Advances 8 (52) (2018) 29583–29588.
[30] B. Raju, et al., Establishment of Process model for rapid prototyping technique (Stereolithography) to enhance the part quality by Taguchi method, Procedia Technology 14 (2014) 380–389.
[31] H. Eng, et al., 3D stereolithography of polymer composites reinforced with orientated nanoclay, Procedia Engineering 216 (2017) 1–7.
[32] K. Chockalingam, et al., Influence of layer thickness on mechanical properties in stereolithography, Rapid Prototyping Journal 12 (2006).
[33] P.J. Ross, Taguchi Techniques for Quality Engineering: Loss Function, Orthogonal Experiments, Parameter and Tolerance Design, 1996.
[34] S. Kumanan, J. Dhas, K. Gowthaman, Determination of submerged arc welding process parameters using Taguchi method and regression analysis, Indian Journal of Engineering and Materials Sciences 14 (2007).
[35] S.S. Mahapatra, et al., Benchmarking of rapid prototyping systems using grey relational analysis, International Journal of Services and Operations Management 16 (4) (2013) 460–477.
[36] J. Rajaguru, et al., Development of rapid tooling by rapid prototyping technology and electroless nickel plating for low-volume production of plastic parts, The International Journal of Advanced Manufacturing Technology 78 (1–4) (2015) 31–40.
[37] Y. Li, et al., Isotropic stereolithography resin toughened by core-shell particles, Chemical Engineering Journal 394 (2020) 124873.
[38] L.L. Wang, et al., Mechanical and tribological properties of acrylonitrile–butadiene rubber filled with graphite and carbon black, Materials & Design 39 (2012) 450–457.
[39] Y. Xue, et al., Tribological behaviour of UHMWPE/HDPE blends reinforced with multi-wall carbon nanotubes, Polymer Testing 25 (2) (2006) 221–229.
[40] F. Ibrahim, et al., Evaluation of the compatibility of organosolv lignin-graphene nanoplatelets with photo-curable polyurethane in stereolithography 3d printing, Polymers 11 (10) (2019) 1544.

[40a] A.K. Kushwaha, et al., Nanocrystalline materials: synthesis, characterization, properties, and applications, Crystals 11 (11) (2021) 1317.

[41] J.Z. Manapat, et al., High-strength stereolithographic 3D printed nanocomposites: graphene oxide metastability, ACS Applied Materials & Interfaces 9 (11) (2017) 10085−10093.

[42] C. Zhang, et al., Stability, rheological behaviors, and curing properties of $3Y-ZrO_2$ and $3Y-ZrO_2$/GO ceramic suspensions in stereolithography applied for dental implants, Ceramics International 47 (10) (2021) 13344−13350.

[43] M. Dawoud, et al., Effect of processing parameters and graphite content on the tribological behaviour of 3D printed acrylonitrile butadiene styrene: Einfluss von Prozessparametern und Graphitgehalt auf das tribologische Verhalten von 3D-Druck Acrylnitril-Butadien-Styrol Bauteilen, Materials Science and Engineering 46 (12) (2015) 1185−1195.

[44] A.C. Uzcategui, et al., Understanding and improving mechanical properties in 3D printed parts using a dual-cure acrylate-based resin for stereolithography, Advanced Engineering Materials 20 (12) (2018) 1800876.

[45] A. Bens, et al., Non-toxic flexible photopolymers for medical stereolithography technology, Rapid Prototyping Journal 13 (2007).

[46] X. Liu, et al., The preparation of ZrO_2-Al_2O_3 composite ceramic by SLA-3D printing and sintering processing, Ceramics International 46 (1) (2020) 937−944.

[47] D. Miedzińska, et al., Experimental study on influence of curing time on strength behavior of SLA-printed samples loaded with different strain rates, Materials 13 (24) (2020) 5825.

[48] C. Janot, Quasicrystals, in: Neutron and Synchrotron Radiation for Condensed Matter Studies, Springer, 1994, pp. 197−211.

[49] D. Gratias, et al., Discovery of quasicrystals: the early days, Comptes Rendus Physique 20 (7−8) (2019) 803−816.

[50] A. Sakly, et al., A novel quasicrystal-resin composite for stereolithography, Materials & Design 56 (2014) 280−285.

[51] C. Choudhari, V. Patil, Product development and its comparative analysis by SLA, SLS and FDM rapid prototyping processes, in: IOP Conference Series: Materials Science and Engineering, IOP Publishing, 2016.

[52] I. Skawiński, et al., FDM 3D printing method utility assessment in small RC aircraft design, Aircraft Engineering and Aerospace Technology 91 (2019).

[53] R. Dermanaki Farahani, et al., Printing polymer nanocomposites and composites in three dimensions, Advanced Engineering Materials 20 (2) (2018) 1700539.

[54] B. Rooks, Rapid tooling for casting prototypes, Assembly Automation 22 (2002).

[55] J.-Y. Jeng, et al., Mold fabrication and modification using hybrid processes of selective laser cladding and milling, Journal of Materials Processing Technology 110 (1) (2001) 98−103.

[56] M. Soshi, et al., Innovative grid molding and cooling using an additive and subtractive hybrid CNC machine tool, CIRP Annals 66 (1) (2017) 401−404.

[57] T. Schmitz, et al., The application of high-speed CNC machining to prototype production, International Journal of Machine Tools & Manufacture 41 (8) (2001) 1209−1228.

[58] M. Dionigi, et al., Simple high-performance metal-plating procedure for stereolithographically 3-D-printed waveguide components, IEEE Microwave and Wireless Components Letters 27 (11) (2017) 953−955.

[59] K.S. Prakash, et al., Additive manufacturing techniques in manufacturing-an overview, Additive Manufacturing 5 (2) (2018) 3873−3882.

[60] V.A. Lifton, et al., Options for additive rapid prototyping methods (3D printing) in MEMS technology, Rapid Prototyping Journal 20 (2014).
[61] C.S. Favero, et al., Effect of print layer height and printer type on the accuracy of 3-dimensional printed orthodontic models, American Journal of Orthodontics and Dentofacial Orthopedics 152 (4) (2017) 557–565.
[62] B. Berman, 3-D printing: the new industrial revolution, Business Horizons 55 (2) (2012) 155–162.
[63] M. Attaran, The rise of 3-D printing: the advantages of additive manufacturing over traditional manufacturing, Business Horizons 60 (5) (2017) 677–688.
[64] H. Peng, et al., On-the-fly print: incremental printing while modelling, in: Proceedings of the 2016 CHI Conference on Human Factors in Computing Systems, 2016.
[65] A. Ganguli, et al., 3D printing for preoperative planning and surgical training: a review, Biomedical Microdevices 20 (3) (2018) 1–24.
[66] L. Pugliese, et al., The clinical use of 3D printing in surgery, Updates in Surgery 70 (3) (2018) 381–388.
[67] A. Squelch, 3D Printing and Medical Imaging, Wiley Online Library, 2018, pp. 171–172.
[68] G. Coelho, et al., Multimaterial 3D printing preoperative planning for frontoethmoidal meningoencephalocele surgery, Child's Nervous System 34 (4) (2018) 749–756.
[69] G. Wurm, et al., Prospective study on cranioplasty with individual carbon fiber reinforced polymere (CFRP) implants produced by means of stereolithography, Surgical Neurology 62 (6) (2004) 510–521.
[70] F.P. Melchels, et al., A review on stereolithography and its applications in biomedical engineering, Biomaterials 31 (24) (2010) 6121–6130.
[71] A. Kaza, Medical applications of stereolithography: an overview, International Journal of Academic Medicine 4 (2018).
[72] S. Zakeri, et al., A comprehensive review of the photopolymerization of ceramic resins used in stereolithography, Additive Manufacturing 35 (2020) 101177.
[73] M. Carve, et al., 3D-printed chips: compatibility of additive manufacturing photopolymeric substrata with biological applications, Micromachines 9 (2) (2018) 91.
[74] S. Wadnap, et al., Biofabrication of 3D cell-encapsulated tubular constructs using dynamic optical projection stereolithography, Journal of Materials Science: Materials in Medicine 30 (3) (2019) 1–10.
[75] K. Lal, et al., Use of stereolithographic templates for surgical and prosthodontic implant planning and placement. Part I. The concept, Journal of Prosthodontics 15 (1) (2006) 51–58.
[76] D. Fuerst, et al., Foam phantom development for artificial vertebrae used for surgical training, in: 2012 Annual International Conference of the IEEE Engineering in Medicine and Biology Society, IEEE, 2012.
[77] C. Hazelaar, et al., Using 3D printing techniques to create an anthropomorphic thorax phantom for medical imaging purposes, Medical Physics 45 (1) (2018) 92–100.
[78] E. Maneas, et al., Anatomically realistic ultrasound phantoms using gel wax with 3D printed moulds, Physics in Medicine and Biology 63 (1) (2018) 015033.
[79] J.J. Coté, et al., Randomized controlled trial of the effects of 3D-printed models and 3D ultrasonography on maternal–fetal attachment, JOGN Nursing 49 (2) (2020) 190–199.
[80] N. Wake, et al., "Pin the tumor on the kidney:" an evaluation of how surgeons translate CT and MRI data to 3D models, Urology 131 (2019) 255–261.
[81] A. Mohammadi, et al., Flexo-glove: a 3D printed soft exoskeleton robotic glove for impaired hand rehabilitation and assistance, in: 2018 40th Annual International Conference of the IEEE Engineering in Medicine and Biology Society (EMBC), IEEE, 2018.

[82] R. Nicot, et al., Using low-cost 3D-printed models of prenatal ultrasonography for visually-impaired expectant persons, Patient Education and Counseling 104 (2021).
[83] O. Guillaume, et al., Orbital floor repair using patient specific osteoinductive implant made by stereolithography, Biomaterials 233 (2020) 119721.
[84] H. Li, et al., Dental ceramic prostheses by stereolithography-based additive manufacturing: potentials and challenges, Advances in Applied Ceramics 118 (1−2) (2019) 30−36.
[85] L.E. Murr, Frontiers of 3D printing/additive manufacturing: from human organs to aircraft fabrication, Journal of Materials Science & Technology 32 (10) (2016) 987−995.
[86] V.B. Morris, et al., Mechanical properties, cytocompatibility and manufacturability of chitosan: PEGDA hybrid-gel scaffolds by stereolithography, Annals of Biomedical Engineering 45 (1) (2017) 286−296.
[87] F. Zhu, et al., Biological implications of lab-on-a-chip devices fabricated using multi-jet modelling and stereolithography processes, in: Bio-MEMS and Medical Microdevices II, International Society for Optics and Photonics, 2015.
[88] A.A. Yazdi, et al., 3D printing: an emerging tool for novel microfluidics and lab-on-a-chip applications, Microfluidics and Nanofluidics 20 (3) (2016) 50.
[89] H. Zhu, et al., Recent advances in lab-on-a-chip technologies for viral diagnosis, Biosensors and Bioelectronics 153 (2020) 112041.
[90] S.M. Oskui, et al., Assessing and reducing the toxicity of 3D-printed parts, Environmental Science & Technology Letters 3 (1) (2016) 1−6.
[91] S. Kreß, et al., 3D printing of cell culture devices: assessment and prevention of the cytotoxicity of photopolymers for stereolithography, Materials 13 (13) (2020) 3011.
[92] H. Li, et al., Three-dimensional printing: the potential technology widely used in medical fields, Journal of Biomedical Materials Research Part A 108 (11) (2020) 2217−2229.
[93] R. Sodian, et al., Application of stereolithography for scaffold fabrication for tissue engineered heart valves, American Society for Artificial Internal Organs Journal 48 (1) (2002) 12−16.
[94] S.N. Economidou, et al., 3D printed microneedle patches using stereolithography (SLA) for intradermal insulin delivery, Materials Science and Engineering: C 102 (2019) 743−755.
[95] G. Morrison, Advances in the skin trade, Journal of Mechanical Engineering 121 (2) (1999) 40−43.
[96] I. Xenikakis, et al., Fabrication and finite element analysis of stereolithographic 3D printed microneedles for transdermal delivery of model dyes across human skin in vitro, European Journal of Pharmaceutical Sciences 137 (2019) 104976.
[97] S.S. Mahdavi, et al., Stereolithography 3D bioprinting method for fabrication of human corneal stroma equivalent, Annals of Biomedical Engineering 48 (2020) 1955−1970.
[98] S. Shin, et al., Melanin nanoparticle-incorporated silk fibroin hydrogels for the enhancement of printing resolution in 3D-projection stereolithography of poly (ethylene glycol)-tetraacrylate bio-ink, ACS Applied Materials & Interfaces 10 (28) (2018) 23573−23582.

Additively manufactured functionally graded metallic materials

4

Dallas Evans, Md Hafizur Rahman, Mathew Heintzen, Jacob Welty, Joel Leslie, Keith Hall and Pradeep L. Menezes
Department of Mechanical Engineering, University of Nevada, Reno, NV, United States

4.1 Introduction

Recent technical advancement has skyrocketed the popularity of additive manufacturing (AM) techniques [1]. As a result, the high-performance utilization of diverse materials has become evident. Functionally graded metallic materials (FGMMs) are elemental metal compounds or alloy-based materials that have a variation in composition and structure gradually over volume, resulting in changes in the material properties [2]. Since being first proposed in the 1980s in Japan, FGMMs have been more widely utilized in the fields such as aerospace, biomedical engineering, sensors, and energy [3]. Metallic FGMMs facilitate optimum mechanical, chemical properties, thermal, and anticorrosive properties, based on the distribution of the mixed materials [4].

FGMMs could be created through various conventional and nonconventional methods. One example of a conventional method of obtaining an inhomogeneous volume is a process called centrifugal casting [5]. Among the nonconventional methods in practice today, AM is rapidly growing in popularity. Subcategories of AM include material jetting (a process where photosensitive material that droplets solidify under ultraviolet light are dispensed and hardened layer by layer), binder jetting (a binder is applied to a bed of powder, bonding layers of material together into one solid piece one layer at a time), and material extrusion (the most common 3D printing process, a continuous material filament is heated and fed through an extruding nozzle), all of which have been explored as solutions to create FGMMs [6]. These new additive technologies vastly simplify the creation of FGMMs and allow for better parts to be constructed. Traditional manufacturing tends to be time-consuming and requires specialized tools and equipment for assembly. It can also be restrictive depending on the part, such as centrifugal casting only being enabled to create cylindrical parts as opposed to newer AM techniques. AM requires less tooling to make highly complex geometry, which saves valuable time and money. AM generally has several advantages over traditional manufacturing processes. Among them, cost, speed, quality, and impact are the main reasons to pursue AM over traditional manufacturing for FGMMs [7].

Tribology of Additively Manufactured Materials. https://doi.org/10.1016/B978-0-12-821328-5.00004-4

AM is a way of creating metallic FGMMs and is a less expensive way of substituting load-bearing components in chemical, medical, and energy industries. By using the AM techniques, metallic FGMMs can be produced both more quickly and precisely, making them ideal for the previously mentioned industries. In this chapter, various manufacturing methods of metallic FGMMs and the current status of their applications in different industrial sectors will be discussed.

Another topic covered in this chapter is the versatile properties of metallic FGMMs. Because of the metallic FGMM's transition between multiple different materials throughout a structure, a gradient of elastic modulus and tensile properties can be customized according to their applications [8]. Similarly, thermal conductivity gradients within metallic FGMMs aid in controlling heat to a higher precision, allowing both insulative and conductive properties to exist within the same material. In addition, another sought out application of metallic FGMMs is due to the modifiable tribological properties of a body's surface. The anticorrosive properties of particular materials can be utilized for metallic FGMM applications as well. The ability to add multiple desired properties to a single part is highly desired in industries trying to reduce weight and increase the performance of components [9]. FGMMs give the ability to combine desired traits as opposed to prioritizing one over the other.

Because metallic FGMMs commonly combine desired features such as corrosion resistance, wear resistance, and durability, their application areas are almost endless. While the FGMM concept is known to be popular in both the automotive and aerospace industries, other industries also rely on its permanence. In combination with AM, FGMMs can be fabricated to precise and accurate detail, making them more appealing to other industries, such as chemical plants, the medical sector, and energy industries.

4.2 Manufacturing of FGMM

4.2.1 Conventional manufacturing of functionally graded metallic materials

The production of FGMM materials has rapidly improved since its inception. Conventional manufacturing is the most widely used form of creating FGMMs and has been developed and refined for decades [10]. The manufacturing of FGMMs could be categorized by a variety of parameters, such as gradient control, product complexity, FGMM type, etc [11]. For the purposes of description, in this chapter, conventional FGMM manufacturing is split into gas-based processes and liquid-phase processes (Fig. 4.1). These processes are further discussed in several sub-sections, as follows-

4.2.1.1 Gas-based processes

4.2.1.1.1 Chemical vapor deposition

One of the most common forms of conventional manufacturing in FGMMs is through the use of chemical vapor deposition (CVD). CVD uses different sources of energy

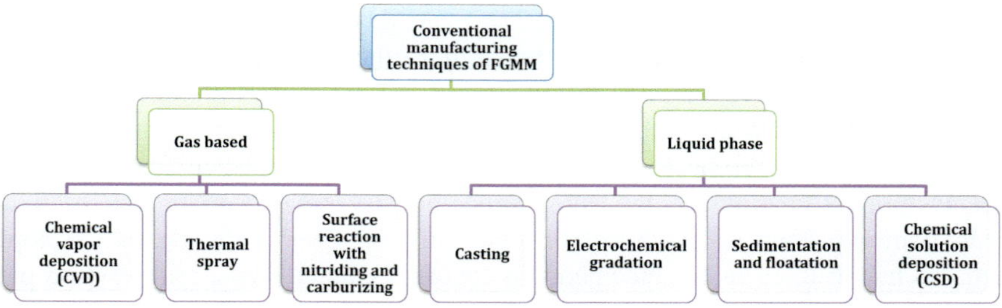

Figure 4.1 Conventional manufacturing techniques of functionally graded metallic materials (FGMMs).

such as heat, plasma, and light to deposit material and coat metals in chemicals such as bromide, hydride, and chloride. By altering the gas ratio, temperature, pressure, flow rate, and gas type, one can alter the FGMM's chemical gradient [12]. The use of CVD permits a direct way of coating metals with materials without the need for more complex machinery and at a relatively low cost.

This technique stems from the incomplete oxidation of firewood, which causes the formation of soot naturally. Throughout the 1970s, CVDs were one of the most attractive processes for fabricating semiconductors and for use in electronics, due to its ability to coat components in protective coatings. This would lead to an expansion in its powers with the ability to coat ceramics, composites, and even solar cells [13]. CVD manufacturing has been a useful method for manufacturing cheap FGMMs that require coatings in various materials that can be vaporized.

CVD results in precise, controlled layers of materials; composition and coating thickness can be adjusted based on application [14]. Micromachining printed circuit boards (PCBs) requires the use of precision routers to cut the material away. Low-quality tools wear quickly and create burrs on the PCB surface. CVD has been used to develop microrouters composed of Ti/TiN/TiCN/a-C:H layers and a diamond coating [15]. The resulting PCB is of higher quality, while the tool life is effectively extended.

4.2.1.1.2 Thermal spray

A similar form of manufacturing could be performed through the use of the thermal spray. This method has been developed since 1882, and the first versions of flame spray were used for tin and lead coatings in 1911. The thermal spray could be broken down into various methods such as wire arc and plasma spray, based on their electric energy sources [16]. The thermal spray could be utilized at temperatures up to 16,000 °C for plasma spray and nearly 4000°C for arc wire spray, as mentioned in the literature [17]. These methods have the capability of creating thermal barriers, which are one of the most important requirements of FGMM from various spraying processes. This has been highly sought after in the aerospace industry for use in jet turbines and remains a significant challenge to get the correct FGMM. Common coating techniques lack the ability to match the thermal expansion coefficients for thermal barriers leading to defects [18–20]. FGMMs utilized for thermal barrier coatings have been seen to make a similar match in the thermal expansion coefficients and have been observed reducing the residual stress in the coatings [21,22].

The aerospace, military, and automobile industries use the thermal spray technique to create desirable thermal barrier coatings and corrosion-resistant surfaces [10,23]. Plasma-sprayed coatings in the automobile industry have been proven to reduce wear, conserve energy, and lower oil consumption. Over a decade of developing methods has resulted in advancements in coating cylinder bores, transmission parts, and piston rings for a wide variety of industrially produced engines [24,25].

4.2.1.1.3 Surface reaction processes

The final common method for fabricating FGMMs utilizing gaseous methods is surface reaction processes. These processes are most common in nitriding and carburizing to steels and other metals for surface treatments [26–28]. Nitriding allows for factors such as temperature and time on microstructure to form phases during the process [29].

Al_2O_3 ceramics are being widely utilized for their strength, hardness, and thermal properties in fields such as ballistic gear and turbine manufacturing [30,31]. Turbines require materials that have both thermal robustness and high fracture toughness. Experiments to produce Al_2O_3/Ti/TiN FGMMs that meet these demands are conducted by nitriding in ammonia salts. This method of creating the FGMM results in an enhanced hardness that meets the properties desired [31].

4.2.1.2 Liquid-phase processes

4.2.1.2.1 Casting

Most recent FGMMs tend to use liquid-phase processes. One of the primary manufacturing methods is casting, which could be carried out in various methods, such as centrifugal, tape, slip, and gel casting. These methods use controls during the solidification process to segregate and reinforce particles.

a) Centrifugal casting: The centrifugal casting probes the advantage of the centrifugal and gravitational forces to coat a solid part in the liquid particles of the additive material and spin it so that it solidifies to the part. This process is helpful for coating the cylindrical parts and could be tuned by changing the rotational speed. For the fabrication of FGMMs, there are two common centrifugal methods, the in-situ technique and the solid particle technique. The centrifugal in-situ technique utilizes centrifugal forces for the solidification step, and the processing temperature is commonly higher than the master alloy temperature [16]. In the solid particle technique, the master alloy receives a higher temperature, and the second phase is solid in the molten metal [32]. A recent study on an Al/Al_2O_3 FGMM for the automobile applications has successfully created FGMMs through the horizontal centrifugal casting method [33]. The result was a component that was stronger and more wear resistant than a homogenous aluminum part [33].

b) Tape casting: Tape casting is another casting method widely used for the creation of the production of ceramic multilayer structures. Here, materials such as ceramics, dispersants, binders, and plasticizers are put onto substrates, which dry and sinter the tape [34]. This method is tested to increase the hardness of components as a laminated casting. A useful application of FGMMs created through tape casting is in the fabrication of armor. Tape casting improves the antipenetration properties of TiB/Ti composites under high-velocity impacts [35].

c) Slip Casting: Slip casting is carried out at 1600°C with high percentages of dispersants. In altering the chemical composition of the FGMM layers, one can gradually increase the grain size of the material. Using these methods, Andertova et al. [36] fabricated slip casting components that had layers with different porosity. They also found that common deformations at high temperatures would cause body deformations due to the material's dilatational transformations.

d) Gel casting: Gel casting is a colloidal casting procedure that includes a relatively short curing time in comparison to other casting methods. This could be used as an FGMM by using monomers' polymerization in a suspension of a free radical initiator [37,38]. In gel casting, ceramic powders are inserted into liquid solutions bearing the monomer and catalyst to create a slurry fluid, which would then be poured into a mold and polymerized [39]. This solution permits a variation in hardness and makes the material smoothed or have a transition from hard to soft without an abrupt seam. For example, Park et al. [40] observed the microstructure of a gel cast FGMM and found smooth variation in hardness from the substrate to the outer layers. This shows that the gel procedure could enhance the transition between materials in FGMMs, unlike nearly any other fabrication method for FGMMs.

4.2.1.2.2 Electrochemical gradation

Electrochemical gradation is the process in which electrochemical treatments are carried out on porous tungsten, and the molten copper is infiltrated into the material. This could be carried out utilizing a wide variety of ceramics and raw metals. This method could alter the density, electric resistivity, porosity, and geometry of an electrolyte within the FGMM [41]. Utilizing this method, FGMMs could be created with similar properties of other manufacturing methods; however, this method is better suited for altering a single component as compared to merging materials at a seam.

Tungsten is generally used as a plasma-facing material for fusion reactors [42]. Because of copper's high thermal conductivity, there has been increasing interest in creating a W/Cu FGMM for use in fusion reactors as diverter assemblies [43]. One way to create an FGMM of tungsten and copper is through electrochemical gradation. W/Cu FGMM gradients can be modified and adjusted through this process by adjusting parameters such as current density, electrolyte resistivity, or tungsten porosity [41]. This method produces a clean gradient at a lower price point.

4.2.1.2.3 Sedimentation and floatation

Sedimentation and floatation are two useful methods for fabricating particle-reinforced FGMMs where there is a density difference between the two components that allow gravity to separate the FGMM into layers [4]. This is a relatively cheap process relying on nature to form the FGMM as opposed to the other methods that require massive amounts of force or power input to fabricate FGMM. The gradient created from sedimentation could define the morphology, size, and mass diffusion of the FGMM [16].

Lucignano and Quadrini [44] utilized the sedimentation process to create a functionally graded polymers using glass beads, silica particles, and UP resin. These samples were then placed through flat punch indentation tests to gauge the mechanical properties of the FGMM. The purpose of the study was to test whether the indentation process could be applied to other FGMMs, such as those containing metals.

4.2.1.2.4 Chemical solution deposition

The final common liquid manufacturing method would be chemical solution deposition (CSD). This method commonly creates a thin electronic oxide film around the object. This film could alter the microstructure of the surface as well as the material's ferroelectric properties [45]. This, in turn, means one can alter the roughness of the surface. It could be created using a liquid deposition dissolved in a solvent, which will most often be organic. This method will create precise crystalline structures at a relatively low cost for a product.

The CSD method is currently being utilized in the creation of functionally graded thermogenerators and sensors. Graded dielectric films for capacitors have been manufactured through the low-cost process of CSD. The thin films can be customized depending on the predicted temperature requirements. Current studies focus on the effectiveness of $BaTiO_3$ to $SrTiO_3$ gradients in these applications [46].

Conventional methods for fabricating FGMM have been rapidly expanding in the last few decades with an extreme need for better functioning parts that have all the qualities desired in less weight. While many of these methods are still in their infancy, the potentials are rapidly allowing for better products with far more affordable prices.

4.2.2 Additive manufacturing process of functionally graded metallic materials

Additive manufacturing is a solid freeform technology, allowing for the direct creation of parts by placing material at set positions. Originally, AM processes were developed and were difficult to scale. This means that many one-off prototypes were constructed and would need to be altered to mass-produce at scale. Today's technologies and newer methods for producing FGMMs in the realms of AM allow for faster production of parts in fewer steps. There are two main compositions of AM FGMMs, which are homogeneous compositions and heterogeneous compositions.

4.2.2.1 Homogeneous FGMMs

Homogeneous FGMMs could be created by varying the density gradients of the material in strategic locations to alter the properties and get desired outcomes. An example of such a process would be mimicking the design of a palm tree with a solid exterior of a part and gradually decreasing the density until it becomes porous in the center. This allows for a lighter part with a higher strength to weight ratios. Fig. 4.2 shows the varied density of an FGMM produced at 15%, 60%, and 65%, where the entire structure is the same metal.

A varying density FGMM was produced by Qiang et al. [48] through a spark plasma sintering (SPS) technique. The resulting FGMM was an alloy with a density varying from 1.74 g/cm^3 to 3.23 g/cm^3.

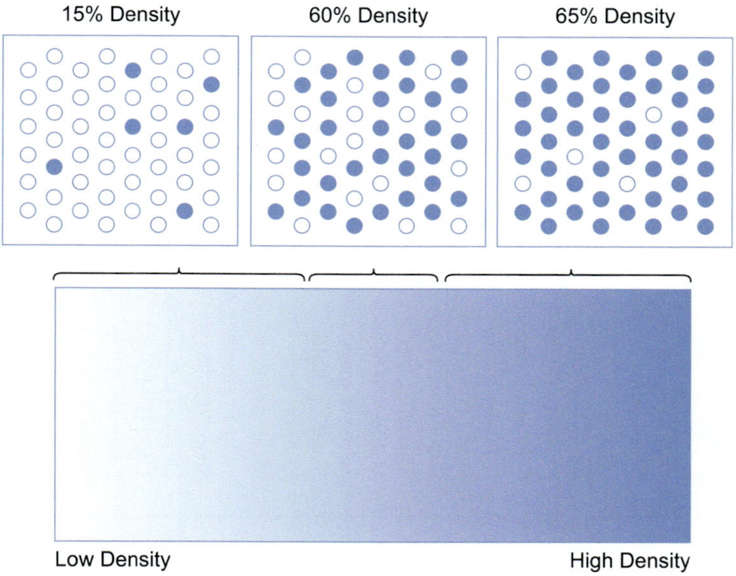

Figure 4.2 Varied densification FGMM produced by AM [47].

4.2.2.2 Heterogeneous FGMMs

Heterogeneous compositions allow for multimaterial through dynamic gradients or complex morphology. By selecting different geometric and material arrangements, one can control the overall function of the FGMM component [47]. Using AM, the bond between two different materials could be improved. This can be carried out with a heterogeneous transition layering compositional parameters and gradients that help to create better bonds to avoid common failures such as cracks and delamination. Additive components have the advantage of being fused to the added materials, gaining the optimum qualities of both materials in the desired location with complex geometry that would otherwise not be capable of being created using traditional manufacturing. Natural examples of heterogeneous FGMMs include, but are not limited to, bones, teeth, and skin due to varying hardnesses, densities, and compositions [49]. Other examples of heterogeneous FGMMs include prosthetic joints, lathe cutting tools, and piezoelectric sensors [50].

4.2.2.3 Advanced additive manufacturing techniques for FGMMs

Currently, not all AM techniques are available for use in FGMMs [51]. Table 4.1 shows the reported types of manufacturing of FGMM components that have been successful. Among them, material extrusions, powdered bed fusions, laminations, and polyjet technology have been investigated in the recent literature [52].

Table 4.1 Reported forms of additive manufacturing for FGMM [47].

Sr.	AM process	Power source	Description	Supporting techniques for FGAM	Materials
1.	Material extrusion	Thermal energy	Material selectively is typically dispensed through a nozzle or extruder	- Freeze-form extrusion fabrication (FEF) - Fused deposition modeling (FDM)	Ceramic slurries, metal pastes, thermoplastics,
2.	Powder bed fusion	High-powdered laser beam electron beams	Feedstock is deposited and selectively fused using a heat source or bonded by means of an adhesive with a view to building up parts.	- Selective Laser Sintering (SLS) - Direct metal Laser Sintering (DMLS) - Selective Laser Melting (SLM) - Selective mask Sintering (SMS) - Electron Beam Melting (EBM)	Atomic metal powder, ceramic powder, polyamide, or polymers
3.	Directed energy deposition	Laser beam	Materials are fused by melting using thermal energy as they are being deposited.	- Laser Engineering net Shape (LENS) - Directed metal Deposition (DMD)	Molten metal powders
4.	Sheet lamination	Laser beam	Material sheets are bonded together and cut selectively in each layer to create desired 3D objects	- Laminated Object Material (LOM) - Ultrasonic Consolidation (UC)	Ceramic tape, Plastic film, metallic sheet
5.	Material jetting	Photo curing	Droplets of build materials are selectively deposited on a layer-by-layer basis	- Polyjet technology (PJT)	Photo polymer digital materials

4.2.2.3.1 Extrusion with 3D printing

Currently, one of the most well-known AM techniques is 3D printing. While there are multiple types of 3D printing, such as fused deposition modeling (FDM) and stereolithography, the most common would be FDM, where materials are extruded to create parts in layers.

a) Fused deposition modeling: FGMM could be manufactured using FDM as well as freeze-form extrusion fabrication (FEF), which are used to make thermoplastics, ceramics, and metal pastes [47]. Having a relatively common form of AM to make FGMMs is extremely efficient and allows for the technology to grow, helping one another.

b) Wire arc additive manufacturing (WAAM): Another form of AM utilizing extrusion is the use of WAAM. WAAM uses a feed wire and utilizes electric potential to create an arc across from the feed to the object, melting the wire into the desired shape. While this is not a traditional form of extrusion, it allows for less porous products to be created, allowing for stronger FGMMs to be created. By utilizing extrusions, multiple materials can be extruded via multi-textured machines that lay individual layers of materials together, allowing for the complex geometries to still be created and by fusing the materials layer by layer together.

Extrusion machines tend to fall short on complex parts that require overlaps and overhangs. This issue arises from the process itself, which requires layers to be laid out one at a time, increasing the height of layers at a time. This means to extrude on an unsupported area, which will most likely be deformed, and the tolerances could be decreased. Therefore, the support material is added to minimize this issue. These supports could increase part cost, require finishing to remove, and even damage the parts if they fail to bond as desired. Although the extrusion processes require support and have shortcomings, the technology is rapidly advancing with superior support materials such as dissolvable supports, which could be dissolved after the process.

A proposed application of FGMM manufacturing via extrusion is seen through an experimental system by Craviero et al. [53] called "rapid construction." The system takes advantage of three multideposition heads to create layered walls of graded composite meshes. This system is unique in the idea that can build structures through an extrusion process that automatically adjusts material composition based on the desired outcome.

4.2.2.3.2 Spark plasma sintering using powder bed fusion

Solid-phase processes have recently become a more widely used and well-established technique for the creation of ceramics, metals, and hard materials. These are mainly created using either SPS of powder metallurgy. SPS refers to a sintering process for modifying ceramic and metals, which stems from a method created in 1910 that used electric energy to consolidate a powdered material [16]. For the following decades, the technique was improved upon where applications ranged from brass and bronze to other materials. This process is like that of powder bed fusion with high-powered laser beams or electron beams. These solid-phase processes utilize a bed of powder, which is used to coat the build volume in a material to be worked. These materials are then targeted by lasers or beams, which fuse the particles together, making a solid part. These parts, while solid, are not always complete and require more postprocessing. This could include heating the parts to a transition temperature and further fusing the parts together. This process is expanded to create FGMMs by including a separate feeder for a second material [16]. This means that multiple materials could be laid out in layers and fused together to create FGMMs.

Due to its high thermal conductivity and low coefficient of thermal expansion, vanadium alloys are often used in fission reactors [49]. However, the cost and corrosion concerns have made it necessary to find a superior solution. SPS has been explored as a method to create a W/V FGMM in order to create a solution superior to just vanadium alone [54].

4.2.2.3.3 Directed energy deposition

Directed energy deposition (DED) involves technologies such as laser engineering net shape (LENS) or directed metal deposition (DMD), where materials are fused by using laser energy and melt to form molten metal powders [32]. One studied application of DED within the manufacturing industry is through the repair of die and molds. Tools that wear quickly create inconsistent parts and need to be replaced. Tools can be repaired through the DED method and reused, saving both time and money [24].

4.2.2.3.4 Sheet lamination

Sheet lamination could be achieved through ultrasonic consolidation using stainless steel as demonstrated by Kumar [55]. This allows the coatings to act as a foil and join utilizing ultrasonic welding techniques. Ultrasonic consolidation allows for the fabrication of various FGMM that could be fitted with laminate sheets to protect or be used as a thermal barrier for the main component. Current applications of the sheet lamination method primarily focus on embedded electrical components and structures [56,57]. Ultrasonic consolidation has the advantage of low-temperature solid-state bonding to fabricate similar components [55].

4.2.2.3.5 Material jetting

Material jetting or commonly known as polyjet, could utilize the widest array of materials in a single print of all AM technologies [47]. It allows for droplets of material to be placed layer by layer, allowing for flexible customization of parts that could take advantage of both homogeneous and heterogeneous AM techniques and give the designer incredible flexibility. This can be especially important to create parts that vary the rigidity of an object and incorporate flexible components that will transform into solid components. A proposed use of an FGMM created by material jetting is a living hinge. Living hinges eliminate complex rotational joints and replace them with a flexible material. The combination of rigid and flexible materials that can be manufactured by material jetting creates a large market for the use of FGMMs [58].

The issues with AM carry over to FGMMs with many defects and poor control being widespread. With limited ability to control the tolerances of parts, mass production surfaces are prone to having a variety of quality among batches of parts. This becomes more common with complex internal geometries and structures that rely on tolerances and machine precision. Tolerance issues, while not always as perfect as a mold, can be dismissed as the parts made via AM techniques may require multiple failed part productions to have a singular successful component; however, these parts have the capabilities and fewer limitations than those that can be made traditionally.

With all the inefficiencies of AM, the mass production of FGMMs can be greatly improved with the newer methods of AM methods. AM allows for the rapid prototyping and the production of complex geometries that would require expendable molds and extremely complex manufacturing capabilities otherwise. AM methods also allow for the capability to repair pre-existing parts and structures by adding on to the parts utilizing the same methods as specified in the CAD software. This allows for a greater ability to fabricate and maintain parts with less machinery and complexity.

4.3 Properties of FGMM

FGMMs exhibit unique properties, which helped them to be implemented in diversified applications. In this chapter, the mechanical, tribological, and corrosive properties of FGMM will be discussed to better understand their functionalities.

4.3.1 Mechanical properties

Functionally graded materials were initially created to decrease thermal stresses at the interface between two materials. With a metal bonded to ceramic, the varying thermal expansion coefficients can create large spikes in stress at the interface between the materials. When the two materials are combined in a metal matrix composite, the metal and ceramic mix at the interface creates a thermal expansion gradient, drastically decreasing the internal thermal stress at the interface. This can eliminate the potential for delamination when temperature drastically increases [59]. These advantages can be seen below in Fig. 4.3. Metal matrix composites are a very practical and easy way to deal with temperature problems that occur at the material interface [60].

Additionally, graded microstructure within a material offers the ability to create a strong and stiff component with less material and less weight. These graded microstructures could be found throughout nature in things such as bamboo as well as human and animal bones. With this advantage, FGMMs can be manufactured to be able to withstand specific repeated stress without sacrificing the weight that would be required with a pure material. In these materials, the elastic modulus is a function of the depth of the material. In doing this, the material is also significantly lighter [61].

WAAM uses metallic feedstock and an electric arc heating source to form the material into its shape. Although there are defects in certain materials when the wire arc manufacturing process is used, it is a process that is ideal for efficiency and could be implemented on various metals [62]. Unlike conventional manufactured FGMMs using the powder bed method, WAAM methods are able to achieve full density materials with little to no porosity problems. This opens up the ability to make much stronger FGMMs for broader applications. Additionally, the opportunities for WAAM are limitless due to the ability to easily combine two materials while eliminating problems at

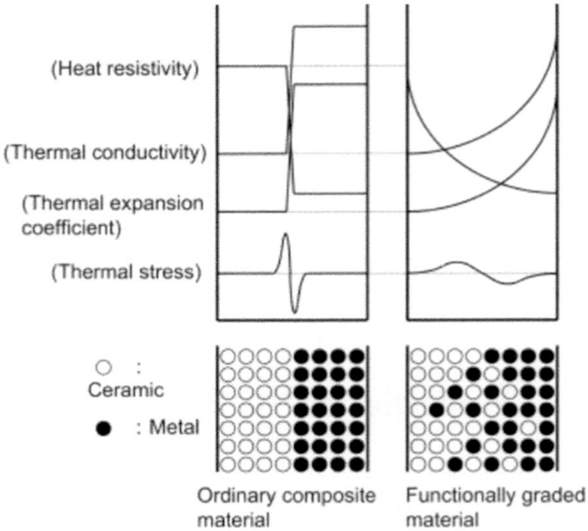

Figure 4.3 Metal matrix composites blend the metal and ceramic at the interface providing many thermomechanical advantages [59].

the interface. This is exemplified by a group of researchers, where Fe—Al intermetallics experimented, and the issues of brittleness were observed at the room temperature [63]. As a possible solution, they used WAAM to manufacture a Fe—FeAl functionally graded material that combines a FeAl intermetallic's traditional properties and a more ductile Fe-based alloy such as steel [62,63]. The graded microstructure of aluminum composition consisted of 15%—50% in 5% increments. After tensile strength and hardness tests were completed, they concluded that the Fe—FeAl FGMM produced met the preferred room temperature ductility. Although they claimed more testing would be needed to prove the practicality before large-scale use, this experiment shows how additive-manufactured FGMM could create much more dynamic metals with multiple preferred mechanical properties that one could not have with a traditionally manufactured material [63].

Through Arc additive manufacturing, researchers have also been able to maintain sufficient plastic deformation and higher tensile strength of SS904L/Hastelloy C-276 than SS904L wrought alloy [64]. During the tensile strength test, the FGMM fractured at the SS904L level with a tensile strength of 680.73 MPa, which is significantly higher than the tensile strength of SS904L wrought alloys (490 MPa). The WAAM process creates a complex cyclic thermal history due to the accumulation of heat and the partial remelting of the previous layer of the metal. This creates variations in the microstructure, such as random dendritic grains. Because of this, the fatigue strength was 28%—35% lower in the FGMM than in the SS904L wrought alloys. This shows that though the WAAM process brought the material to full density and was shown to increase tensile strength, it has its downfalls like all other manufacturing processes. Yet still, with the significant increase in tensile strength, this manufacturing process has much promise of further application. The increased mechanical property found in the research shows that wire arc manufacturing is an ideal process to create materials where a sudden change is needed in the metallic material [64].

4.3.2 Tribological properties

When talking about the tribological properties of a substance or material, one is referring to the different interactions the surface of material undergoes while in motion and in contact with another substance. Tribology is a science that deals with friction, wear, and lubrication [65]. Since there is no such thing as a frictionless surface, the tribological properties must be considered while choosing the material. However, the problem with choosing a traditional composite, which is mostly a homogeneous mixture of constitutes, is that a compromise has to be made between the properties of the constituent material and the application's requirement due to the loss of the overall identity of the constituent while in the composite mixture. Because of this loss of identity and property compromise most composites have to make, most mechanical failures can be tied back to one of these traits. Functionally graded materials (i.e., FGMMs), on the other hand, comprise two or more different layers of materials with a graded interlayer, meaning there is no compromise that most composites have to make since FGMMs can exhibit the same characteristics of each individual material it is composed of.

The benefit of using functionally graded materials is that almost all of their tribological properties could be specifically tailored to meet the requirements of the intended applications, such as better hardness or better resistance to wear. For example, a nearly wear-resistant functionally graded material could be a ceramic–metal composite. A ceramic face could be used on the exposed surface to benefit from its high wear resistance, while a metallic body could be used as a durable and reliable internal body.

A material similar to the above example was created and tested by Srinivas et al. [66]. They made an aluminum-based functionally graded material with silicon-carbide and magnesium peroxide constituents. Four layers of the FGMM were formulated, such as 100% pure aluminum, 90%Al + 10%SiC, 90%Al + 5%SiC +5%MgO_2, and 85%Al + 5%SiC +10%MgO_2, respectively [66]. They investigated the tribological properties of their functionally graded material by the pin-on-disc test, which is widely used to test the wear and friction characteristics of a material. Depending on what kind of tests the user wants to run, the material for the pin and the disc can vary. A typical pin-on-disc setup is shown in Fig. 4.4.

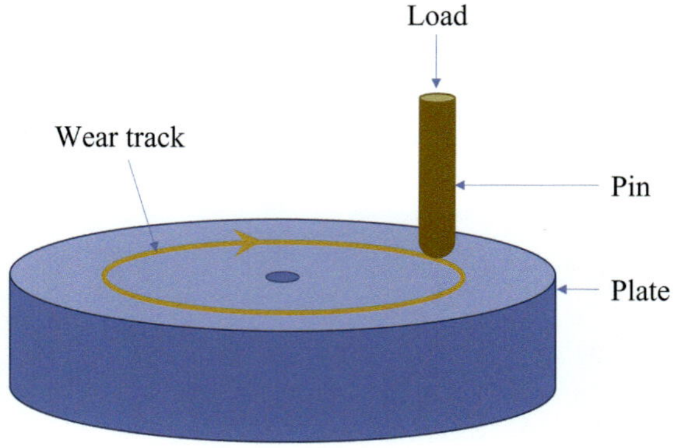

Figure 4.4 Pin-on-disc test apparatus.

The test that Srinivas et al. [66] conducted for stainless steel pin at 15 N to a disc made of FGMM observed less wear for the FGMM disc compared to that of aluminum. The FGMM performed much better due to its increased hardness from the presence of silicon carbide. They also analyzed that the further they went down into the FGMM layer-wise when they got to the layers with less silicon carbide, the wear started to resemble that of the pure aluminum due to magnesium's low hardness and high brittleness [66].

A rotating pin-on-disc test is very useful when studying the tribological properties of functionally graded materials; however, some applications in which the materials may be used for such as in automobile components, which could include engine blocks, piston rings, and cylinder liners, to name a few, are subjected to reciprocating forces rather than rotating forces. Due to the parts being under different wear forces, different tests could be carried out to replicate the environment in which the components will be subjected to everyday use and wear.

Different from the pin-on-disc test, a reciprocating wear test has the pin held stationary with a load acting perpendicular to the surface of the testing material [67]. The material then undergoes a reciprocating action for a set number of cycles (Fig. 4.5). As Karun et al. [67] observed, the rotating as well as the reciprocating wear performance of aluminum alloys and metal matrix composites (MMC)/FGMM not only depends on the service conditions (load, sliding distance, sliding speed, countersurface temperature, and roughness of the surface) but also depends upon various material-related mechanical properties (hardness, ductility, and toughness) and microstructure [67]. The objective of their investigation was about processing of A356 alloy and 10 and 20 wt.% SiCp-reinforced 356 Al MMC and FGMM using liquid metal stir casting as well as centrifugal casting techniques and evaluation of microstructure, mechanical properties, rotary, and reciprocating wear characteristics [67].

Karun et al. [67] conducted both the pin-on-disc test and the reciprocating wear test with dry sliding conditions during their tests. The material they used for the pin-on-disc test was a 165 mm-diameter high-carbon alloy steel with a hardness of 63 HRC with compressive strength and abrasion resistance and material composition of Cu 1.00%, Mg 0.50%, Cr 1.40%, and Si 0.20%, with the test parameters being a sliding

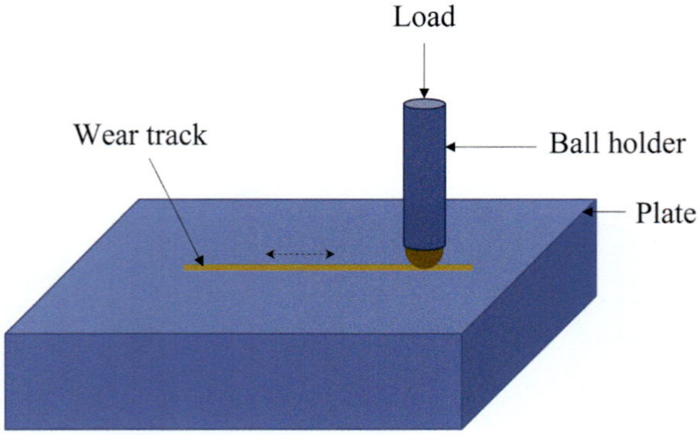

Figure 4.5 Reciprocating wear test apparatus.

distance of 330 m and a maintained constant velocity of 1 m/s and four different loads of 1, 4, 8, and 12 kg [67]. As for the reciprocating tests, they set them up using heat-treated parts, which were 6 mm in diameter and 30 mm long, and a cantilevered loaded arm. They also maintained a constant sliding distance of 330 m and with seven total load variations: 1, 4, 6, 8, 10, and 12 kg. They determined that the reciprocating wear rate depended on various parameters such as the mean reciprocating velocity, applied load, the material used, and the countersurface temperature as a result of continuous sliding of the pin over it [67].

The tests that Karun et al. [67] were running were testing the differences in the hardness and tribological properties of "as-cast" and "heat-treated" samples of their composites. The results confirmed that their heat-treated casting of SiC-reinforced aluminum composites was by far stronger than their "as-cast" version. A minimum hardness of 54 BHN and maximum hardness of 104 BHN was observed for the as-cast and heat-treated samples, respectively. Metal matrix composites were enhanced with a maximum hardness of 122 and 114 BHN for the heat-treated samples of A356-20 wt.% SiCp and A356-10 wt.% SiCp, and a minimum BHN of 76 and 75 for the same samples. The hardness difference of composites seems to be less because of the composite with 20 wt.% SiCp, having an agglomeration of particles, which leads to a reduction in hardness values. Hardness data disclose the fact that the hardness values are greater for heat-treated than the as-casts samples [67].

The increase in hardness that Karun et al. [67] achieved with their functionally graded material alloy is just an example of the kind of tailoring this technology could lead to. A common and economical way of producing functionally graded materials is by centrifugal casting. That is because the centrifugal effect can produce different layers with different specifications, all within the same component. A specific example would be the functionally graded aluminum/zirconia metal matrix. The hardness of the composite was higher at the outer surface compared to the rest of the composite, and the tensile strength was higher on the upper section due to the higher segregation of reinforcement particles [68]. With AM technology developing and advancing every day, soon the same, if not better, properties could be achieved via this method compared to centrifugal casting or other methods.

4.3.3 Corrosion and tribocorrosion properties

Metallic FGMMs have the ability to possess exceptional corrosion or tribocorrosion resistance properties, which make them an excellent option for certain toxic environment requirements [69]. Tribocorrosion is defined as a metallic material degradation process due to the synergistic effect of wear and corrosion [65]. Fig. 4.6 shows a typical schematic for tribocorrosion testing. Here, a corrosion testing setup is integrated with the reciprocating wear testing setup [70]. The metallic FGMM's come in different forms to satisfy the need for corrosion resistance, such as composite metallic materials or composite metallic coatings [66]. These metallic FGMMs are produced by evaluating certain elements or alloys and incorporating them into another material to fit the requirements of the environment (toxic). FGMMs could be produced

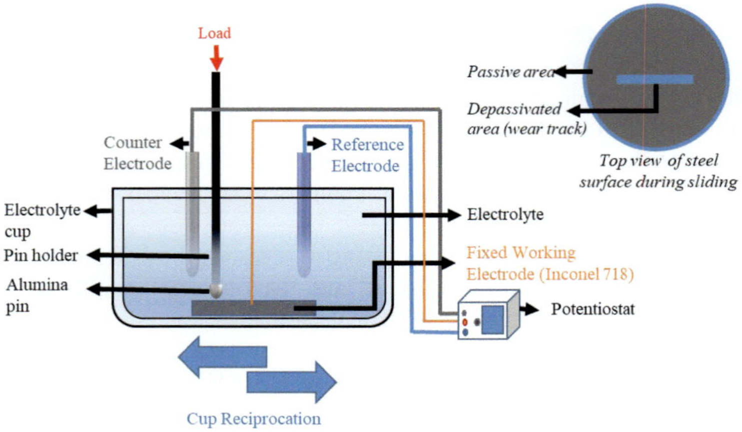

Figure 4.6 Linear reciprocating ball-on-plate tribometer [73].

by a number of materials with excellent tribological behaviors, such as aluminum, silicons, carbides, and magnesium. Some examples of different types of metallic FGMM's and their corrosive properties and characteristics are provided below.

Composite metallic coatings are capable of changing a material surface's mechanical and chemical properties to address certain material requirements [71]. These composite coatings often have higher corrosion resistance than basic (pure) metallic surfaces [72]. Specifically, nanocomposite coatings show superior anticorrosive properties and could be created using metallic FGMMs. Lajevardi et al. [73] evaluated functionally graded $Ni-Al_2O_3$ nanocomposite coating in 2017. $Ni-Al_2O_3$ is a nickel−alumina FGMM that is produced by homogenous metal distribution (electrodeposited inert particles in metal-matrix) and has superior tribocorrosion characteristics [73]. Additionally, $Ni-Al_2O_3$ is abrasive resistant to metallic materials and has high wear resistance and hardness, making it an exceptionally rounded material. The anticorrosion property was evaluated using an ultrasonic agitation experiment, which was conducted at 3.5%wt. NaCl solution, with different cycle loads at different frequencies. Additionally, the use of mechanical (100 RPM) and magnetic (300 RPM) stirring with the 20 kHz, 20W agitation kept the particles uniform throughout the evaluation. The tribocorrosion test was performed using a linear reciprocating ball-on-plate tribometer. The coated surface was immersed in the 3.5%wt. NaCl solution and a 4.75 mm alumina ball rubbed against the coating surface at a frequency of 1 Hz. The stroke length was measured at 5 mm for 60 min (total length: 144 m) with a 20 N load.

The roughness of the surface was measured using two load cells, which could measure the tangential forces simultaneously. Each examination used five samples, which had different deposited load cycles (A: 90%, B: 75%, C: 50%, D: 25%, and E: 10%). Due to the different methods in which the sample was prepared, the samples have different thicknesses and nanoparticle structures. The coated sample served as the electrode, and its potential was controlled using a Gamry potentiostat.

The results of the examinations included that the corrosion resistance of the samples decreased as the volume increased of alumina particles in low load cycle settings. This also resulted in higher hardness readings of the soft composite coating, resulting in excellent tribocorrosion properties [73]. Additionally, morphological investigations were conducted and found that adhesive wear was the primary cause of wear and dominated the samples with softer textures.

The corrosion examination was obtained by centrifugally producing both Al/Al_3Ti and Al/Al_3Zr and dipping them in a 0.6 M NaCl solution and scanned for potential (−1.4 to 0.2 V at a scan rate of 2 mV s^{-1}) with a PGP201 potentiostat/galvanostat controlled by the VoltMaster-1 software. The tribocorrosion properties were evaluated in the outer regions of the materials with a reciprocating tribometer (Plint TE67/R). The results of the Al/Al_3Ti FGMM samples show an increase of current density at the beginning of sliding, which corresponds to the fresh material becoming exposed. Eventually, a small decrease in current density until steady state occurs around 100s for 30G and 120G, while 60G sample took much more time. Although all materials show benefiting behavior, 120G presents a relatively more beneficial tribocorrosion behavior with a lower friction coefficient and current density. Fig. 4.7 below shows the results of the evaluation. The results of Al/Al_3Zr FGMM samples, in comparison, showed higher current density and friction coefficient, portraying worse tribocorrosion performance compared to that of Al/Al_3Ti [74].

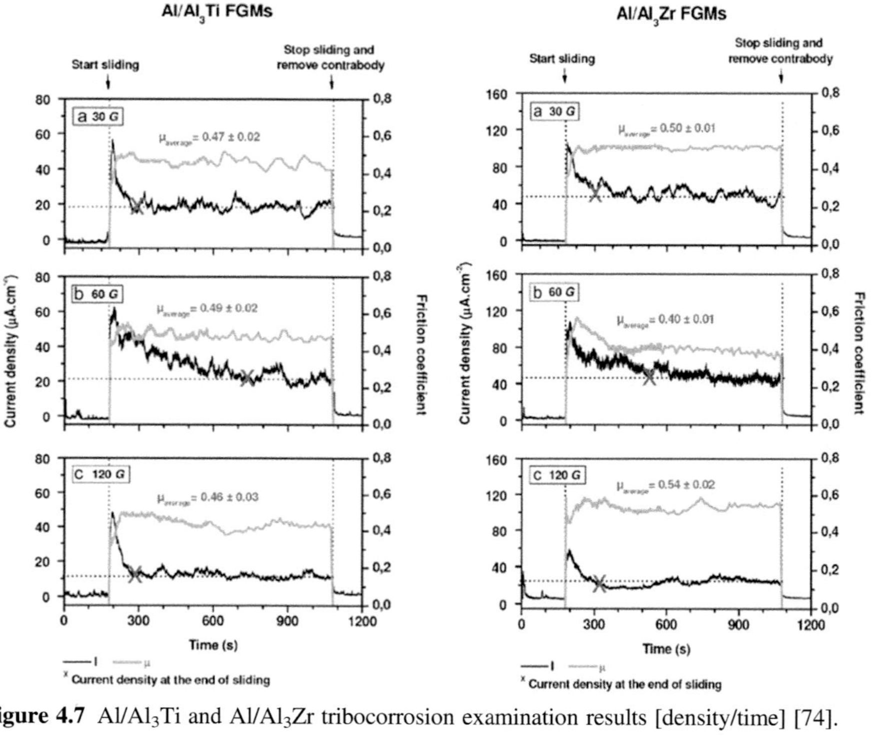

Figure 4.7 Al/Al_3Ti and Al/Al_3Zr tribocorrosion examination results [density/time] [74].

This particular examination showed the benefits metallic FGMMs could bring to a variety of fields. Therefore, the accelerated damage and unwanted material removal could be minimized by using FGMMs in different applications. Specifically, chemical plants, the medical sector, and the energy industry have been benefitted significantly from the FGMMs. In the following section, different potential applications of FGMMs are described.

4.4 Applications

FGMM has wide potential to aid different industries. The medical, aerospace, and energy industries have already observed various successful applications of the metallic FGMM in both high-temperature and corrosive environments. Current applications of metallic FGMMs in these industries include, but are not limited to, bio-implants, inner walls of nuclear reactors, composite piping systems, thermal barrier coatings, and the graded electrode for the production of solid oxide fuel [75].

4.4.1 Medical sector

The fields of medicine and biomedical engineering have multiple applications for FGMMs, such as bone and tooth implants [76,77]. The specific metallic FGMMs used for the medical sector are known as "natural FGMMs" due to their structure, which is commonly found in nature. Additionally, AM is growing in terms of producing medical devices and various implants. Medical devices could be created (modeled) using computer-aided design (CAD) programs and produce g.code for process manufacturing. The devices (or implants) are required to be made of various materials based on application and could include polymer—ceramics, metal—ceramics, composites/resins, and various other FGMMs.

Implant operations include skeletal, orthopedic, and dental applications, where the properties of FGMMs used vary for the different types of implants [78]. The word "implant" may be classified as an artificial bone for medical use and "dental implant" as an artificial tooth for dental use. Originally, many implants were produced using composites that could potentially fail under extreme working conditions, which requires the need to produce metallic FGMMs [78]. Implants for orthopedics are used mostly as structurally enforced artificial support, which is inserted inside the area needing support. Medical implants lay more weight on strength, torque, and toughness in mechanical properties and tribological abrasion resistance (artificial joints) [78]. Dental implants are usually much smaller and are used to restore the masticatory function (absent when a tooth is lost or removed). FDMMs are useful as dental implants since natural teeth have a "gradient" structure [79]. In fact, dental implants made with FDMMs demonstrate superior properties compared to all-ceramic restorations [80]. Implant materials must all satisfy the requirements of biocompatibility, strength, and corrosion resistance. Fig. 4.8 shows some natural examples of FGMMs that contained graded properties, such as in human teeth, human bones, palm stem, and

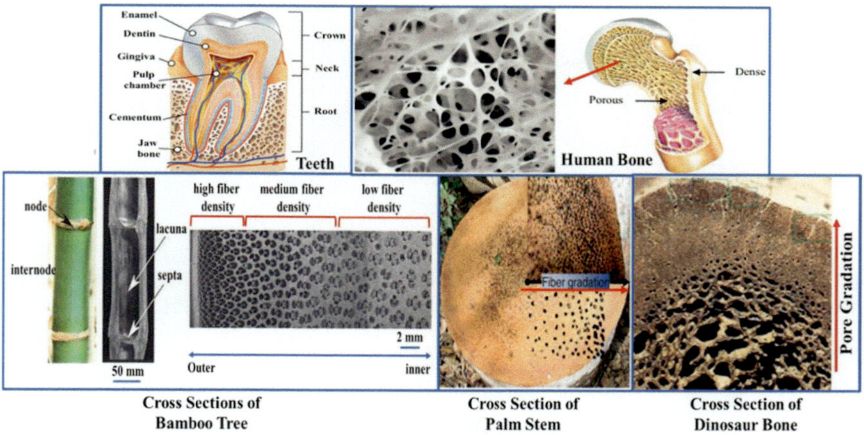

Figure 4.8 Some natural examples of FGMMs contain structures with graded properties [11,81,82].

dinosaur bone [11]. It is also observed that bamboo has three density regions: high fiber, medium fiber, and low fiber [81]. These graded properties provide their strength, and therefore, bamboo is used as a structural material in many parts of our world. Therefore, researchers not only tried to replicate those graded properties, as in human or dinosaur bones or teeth, but also alike bamboo structures [11].

Researchers at Hokkaido University Graduate School and Institute of Material Research examined the FGMM titanium/hydroxyapatite (Ti/HAP) due to its longitudinal gradient structure and cylindrical shape produced from the AM process of powder metallurgy [83]. Titanium is one of the best metallic biocompatible materials and is commonly used in implants due to its desirable properties, such as high strength, low density, inertness to the environment of the body, low Young's modulus, and corrosion resistance [84]. Hydroxyapatite (HAP) is the main component in bone and teeth and possesses bioactive properties that cause new bone formation [85]. The powder metallurgy also optimized both biocompatible and mechanical properties of each material during formation. In general, Ti-based alloys are finding more applications biomedically due to their excellent mechanical, physical, and biological performance [83]. Therefore, the FGMMs from Ti are of significant interest in this sector.

The Ti/HAP FGMM was evaluated using the Brinell Hardness examination based on different types of Ti powders (<45 μm and <150 μm) and HAP powders (<20 μm and 20–45 μm) and different regions inside the Ti/HAP FGMM, as shown in Fig. 4.9. The examinations were conducted on sample sizes of $3 \times 3 \times 7$ mm and $2 \times 2 \times 14$ mm on the Instron 4204 universal testing machine. Additionally, imaging examinations were conducted, such as visual inspection (VI), optical microscope (OM), scanning electron microscopy (SEM), and X-ray scanning analytical microscopy (XSAM). Biocompatibility was assessed using an animal implementation test in which uniform sample implants were inserted into the bone marrow of the femora in rats [86].

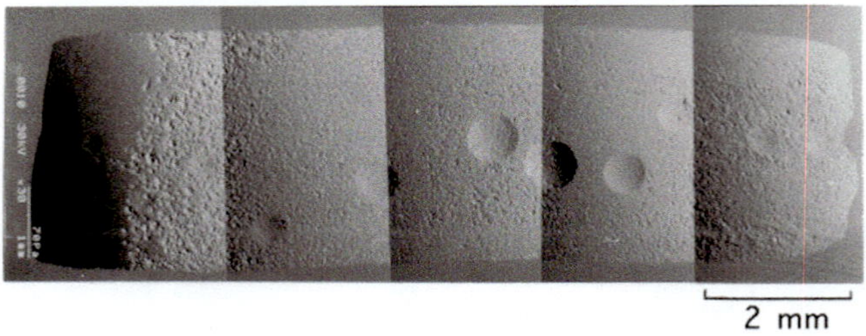

Figure 4.9 Ti/HAP FGMM after the Brinell Hardness test observed by reflective SEM imaging where the round indentations indicate the hardness test in each region [86].

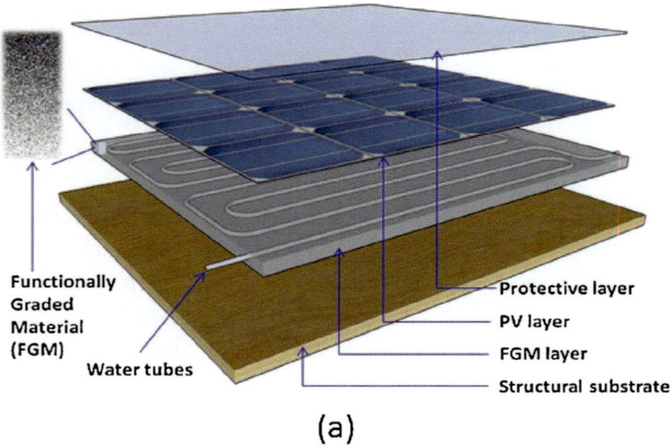

Figure 4.10 The functionally graded microstructure provides insulation to the structural substrate while exchanging the heat from the photovoltaic (PV) layer into the water tubing [87].

The results of the Brinell Hardness examination were analyzed with SEM and indicated a hardness change with the HAP content, which is shown in Fig. 4.10 above [86]. The Brinell test is a long preferred method of assessing metals' hardness during the forming process [88]. The hardness was low in the center region of the material (where Ti and HAP were mixed comparably), which could be due to friction between particles during the AM process. The addition of Ti to HAP for this FGMM greatly increased the fracture toughness and the maximum compression strength of the bone (which is now 150 MPa). Additionally, the decreasing tendency of hardness in the jaw direction contributes to stress relief and relaxation, which can alleviate the damage by any imposition on the implant. The only issue comes with the binding force between Ti and HAP, which is still compromised and is not enough for practical use. Hokkaido University of Dental and Institute for Material Research Ti/HAP optimized each of the mechanical and biocompatibility properties of each material and the maturation of newly formed bone in animal specimens (HAP). The study also demonstrated tissue reaction to gradually modified materials such as FGMMs, which implies the further development of Ti/HAP [86].

4.4.2 Energy industry

As humanity's demands for energy increase, research and development in creating new energy harvesting methods and improvement on the current methods is growing exponentially. FGMMs have made their way into this field due to their advantages in their thermomechanical properties and increased strength and durability. In the renewable energy industry, metallic FGMMs are being implemented to increase the efficiency of building integrated photovoltaics through enhanced thermomechanical properties. FGMMs can also be seen being researched in the offshore oil drilling industry in the search for strong and durable materials for marine risers.

A current problem with solar panels is their inability to regulate temperature [89,90]. As the temperature of the solar panel increases, the efficiency of energy absorbed decreases [89]. The extra heat can also be detrimental to the operational life of the panel [89]. The thermo-mechanical properties of metallic FGMMs make them a viable solution to increase the efficiency and lifespan of building-integrated solar panels. Researchers at Columbia University have manufactured a heat exchanging layer that goes within a photovoltaic that insulates the building from the large amounts of heat absorbed. Meanwhile, it can easily dissipate the heat to water flowing through an integrated piping system [87]. This concept can be seen in the labeled diagram in Fig. 4.10. This is possible because the thermal conductivity gradients within metallic FGMM aides in controlling heat to a higher precision allowing both insulative and conductive properties to exist within the same material.

The manufacturing process of creating this heat exchanging layer consists of mixing aluminum (Al) powder with high density polyethylene (HDPE). Because Al and HDPE have very different densities when a liquid such as ethanol is introduced, the particles will fall through the mixture at different rates. Once the desired microstructure has formed, the liquid is filtered out, and the mixture is melted to form a metal matrix composite with a graded microstructure. The layer is manufactured with water tubing near the top of the material where it is most conductive. The higher thermal conductivity increases the heat exchange from the PV layer to the cooler water flowing through the tubing. Additional to more efficient solar panels, the FGMM allows the solar panel to create heated water for the building during the day to save more energy. The functionally graded Al layer is necessary to achieve the added efficiency due to the PV layer operating at a lower temperature. Otherwise, significantly more energy would have to be added to the system for heat exchange [87]. In addition, the FGMM increases heat exchange to the water flowing through the system, which leads to less energy used in the water heating systems in the building. After confirming the advantages of using this FGMM, more research must be conducted to achieve large-scale manufacturing.

In another sector of the energy industry, FGMMs are being experimented within marine risers used for offshore drilling [91,92]. Marine risers are used in the transportation of oil from below the ocean floor up to the drilling platform. These cylindrical conduits perform a critical operation as a failure in these risers can lead to oil spewing out into the ocean, causing environmental catastrophe [91]. Taking into consideration the corrosiveness of the ocean water and the large magnitude hydrodynamic forces that the marine riser is exposed to, material selection is imperative to the resistance against these aspects of the ocean. The article *WAAM of functionally graded material for marine risers* does an in depth analysis on an R 70S-6 Carbon Manganese steel and ER 2209 duplex stainless steel FGMM [93]. A concept design of the FGMM marine riser can be seen in Fig. 4.11. This FGMM would replace the current material used in marine risers. The manufactured FGMM had a carbon manganese steel layer of 14.47 mm and ER 2209 duplex stainless steel layer of 3 mm. The carbon manganese layer is used for its anticorrosive properties, while the duplex steel is used for its strength. The hardness tests completed on the FGMM showed that the interface has the highest micro hardness at 307–320 UV. The corrosion tests showed the superior corrosion properties in the carbon manganese steel with a corrosion rate of 823.35 g/cm versus the 70.77 g/cm rate of the duplex steel [93].

The FGMM is comparable to the X-52 carbon manganese steel that is used in the current marine risers [93]. This study ultimately found that tensile strength, percent elongation, and yield strength are similar to that of X-52 but has slightly increased ultimate strength. Corrosion comparisons show that the FGMM is 12 times more corrosion resistant than that of the primary materials of X-52. The analysis shows that the functionally generated marine riser material is significantly more capable than the current materials used in marine risers and could help prevent failures in the marine risers and furthermore environmental devastation.

Additive-manufactured FGMMs are also beginning to be implemented in electric transference in high-voltage contact parts [94]. Electrical contacts are used as a controller to complete or disrupt high-load circuits. These contacts are subjected to extreme temperatures during arcing. Traditional tungsten copper contacts are manufactured with an infiltration or mixed powder sintering method [94]. This creates a product that has poor antielectrical corrosive properties and creates a short usability life for the contacts [94]. The copper tungsten FGMM is made of powder-fed laser AM, where the copper feeder adjusts from 0.2 to 15 g/min and the tungsten from 0 to 25 g/min. The mass volume of tungsten goes from 0% to 99%. The gradient layering of the copper tungsten FGMM can be seen below in Fig. 4.12. Studies show that the FGMM copper tungsten has a higher arc erosion resistance and a longer operational life than that of the traditionally manufactured copper tungsten contact [94].

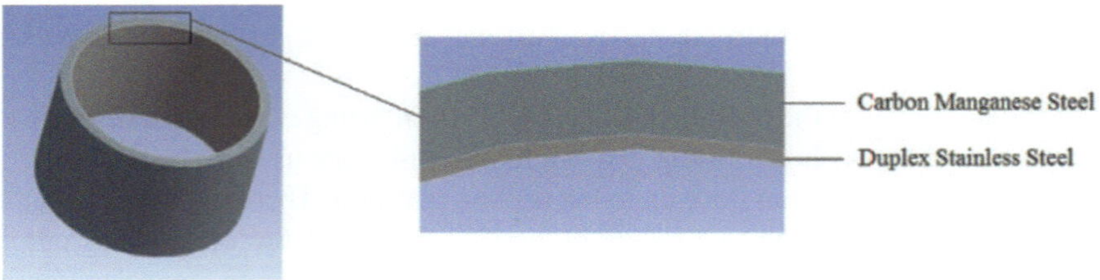

Figure 4.11 Marine riser FGMM concept that shows the duplex stainless and carbon manganese steel layers of the piping [93].

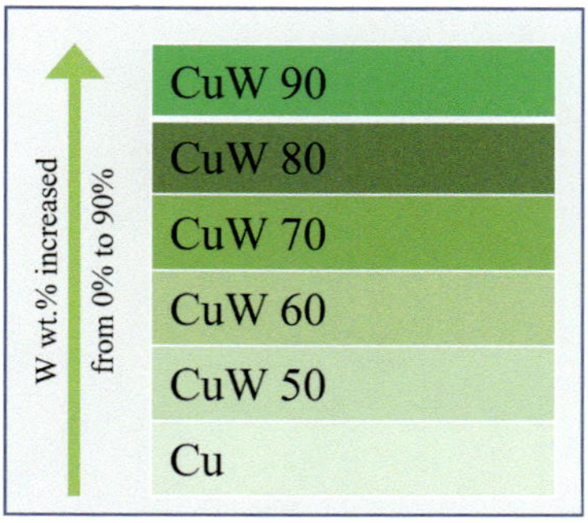

Figure 4.12 The functionally graded copper tungsten gradient layering [94].

4.4.3 Future prospects in the chemical industry

While FGMMs are not widely utilized in the chemical industry at this time, the prospective use could dramatically change the way chemicals are stored and produced.

As AM continues to make proportionate technical advances, so does the possibility for FGMMs in the chemical sector. Due to the volatile nature of the materials being handled, components need to be able to withstand contradictory requirements. The ideal material needs to be able to handle thermal stresses, mechanical stresses, and strains, as well as the corrosive or erosive properties of which the material is in constant contact. Metallic FGMMs are a unique solution as the resulting material gradient can withstand these requirements more than a singular material. By utilizing FGMMs over composite structures, one reaps the benefit of distributed thermal and mechanical properties that would otherwise be separated at a material boundary, usually resulting in breakage or failure along said boundary.

Components of a chemical plant that already benefit from the use of FGMMs include pressure vessels, heat exchanger tubes, and piping systems [75]. Exhaust valves in combustion engines are subjected to thermal stress, mechanical stresses, and corrosive chemicals. Depending on the materials used, the valve can experience significant friction and wear [95]. Using $NiAl-Al_2O_3$ (an FGMM) for this application would result in a lighter, longer-lasting component. When in a gradient, $NiAl-Al_2O_3$ has high mechanical strength, hardness, and resistance to corrosion and abrasive wear [96].

Another viable use for additive-manufactured FGMMs in the chemical industry is in regard to heat sinks and heat exchangers. Shell-and-tube heat exchangers, which can be seen in Fig. 4.13, are universally the most common heat exchangers for industrial use as they are reliable and easily maintainable [98]. Traditionally, copper and aluminum are the first choices in heat exchanger tube design due to their high thermal conductivity, which is 398 (W/m K) and 247 (W/m K), respectively [99,100]. Due to the temperature difference between the exchanger's shell side and tube side, thermal stresses play a major role in material selection for the heat exchanger tubes. Both thermal fatigue and thermal expansion failures are problematic with aluminum and copper exchanger tubes. Thermal fatigue failures are a result of cyclic thermal stresses

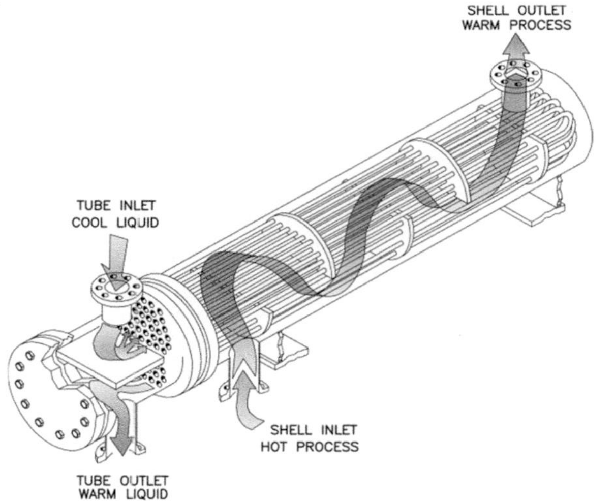

Figure 4.13 An internal view of a shell-and-tube exchanger commonly used in chemical manufacturing plants [97].

exceeding the tensile strength of a material and usually occur at the U-bend of a heat exchanger tube [101]. Thermal expansion failures result in increased pressure, which then exceeds a material's yielding point [102]. In addition to these two failure types related to temperature, the tubes need to be able to withstand any stress due to fluid speed, fluid pressure, and vibrational effects. A proposed alternative to these copper and aluminum heat exchanger tubes is the introduction of aluminum nitride/tungsten, or AIN/W, due to its high thermal conductivity and strength [103]. Although aluminum nitride by itself is extremely strong and has a high compressive strength, its tensile strength is still relatively low since it is a ceramic [104]. The addition of tungsten creates a FGMM that is able to withstand very high stress due to very high tensile strength under high temperatures [105].

4.5 Discussion and conclusion

As AM continues to advance, so does the development of FGMMs. By using AM as a means for creating FGMMs, less waste is produced in the creation process, and the piece is more affordable to produce as compared to traditional practices. AM of FGMMs allows for the manufacturing process to take less time, increasing speed and efficiency. Because the mechanical properties of metallic FGMMs are highly customizable due to the gradient of materials input, their resistance to thermal and bending stresses makes them highly desirable. Enhanced tribological and corrosion resistance properties aid in implementing FGMMs in chemical, medical, and energy industries.

Conventional methods of manufacturing have evolved since their conception. There are two main methods available for producing FGMMs: gas-based methods and liquid-phase methods. Gas-based methods include techniques such as CVD, thermal spraying, and wire-arcing. While some gas-based methods are highly desirable (such as thermal spraying), the results tend to be inconsistent and have a large potential for creating defects in the manufactured FGMM. Liquid-phase methods include various forms of casting, electrochemical gradation, and sedimentation processes. Liquid-phase methods are generally the most commonly used for creating FGMMs.

AM is gaining popularity due to its cost, accuracy, speed, and customizability of resulting workpieces. AM also provides a way for FGMM pieces to be fabricated with consistency in mind. Computer-generated and controlled AM methods decrease the error between components and accurately distribute the desired materials throughout the volume. Unfortunately, not all methods of AM can produce FGMMs. Extrusion processes such as freeze form extrusion and wire arc AM are well-known forms of AM for FGMMs. Extrusion processes are quick and efficient; however, these methods result in weak pieces due to the weak bonding between layers of material. The extrusion also falls short when a piece requires overhangs. Supports need to be implemented into the design, which leads to complications when removing them from the part itself. For this reason, solid-phase processes are preferred over extrusion processes.

It is already known that the medical sector takes advantage of both AM and FGMMs in the form of implants and prosthetics. AM is essential to rapid product development and provides a boundless opportunity for customization. Furthermore, the advances of metallic FGMMs have only increased the tailored facet of the medical field. Prosthetics and implants made of FGMMs were created to have enhanced biocompatibility and are considered an upgrade compared to conventional methods. FGMMs have allowed the energy industry to make advancements in harnessing solar power due to their resistance to stresses caused by temperature. Previous models of solar panels were prone to rapid degradation. Solar panels can take advantage of the thermoresistant properties of metallic FGMMs, and as a result, the degradation is minimized.

While the use of FGMMs has not been documented extensively in the chemical industry, the potential for their use would prove to be beneficial. Chemical plants need equipment that can handle abrasives, corrosives, and high-temperature products running through them constantly. Exhaust valves and heat exchangers are just two examples of how the chemical industry can benefit greatly from FGMMs.

Both AM and FGMMs have advanced exponentially since they were both conceived. The use of AM to create FGMMs can and will continue to progress in many fields, including the chemical, medical, and energy sectors. The advanced features and properties of the FGMMs are very desirable for versatile industries, especially due to the overall progress in the AM field over the last decade.

Acknowledgments

The authors would like to acknowledge the help from Soumya Sikdar and Harmony Werth in finding some articles relevant to functionally graded metallic materials.

References

[1] M.K. Thompson, et al., Design for additive manufacturing: trends, opportunities, considerations, and constraints, CIRP Annals 65 (2) (2016) 737−760.
[2] Y. Miyamoto, et al., Functionally Graded Materials: Design, Processing and Applications, vol. 5, Springer Science & Business Media, 2013.
[3] S.K. Bohidar, R. Sharma, P.R. Mishra, Functionally graded materials: a critical review, International Journal of Research 1 (4) (2014) 289−301.
[4] J.J. Sobczak, L. Drenchev, Metallic functionally graded materials: a specific class of advanced composites, Journal of Materials Science & Technology 29 (4) (2013) 297−316.
[5] W.S. Ebhota, T.-C. Jen, Casting and applications of functionally graded metal matrix composites, in: Advanced Casting Technologies, 2018, pp. 60−86.
[6] J. Gardan, Additive manufacturing technologies: state of the art and trends, International Journal of Production Research 54 (10) (2016) 3118−3132.
[7] M. Attaran, The rise of 3-D printing: the advantages of additive manufacturing over traditional manufacturing, Business Horizons 60 (5) (2017) 677−688.
[8] L.-L. Ke, Y.-S. Wang, Two-dimensional contact mechanics of functionally graded materials with arbitrary spatial variations of material properties, International Journal of Solids and Structures 43 (18−19) (2006) 5779−5798.
[9] S. Jing, et al., Optimum weight design of functionally graded material gears, Chinese Journal of Mechanical Engineering 28 (6) (2015) 1186−1193.
[10] Y. Li, et al., A review on functionally graded materials and structures via additive manufacturing: from multi-scale design to versatile functional properties, Advanced Materials Technologies 5 (6) (2020) 1900981.
[11] B. Saleh, et al., 30 years of functionally graded materials: an overview of manufacturing methods, applications and future challenges, Composites Part B: Engineering (2020) 108376.
[12] T. Hirai, CVD processing, MRS Bulletin 20 (1) (1995) 45−47.
[13] K. Choy, Chemical vapour deposition of coatings, Progress in Materials Science 48 (2) (2003) 57−170.
[14] R.S. Parihar, S.G. Setti, R.K. Sahu, Recent advances in the manufacturing processes of functionally graded materials: a review, Science and Engineering of Composite Materials 25 (2) (2018) 309−336.
[15] C. Guo, Amorphous hydrogenated carbon coatings on microrouters by ECR-CVD system, Surface and Coatings Technology 201 (16−17) (2007) 7122−7129.
[16] M. Naebe, K. Shirvanimoghaddam, Functionally graded materials: a review of fabrication and properties, Applied Materials Today 5 (2016) 223−245.
[17] M.R. Dorfman, Thermal spray applications, Advanced Materials & Processes 160 (10) (2002) 66−68.
[18] J. Kim, M. Kim, C. Park, Evaluation of functionally graded thermal barrier coatings fabricated by detonation gun spray technique, Surface and Coatings Technology 168 (2−3) (2003) 275−280.
[19] A. Kawasaki, R. Watanabe, Evaluation of thermomechanical performance for thermal barrier type of sintered functionally graded materials, Composites Part B: Engineering 28 (1−2) (1997) 29−35.
[20] R.A. Miller, C.E. Lowell, Failure mechanisms of thermal barrier coatings exposed to elevated temperatures, Thin Solid Films 95 (3) (1982) 265−273.
[21] C. Pan, B. Chen, Formation of the deformation twinning in austenitic stainless steel weld metal, Journal of Materials Science Letters 14 (24) (1995) 1798−1800.

[22] E. Pan, Exact solution for functionally graded anisotropic elastic composite laminates, Journal of Composite Materials 37 (21) (2003) 1903–1920.

[23] S. Seifert, J.I. Kleiman, R.B. Heimann, Thermal optical properties of plasma-sprayed mullite coatings for space launch vehicles, Journal of Spacecraft and Rockets 43 (2) (2006) 439–442.

[24] G. Wuest, G. Barbezat, S. Keller, The key advantages of the plasma-powder spray process for the thermal spray coating of cylinder bores in automotive industry, SAE Transactions (1997) 37–47.

[25] B. Gérard, Application of thermal spraying in the automobile industry, Surface and Coatings Technology 201 (5) (2006) 2028–2031.

[26] E. Menthe, et al., Structure and properties of plasma-nitrided stainless steel, Surface and Coatings Technology 74 (1995) 412–416.

[27] B. Larisch, U. Brusky, H.-J. Spies, Plasma nitriding of stainless steels at low temperatures, Surface and Coatings Technology 116 (1999) 205–211.

[28] T. Christiansen, M.A. Somers, Low temperature gaseous nitriding and carburising of stainless steel, Surface Engineering 21 (5–6) (2005) 445–455.

[29] A. Zhecheva, et al., Enhancing the microstructure and properties of titanium alloys through nitriding and other surface engineering methods, Surface and Coatings Technology 200 (7) (2005) 2192–2207.

[30] A. Krell, J. Klimke, T. Hutzler, Advanced spinel and sub-μm Al_2O_3 for transparent armour applications, Journal of the European Ceramic Society 29 (2) (2009) 275–281.

[31] J.G. Miranda-Hernández, et al., Synthesis of Al_2O_3/Ti/TiN functional graded materials by means of nitriding in salts of Al_2O_3/Ti composites, Materials Science Forum 691 (2011) 58–62, https://doi.org/10.4028/www.scientific.net/MSF.691.58 (Trans Tech Publ).

[32] L. Drenchev, et al., Numerical simulation of macrostructure formation in centrifugal casting of particle reinforced metal matrix composites. Part 1: model description, Modelling and Simulation in Materials Science and Engineering 11 (4) (2003) 635.

[33] B.I. Saleh, M.H. Ahmed, Development of functionally graded tubes based on pure Al/Al 2 O 3 metal matrix composites manufactured by centrifugal casting for automotive applications, Metals and Materials International 26 (9) (2020) 1430–1440.

[34] D. Hotza, P. Greil, Aqueous tape casting of ceramic powders, Materials Science and Engineering: A 202 (1–2) (1995) 206–217.

[35] Z. Zhong, et al., Design and anti-penetration performance of TiB/Ti system functionally graded material armor fabricated by SPS combined with tape casting, Ceramics International 46 (18) (2020) 28244–28249.

[36] J. Andertová, et al., Functional gradient alumina ceramic materials—heat treatment of bodies prepared by slip casting method, Journal of the European Ceramic Society 27 (2–3) (2007) 1325–1331.

[37] C. Tallon, G.V. Franks, Recent trends in shape forming from colloidal processing: a review, Journal of the Ceramic Society of Japan 119 (1387) (2011) 147–160.

[38] A.C. Young, et al., Gelcasting of alumina, Journal of the American Ceramic Society 74 (3) (1991) 612–618.

[39] J. Yang, J. Yu, Y. Huang, Recent developments in gelcasting of ceramics, Journal of the European Ceramic Society 31 (14) (2011) 2569–2591.

[40] S.H. Park, et al., Fabrication and mechanical characterization of Al 2 O 3/ZrO 2 layered composites with graded microstructure, in: AIP Conference Proceedings., American Institute of Physics, 2008.

[41] R. Jedamzik, A. Neubrand, J. Rödel, Functionally graded materials by electrochemical processing and infiltration: application to tungsten/copper composites, Journal of Materials Science 35 (2) (2000) 477–486.

[42] B. Khripunov, et al., Study of tungsten as a plasma-facing material for a fusion reactor, Physics Procedia 71 (2015) 63–67.
[43] Z.-J. Zhou, et al., Performance of W/Cu FGM based plasma facing components under high heat load test, Journal of Nuclear Materials 363 (2007) 1309–1314.
[44] C. Lucignano, F. Quadrini, Indentation of functionally graded polyester composites, Measurement 42 (6) (2009) 894–902.
[45] G. He, et al., Surface morphology and ferroelectric properties of compositional gradient PZT thin films prepared by chemical solution deposition process, Applied Surface Science 283 (2013) 532–536.
[46] E. Mueller, et al., Functionally graded materials for sensor and energy applications, Materials Science and Engineering: A 362 (1–2) (2003) 17–39.
[47] G.H. Loh, et al., An overview of functionally graded additive manufacturing, Additive Manufacturing 23 (2018) 34–44.
[48] S. Qiang, et al., Preparation of Mg– Ti system alloy and FGM with density gradient by spark plasma sintering technique, Journal of Wuhan University of Technology - Materials Science Edition 19 (1) (2004) 58–60.
[49] D. Smith, B. Loomis, D. Diercks, Vanadium-base alloys for fusion reactor applications—a review, Journal of Nuclear Materials 135 (2–3) (1985) 125–139.
[50] V. Birman, Functionally graded materials and structures, in: Encyclopedia of Thermal Stresses, 2014, pp. 1858–1865.
[51] C. Zhang, et al., Additive manufacturing of functionally graded materials: a review, Materials Science and Engineering: A 764 (2019) 138209.
[52] I. Astm, ASTM52900-15 standard terminology for additive manufacturing—general principles—terminology, ASTM International, West Conshohocken, PA 3 (4) (2015) 5.
[53] F. Craveiro, et al., A design tool for resource-efficient fabrication of 3d-graded structural building components using additive manufacturing, Automation in Construction 82 (2017) 75–83.
[54] Y. Tang, et al., Preparation of W–V functionally gradient material by spark plasma sintering, Nuclear Engineering and Technology 52 (8) (2020) 1706–1713.
[55] S. Kumar, Development of functionally graded materials by ultrasonic consolidation, CIRP Journal of Manufacturing Science and Technology 3 (1) (2010) 85–87.
[56] G.J. Ram, et al., Use of ultrasonic consolidation for fabrication of multi-material structures, Rapid Prototyping Journal 13 (4) (2007) 226–235, https://doi.org/10.1108/13552540710776179.
[57] K. Johnson, et al., New discoveries in ultrasonic consolidation nano-structures using emerging analysis techniques, Proceedings of the Institution of Mechanical Engineers, Part L: Journal of Materials: Design and Applications 225 (4) (2011) 277–287.
[58] D.V. Kaweesa, D.R. Spillane, N.A. Meisel, Investigating the impact of functionally graded materials on fatigue life of material jetted specimens, in: Solid Free. Fabr. Symp., 2017.
[59] Y. Shinohara, Functionally graded materials, in: Handbook of Advanced Ceramics, 2013.
[60] A.P. Mouritz, Introduction to Aerospace Materials, Elsevier, 2012.
[61] H.Q. Wang, Qing-Hua, Meshless analysis for two-dimensional elastic problems, in: Methods of Fundamental Solutions in Solid Mechanics, Elsevier, 2019, pp. 143–210.
[62] A. Horgar, et al., Additive manufacturing using WAAM with AA5183 wire, Journal of Materials Processing Technology 259 (2018) 68–74.
[63] C. Shen, et al., Fabrication of Fe-FeAl functionally graded material using the wire-arc additive manufacturing process, Metallurgical and Materials Transactions B 47 (1) (2016) 763–772.

[64] A.R. Kannan, et al., Process-microstructural features for tailoring fatigue strength of wire arc additive manufactured functionally graded material of SS904L and Hastelloy C-276, Materials Letters 274 (2020) 127968.
[65] P.L. Menezes, et al., Tribology for Scientists and Engineers, Springer, 2013.
[66] P.N.S. Srinivas, B. Balakrishna, Microstructural, mechanical and tribological characterization on the Al based functionally graded material fabricated powder metallurgy, Materials Research Express 7 (2) (2020) 026513.
[67] A.S. Karun, et al., Enhancement in tribological behaviour of functionally graded SiC reinforced aluminium composites by centrifugal casting, Journal of Composite Materials 50 (16) (2016) 2255−2269.
[68] N. Radhika, R. Raghu, Mechanical and tribological properties of functionally graded aluminium/zirconia metal matrix composite synthesized by centrifugal casting, International Journal of Materials Research 106 (11) (2015) 1174−1181.
[69] S. Khan, Analysis of tribological applications of functionally graded materials in mobility engineering, International Journal of Scientific and Engineering Research 6 (3) (2015) 1150−1160.
[70] A. Siddaiah, A. Kasar, P. Kumar, J. Akram, M. Misra, P.L. Menezes, Tribocorrosion behavior of inconel 718 fabricated by laser powder bed fusion-based additive manufacturing, Coatings 11 (2) (2021) 195, https://doi.org/10.3390/coatings11020195.
[71] A. Nouri, C. Wen, Introduction to surface coating and modification for metallic biomaterials, in: Surface Coating and Modification of Metallic Biomaterials, 2015, pp. 3−60.
[72] M. Lekka, Electrochemical Deposition of Composite Coatings, 2018.
[73] S. Lajevardi, T. Shahrabi, J. Szpunar, Tribological properties of functionally graded Ni-Al_2O_3 nanocomposite coating, Journal of the Electrochemical Society 164 (6) (2017) D275.
[74] S. Ferreira, et al., Microstructural characterization and tribocorrosion behaviour of Al/Al_3Ti and Al/Al_3Zr FGMs, Wear 270 (11−12) (2011) 806−814.
[75] R.M. Mahamood, E.T. Akinlabi, Types of functionally graded materials and their areas of application, in: Functionally Graded Materials, Springer, 2017, pp. 9−21.
[76] I. Bharti, N. Gupta, K. Gupta, Novel applications of functionally graded nano, optoelectronic and thermoelectric materials, International Journal of Materials, Mechanics and Manufacturing 1 (3) (2013) 221−224.
[77] H. Shi, et al., Functional gradient metallic biomaterials: techniques, current scenery, and future prospects in the biomedical field, Frontiers in Bioengineering and Biotechnology 8 (2021) 1510.
[78] W. Pompe, et al., Functionally graded materials for biomedical applications, Materials Science and Engineering: A 362 (1−2) (2003) 40−60.
[79] A.R. Studart, Biological and bioinspired composites with spatially tunable heterogeneous architectures, Advanced Functional Materials 23 (36) (2013) 4423−4436.
[80] B. Henriques, Inhomogeneous materials perform better: functionally graded materials for biomedical application, Journal of Powder Metallurgy and Mining 2 (2013) 1−2.
[81] S. Mannan, J. Paul Knox, S. Basu, Correlations between axial stiffness and microstructure of a species of bamboo, Royal Society Open Science 4 (1) (2017) 160412.
[82] F. Ramírez-Gil, et al., Optimization of functionally graded materials considering dynamical analysis, in: Computational Modeling, Optimization and Manufacturing Simulation of Advanced Engineering Materials, Springer, 2016, pp. 205−237.
[83] Y. Li, et al., New developments of Ti-based alloys for biomedical applications, Materials 7 (3) (2014) 1709−1800.
[84] V.S. de Viteri, E. Fuentes, Titanium and titanium alloys as biomaterials, in: Tribology-Fundamentals and Advancements, 2013, pp. 155−181.

[85] C. Petit, L. Montanaro, P. Palmero, Functionally graded ceramics for biomedical application: concept, manufacturing, and properties, International Journal of Applied Ceramic Technology 15 (4) (2018) 820–840.

[86] F. Watari, et al., Biocompatibility of materials and development to functionally graded implant for bio-medical application, Composites Science and Technology 64 (6) (2004) 893–908.

[87] F. Chen, X. He, H. Yin, Manufacture and multi-physical characterization of aluminum/high-density polyethylene functionally graded materials for green energy building envelope applications, Energy and Buildings 116 (2016) 307–317.

[88] R. Hill, B. Storåkers, A. Zdunek, A theoretical study of the Brinell hardness test, Proceedings of the Royal Society of London. A. Mathematical and Physical Sciences 423 (1865) (1989) 301–330.

[89] S. Bhatia, Advanced Renewable Energy systems, (Part 1 and 2), CRC Press, 2014.

[90] A. Amelia, et al., Investigation of the effect temperature on photovoltaic (PV) panel output performance, International Journal on Advanced Science Engineering Information Technology 6 (5) (2016) 682–688.

[91] H. Crumpton, Well Control for Completions and Interventions, Gulf Professional Publishing, 2018.

[92] S. Chandrasekaran, Design of Marine Risers with Functionally Graded Materials, Woodhead Publishing, 2020.

[93] S. Chandrasekaran, S. Hari, M. Amirthalingam, Wire arc additive manufacturing of functionally graded material for marine risers, Materials Science and Engineering: A 792 (2020) 139530.

[94] 李涤尘张安峰严深平姚金刚曹伟产, Additive Manufacturing Method for Electrical Contact Made of Copper-Tungsten Functionally Gradient Material, 2016.

[95] P. Forsberg, P. Hollman, S. Jacobson, Wear mechanism study of exhaust valve system in modern heavy duty combustion engines, Wear 271 (9–10) (2011) 2477–2484.

[96] M. Chmielewski, K. Pietrzak, Metal-ceramic functionally graded materials–manufacturing, characterization, application, Bulletin of the Polish Academy of Sciences. Technical Sciences 64 (1) (2016) 151–160.

[97] R.A. Parisher, R.A. Rhea, Chapter 6 - Mechanical Equipment, in: R.A. Parisher, R.A. Rhea (Eds.), Pipe Drafting and Design, Third Edition, Gulf Professional Publishing, 2012, pp. 112–133, https://doi.org/10.1016/B978-0-12-384700-3.00006-2. ISBN 9780123847003.

[98] B.I. Master, et al., Most frequently used heat exchangers from pioneering research to worldwide applications, Heat Transfer Engineering 27 (6) (2006) 4–11.

[99] P. Rodriguez, Selection of Materials for Heat Exchangers, 1997.

[100] D. Chung, Materials for thermal conduction, Applied Thermal Engineering 21 (16) (2001) 1593–1605.

[101] K.B.S. Rao, B. Raj, A. Nagesha, Fatigue Testing: Thermal and Thermomechanical, 2017.

[102] M.P. Schwartz, Four Types of Heat Exchanger Failures, ITT Bell & Gosset, 1982.

[103] Y. Miyamoto, The applications of functionally graded materials in Japan, Materials Technology 11 (6) (1996) 230–236.

[104] T.J. Holmquist, D.W. Templeton, K.D. Bishnoi, Constitutive modeling of aluminum nitride for large strain, high-strain rate, and high-pressure applications, International Journal of Impact Engineering 25 (3) (2001) 211–231.

[105] R. David, CRC Handbook of Chemistry and Physics. 2000–2001, CRC Press, Boca Raton, 2000, p. 4.

Fused deposition modeling (FDM): processes, material properties, and applications

5

Matthew Montez [1], *Keegan Willis* [1], *Henry Rendler* [1], *Connor Marshall* [1], *Enrique Rubio* [1], *Dipen Kumar Rajak* [2], *Md Hafizur Rahman* [1] *and Pradeep L. Menezes* [1]
[1]Department of Mechanical Engineering, University of Nevada, Reno, NV, United States; [2]Department of Mechanical Engineering, Sandip Institute of Technology & Research Centre, Nashik, Maharashtra, India

5.1 Introduction

Additive manufacturing (AM) is the process of creating an object by depositing materials in layer by layer to create a 3D object directly from a computer-aided design (CAD) model [1]. AM has several benefits, including saving money, reducing idle time, and creating complex parts. Enormous reductions in waste and overall carbon footprint from these processes eliminate the need to consume a lot of materials. Furthermore, these products are made in less time, still in fine quality [2]. Therefore, cheaper but superior replacement parts, whole products, and structures could be available to the everyday consumer at a cost drastically less than conventional manufacturing methods used today [3].

One of the widely used AM methods is fused deposition modeling (FDM). FDM printers typically use a thermoplastic filament, which is heated until it melts and then extruded through a nozzle onto a platform [4]. FDM is an important manufacturing process because it can create parts with good thermal and chemical resistance and strength to weight ratios, making it ideal for demanding applications [5]. It is one of the most common methods and usually the cheapest, depending on the model. Additionally, FDM is an excellent option when designing a custom or complex part [6]. Apart from FDM, other additive manufacturing processes, such as selective laser sintering, binder jetting, laminated object manufacturing have been used for high resolution prototyping. These processes are each great option in their own realm, but none of the processes can be used unilaterally in the form of a "one-size-fits-all" practice [7]. The tribological properties of the additives used can play a large part in the process selection; this depends on the application and whether or not the part needs to be wear-resistant or if it is not a priority. The processes can also be chosen based on how the manufactured part will corrode and what limits are acceptable. The tribocorrosion properties vary depending on the manufacturing process as well as the subset of additive that is used to create the part. As mentioned above, FDM commonly uses ABS

Tribology of Additively Manufactured Materials. https://doi.org/10.1016/B978-0-12-821328-5.00005-6

filaments, but there are also other filaments such as polyethylene terephthalate (PETG) and polylactic acid (PLA) that offer different physio-chemical as well as tribological properties. Knowing the options for each process allows the engineer to select the best process that will fit the needs based on the properties of materials that can be used.

5.2 FDM processes

5.2.1 Material selection

While selecting which material to use when using an FDM printer, it is crucial to understand that there are many different materials to choose from. Some of the popular polymeric filament materials are the following: Polylactic acid (PLA), acrylonitrile butadiene styrene (ABS), nylon, acrylonitrile styrene acrylate (ASA), high-impact polystyrene (HIPS), polyethyleneterephthalate (PET), polyethyleneterephthalate glycol (PETG), polycarbonate (PC), thermoplastic polyurethane (TPU), thermoplastic elastomer (TPE), polyvinyl alcohol (PVA), polypropylene (PP), and composite [8]. PLA is a thermoplastic, which is a common material when using FDM. PLA is a semi-crystalline polymer with a melting temperature of 180°C and is easy to work with. ABS is an amorphous thermoplastic that is commonly used due to being a hard-strong plastic with some flexibility when compared to PLA [9]. ABS has a extruder temperature between 200°C and 260°C [10]. Nylon is a mostly semi-crystalline thermoplastic material [11]. Nylon is a tough material with a melting point of around 230−300°C [12]. ASA is an amorphous plastic that has similar properties when compared with ABS. ASA has a high impact and wear resistance, but it is commonly used due to its strong UV resistance; it's printing temperature has a range from 220 to 245°C [13]. HIPS is a dissolvable support material typically used with ABS. Compared to ABS, HIPS is lighter and dimensionally more stable; it is also dissolvable by D-Limonene and has a extruder temperature with a range of 230−245°C [14]. PET and the glycol modified version (PETG) are semi-rigid materials with a smooth surface and have impact resistance. PET and PETG have thermal characteristics that prevent warping when cooled and the extruder temperature typically ranges between 230 and 250°C [15]. PC is an extremely durable amorphous material; it has impact resistance, heat resistance, and flexibility. The extruder temperature of PC ranges between 260−310°C [16]. TPEs and TPU are flexible materials that can be printed through FDM; they have good impact resistance and are soft compared to the materials listed before. TPE and TPU have a extruder temperature with a range of 225−245°C [17]. PVA is a soft material that dissolves in water and is used as support similar to HIPS. PVA has a melting point ranging from 180 to 200°C [18]. PP is a semi-crystalline material similar to PET and PETG with a smooth surface and fatigue resistance. PP is typically used for low-strength applications and has a extrusion temperature with a range from 220 to 250°C [19]. Composite material typically has components of a core polymer material and a reinforcing material. Composite materials have a much higher strength and stiffness when compared to the other polymers and can even compete in strength with some metals like aluminum [20].

There are many different variations of the materials listed, which makes the material stronger, easier to print, more flexible, more heat resistant, and many other desired attributes that ensure that there is a material for every use and application. ABS-M30 is about 70% stronger than standard ABS and has high resolutions for more intricate details or larger models; it also comes in many different colors [21]. ABS-ESD7 is compounded with carbon, which has static dissipative properties, which work well to encase electronic components. ABS-ESD7 has superior strength compared to standard ABS. ABS-M30i is a sterilizable material that has tensile, flexural, and impact strength; it is also certified to contact food [21]. PC-ABS is a blend of PC and ABS that provides the strong points of each material, including the strength and heat resistance of PC with the flexibility of ABS. Another certified material for food packaging is PC-ISO, a bio-compatible transparent material with a high tensile and flexural strength with heat resistance. Polyphenylsulfone (PPSF, also known as PPSU) is a material that has strong mechanical properties and a high temperature and chemical resistance, making it an ideal material for harsh conditions [21]. Nylon 12 has a high fatigue resistance, which makes it ideal for an environment that is exposed to high vibration and shock. Nylon 12CF is a composite material composed of Nylon 12 and chipped carbon fiber; it is similar to Nylon 12, but it has an improved strength and stiffness to weight ratio, and it also has electrostatic discharge properties. TPU 92A elastomer increases the flexibility and elongation of TPU while providing abrasion resistance perfect for a vibration dampener [21]. When using FDM it is common to select a composite material to gain the desired attributes when compared to the pure polymer material. A common type of composite material is fiber-reinforced composites, a combination of polymer material and continuous or short reinforcing fibers [22]. Some of the main fiber reinforcements are carbon fiber, glass fiber, and kevlar. By adding these fibers to the polymer, new mechanical properties are formed. Short fibers are added to the polymer to enhance the strength of the desired part and are typically added by mixing the fibers into the molten thermoplastic polymer. Continuous fibers are used by separately supplying the FDM printer. The fiber is then passed through the nozzle to combine with the thermoplastic as shown in Fig. 5.1.

As shown in Table 5.1, additive materials, such as ceramic, copper, carbon fiber, and aluminum can be used with PLA plastics [23]. To determine the effects of these additives, the mechanical properties were tested and analyzed combining these additives to PLA. Pure PLA had a tensile strength of 42.4 MPa, 6% elongation when it broke, a flexural strength of 69.7 MPa, and a 3% deflection at break [23]. PLA and wood had a tensile strength of 29.5 MPa, 5% elongation when it broke, a flexural strength of 40.4 MPa, and a 2% deflection at break. PLA and ceramic had a tensile strength of 43.2 MPa, 5% elongation when it broke, a flexural strength of 57.8 MPa, and a 2% deflection at break [23]. PLA and copper had a tensile strength of 40.3 MPa, 8% elongation when it broke, a flexural strength of 55.5 MPa, and a 3% deflection at break. PLA and aluminum had a tensile strength of 40.2 MPa, 5% elongation when it broke, a flexural strength of 62.8 MPa, and a 4% deflection at break. And PLA with carbon fiber had a tensile strength of 32.8 MPa, 8% elongation when it broke, a flexural strength of 50.3 MPa, and a 4% deflection at break [23].

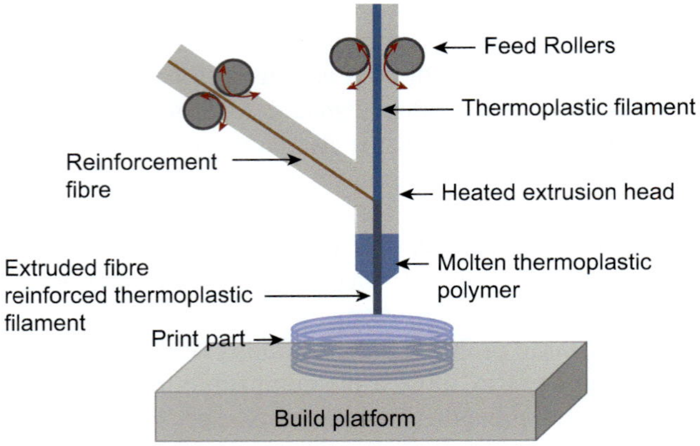

Figure 5.1 Fusion deposition modeling continuous fiber process [22].

Table 5.1 Variation in properties through using different additives with PLA.

Sr.	Materials	Additive	Tensile strength (Mpa)	Flexural stress (Mpa)	% Deflection when breaks	% Elongation when breaks
1	PLA	No additive	42.4	69.7	3	6
2	PLA	Ceramics	43.2	57.8	2	5
3	PLA	Copper	40.3	55.5	3	8
4	PLA	Aluminum	40.2	62.8	4	5
5	PLA	Carbon fiber	32.8	50.3	4	8

When looking at the above information, it is important to note that each of the materials was printed with the same dimensions, so the weight of each piece is different. When selecting a material for FDM printing, it is also important to consider the difficulty of printing and forming the material. PLA is considered the easiest polymer to print, making it a good beginner material [24]. Since ABS has a higher printing temperature, it requires a heating bed and is more likely to warp when it is cooled, making it a more difficult material to print. Nylon has a high melting temperature and requires the use of a heating bed as well. Also, it is prone to warping when cooled, absorbs moisture, which can form bubbles in the printed parts [24]. PET is also an easy material to print compared to ABS and nylon, but its higher melting point is not as easy to print compared to PLA. TPU is widely considered to be the most difficult material to print; this is because of how flexible the material is. Lastly, PC is another difficult material to print due to the high melting point temperatures, which requires a heating bed, making it prone to warping when cooled. PC is also known to ooze while printing and absorbs moisture making an enclosure required [16].

5.2.2 Material processing/extrusion

After the material is selected according to the use case of the final product of the FDM part, the material must be processed from its raw form into useable material. Plastic is an organic chemical compound that is formed from the continuous linking of many small molecules known as monomers. The chains formed in this process are called polymers, and the process of turning monomers into polymers is called polymerization. Plastics are one form of polymers. In the case of FDM manufacturing, thermoplastics are required as thermoset plastics do not melt once they have been set. The raw materials used to obtain the monomers that make up the polymers vary based on the classification of thermoplastic to be manufactured. Because thermoplastics can be melted and remelted without destroying the integrity of the plastic, they can be processed through either new material manufacturing or through recycling. In the case of PLA, one of the most common materials used in FDM, lactic acid is the base raw material. The synthesis of PLA is a multistep process that begins with lactic acid production and ends with polymerization. Lactic acid can be sourced from either chemical synthesis or from carbohydrate fermentation [25]. Because lactic acid comes from natural sources and from fermentation, PLA is considered bioplastic and its sources are considered renewable. Lactic acid is a product of the fermentation of basic sugars by Lactobacilli bacteria. Common sources of these sugars are corn, potato, sugar cane, and beets. Because the lactic acid is produced with other byproducts of lactic acid fermentation, it needs to be isolated. This can be done through ultra-filtration, nano-filtration, ion-exchange processes, and electrodialysis. As per Belgacem et al. [25], the polymerization process can be pursued through the azeotropic condensation technique. Azeotropic condensation polymerization is able to produce high chain lengths without the involvement of chain extenders. For example, Mitsui Chemicals (Japan) has commercialized an azeotropic condensation polymerization process wherein lactic acid and a catalyst are dehydrated azeotropically in a high boiling, refluxing, aprotic solvent at reduced pressures to obtain PLA with high molecular weight. A distillation of lactic acid is then carried out for 2–3 h at 130°C to remove the water condensates. Then, the catalyst and diphenyl ether are added, and a tube filled with molecular sieves is attached to the reaction vessel. The refluxing solvent is taken back to the vessel through the molecular sieves for 30–40 h at 130°C. At the final step, the ensuing PLA is purified [25]. The PLA can then be pelletized, forming nurdles. Nurdles, little beads of plastic, are later used in the extrusion process to form filaments, the standard with other new plastic manufacturing; hydrocarbons such as ethylene or propylene are derived from breaking down molecules obtained in the refinement of crude oil. The monomers obtained in this process are then used in polymerization reactions to create polymer resin. These are then processed into the nurdles used to make the filaments.

In the case of recycling, used plastics obtained from bottles or other forms of waste can be ground up into polymer flakes. Before these bottles are ground up, they need to be cleaned. After they are cleaned, they are ground up and cleaned again. After a few rounds of cleaning the polymer in order to minimize impurities, the flakes can be fed through a hopper, melted, and pushed through an extruder with many streams of extrusion. These extrusions, less precise than would be done with an extruder meant to make filaments, are then pelletized. These pellets are known as nurdles, the basic term for any raw thermoplastic that has yet to be conglomerated and formed. These pellets are only a few millimeters in diameter, and thousands of them are needed to make a sizable roll of filament. Once the nurdles have been formed, they can then go on to be melted down again and then extruded into filaments using a machine called a single screw extruder (SSE) [26]. Fig. 5.2 shows the leveling diagram of a SSE.

An SSE consists of a hopper that feeds into a screw, turned by a motor, controlled by a gearbox. The screw feeds into a die of a specified size. In order to extrude the nurdles into filament, the nurdles are first fed into the hopper. If additives are added to the polymer, they are added to the hopper after first being blended with the nurdles. Many different additives can be mixed with the polymers to give the material various desirable properties. There are colorants, plasticizers, flame retardants, and more. The hopper then feeds the nurdles into the extruder barrel, where the material is melted by the heaters, which heat the polymer uniformly. As the polymer is being heated, it is forced down the barrel using the screw, also known as a plasticator. The screw is turned by a drive motor and a gear box. As the screw turns, the material residing in between the threads is shoved toward the die. The material is then pulled, not pushed, through the die. This process forms a filament. The filament is often checked with laser diameter gages as it is being formed to ensure that the product is within the specified tolerance. The die is a set diameter, and although it may seem that the diameter of the filament would be the same diameter as the die, this is not the case. In the process of pulling the filament through the die, the diameter changes proportionally to the speed at which the plastic is pulled, and thus the speed at which the material is pulled through the die is actually more critical to achieving a

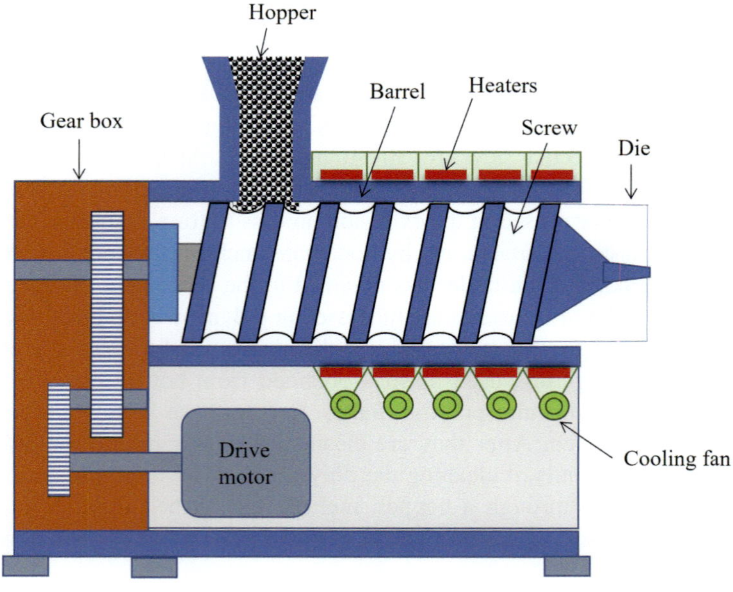

Figure 5.2 Schematic layout of single screw extruder.

proper diameter than the die size itself [27]. After the filament is pulled through the die, it needs to cool down and is fed onto a spool.

5.2.3 Fabrication

Material extrusion is a common process for manufacturing parts in low quantities, allowing easy prototyping while keeping costs to a minimum [28]. Using high flow rates, injection molding was able to utilize the full cavity of the mold, increasing the part's resolution. The issue of dimensionality has been a recurring issue for AM and the use of high flow rates has been able to mitigate this issue through the use of injection printing. The properties of the injection printed part also increased on average by a factor of 3.2 compared to the original material extrusion process [28]. The strength of the parts also showed an increase in stiffness, strength, and strain to failure when loaded in a tensile configuration of 21%, 47%, and 35%, respectively [28]. The injection printing process prints the shell of the cavity and then fills the cavity with material allowing for parts to be created with higher precision than they could be achieved using previous techniques. There is another technique described as Z-pinning in which material is deposited into different layers across the parts. This compartmentalization and filling technique have proven to increase the parts' strength and stiffness by a factor of 3.5 in the z-direction for some materials. This process and technique speed up parts production while decreasing the amount of downtime required to set up and manufacture parts. In addition to "Z-Pinning," engineers have developed another technique that utilizes the properties of epoxy and that utilizes the properties of epoxy and the selected material to increase the overall part properties. This process is a form of "fill compositing" that leaves voids in the 3D printed material to allow for the voids to be filled with an epoxy/resin material afterward. After the voids are filled with material and it is allowed to cure, the parts can be removed from the machine. The results have shown that the overall part characteristics of strength and stiffness can increase by up to 25% and 45%, respectively. The largest issue today with the AM fabrication processes is that there are severe rate limiting factors that hinder its use for small scale manufacturing. The largest issue is that of delivering enough heat to allow the extrusion process to be carried out. One of the methods that are currently being researched involves the use of a plasticating screw that heats and melts the polymer along a channel resembling that of a screw [28]. This technique provides much more surface area than conventional 3D printing, with melting lengths being much longer, allowing for more heat input into the material. This process has currently only been used for large-scale AM, but there are attempts to make it a reasonable retrofit for an off-the-shelf, consumer, 3D printer. New developments in 3D printing filament have allowed for much stronger bonds between layers as well as building in abrasion resistance and impact resistance properties. The 3D printer filament uses milled carbon fiber that is mixed in with common polymers of printer filaments such as PLA, ABS, and PETG. Data showed a 27% increase in the material strength of ABS when it was mixed with carbon fiber [29]. When printing with carbon fiber filament, there is the option of using filament that is infused with the fiber itself or using two separate spools of material, as shown in Fig. 5.3.

A promising solution for the volumetric, rate-limiting factor has been tested by engineers where the length and size of the nozzle are increased, allowing for more heat

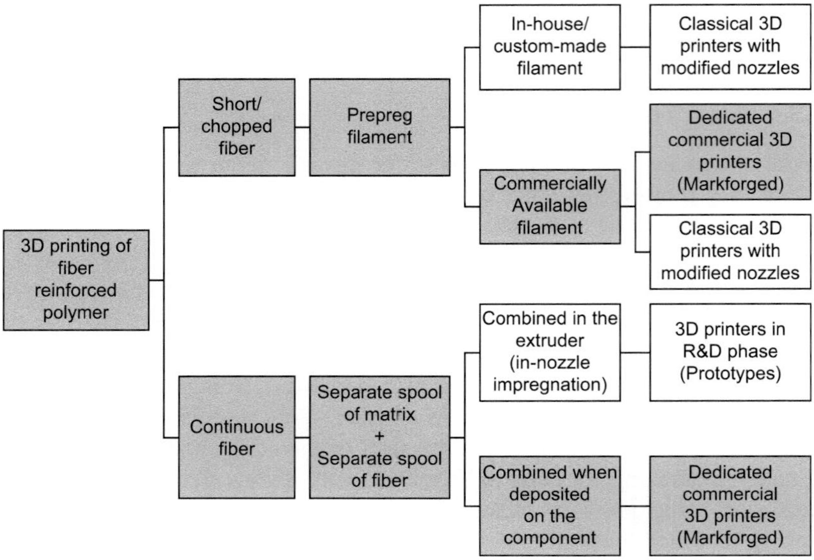

Figure 5.3 Carbon fiber printing options [29].

transfer to the material. The volumetric flow rate shows an increase from 12.5 to 20 mL/s [28]. This process change results in an increased volumetric output of 38%. A new generation of material flow system promises 35−78 mL/s, an average output increase of approximately 4.5 times what is available today. The process uses a nut-feed extruder and laser-heated polymer liquefier to achieve a much higher force when the material is extruded. This process could make better material flow rates available for the high-end consumer and industrial markets as these flow rates are slightly under 7 times higher than what is currently on the market for industrial material extrusion in the part fabrication industry [28]. The ability to rapidly manufacture industrial parts both in large quantities and in prototype situations is a great use for FDM printing. A study was performed to create a "gastric-resistant" tablet through the use of FDM printing with dual extruders [30]. This process utilized thermo scientific's HAAKE MiniCTW hot melt extruder to extrude the combination of filler material, plasticizer, and the talc filler material [30]. This chemical combination was paired with an extruder temperature of 90°C and a 1.25 mm nozzle size. This process allowed for the tablets to be printed successfully. The researchers were also able to modify the print settings and filament types to produce the anti-inflammatory budesonide as well as diclofenac sodium. The shell of the tablet was designed to be caplet. The shell thicknesses ranged from 0.17 to 0.87 mm. The varying shell thicknesses allowed for the tablets to hold onto the contents for varying amounts of time while the body breaks them down. The processes used a dual extruder head as well as a single extruder head in other tests. The results showed that there was frequent blocking of the PVP filament when the dual extruder assembly was used. However, when the single extruder assembly was used, there was little to no blockages in the extruder. This is theorized to be a result of the filament in the second extruder remaining at high temperature when not in use and essentially gluing itself to the inside of the extruder nozzle. After some trial and error, the realization was made that the extruder nozzle can be lubricated, the lubricant of choice was castor oil, oleic acid, and or PEG 400. This allowed the dual extruder to be utilized without clogging occurring [30]. This not only increased the reliability of the extruder nozzles but also decreased downtime of the process. In addition to having personalized drugs converging on the perfect dose for patients, the drugs are also incredibly stable. In the case of ramipril, it does not undergo any sort of degradation when printed even at temperatures as high as 90°C [31]. A breakdown of the process using a screw-driven printer is shown in Fig. 5.4.

Using 3D printing, the capsules were able to be printed quickly and efficiently. After testing, it became apparent that the shell thickness had to be greater than 0.52 mm to achieve core protection when submerged in acid [30]. Therefore, not only can FDM printing be used in the fabrication industry, but it can also be used to create pharmaceuticals. In the near future, it is quite possible that this technology can be used to print and create customized pharmaceutical compounds without the overhead that is typically required to make modern drugs. The ability to make user-specific drug doses is not out of the question with FDM printing, as the doses can vary based on each patient's body mass and metabolic rate. The ability to create doses that meet the needs of the individual will create medicines that are not only more accurate and reduce the risk of side effects, but it will also decrease the cost of manufacturing. This reduction in production cost can then be passed onto the patient [32]. Therefore, the process improvement utilizing FDM printers is not only more cost-effective but also safer for the patient. The ability to customize the type of release mechanism is also an added benefit to the process. Depending on the drug cocktail and the necessary release times, the walls of the drug shell can be altered to meet these requirements. Another amazing process improvement is to create multistage capsules that allow for more than one drug type to be released based on different time requirements [30]. The ability to create

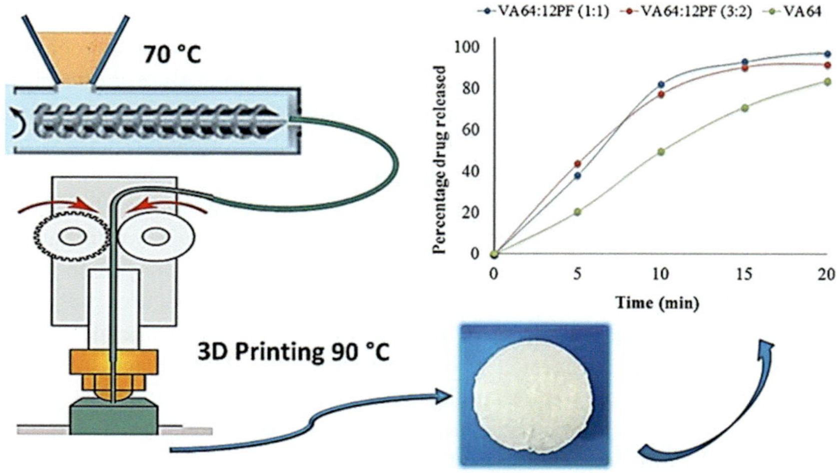

Figure 5.4 3D printing of pharmaceuticals [31].

custom pharmaceuticals that vary per patient is an incredible feat in the engineering and medical world as medicine has never been so personal. FDM processes have been used across the medical industry to create pharmaceuticals products, such as tablets or capsules. It has been used for the rapid manufacturing of prototypes as well as high volume manufacturing as an incredibly efficient and inexpensive method. In addition to both of these fabrication opportunities, it has also proven to be very useful in the fabrication of thermoelectric materials on the nanoscale level. This allows scientists and engineers to create materials that have incredibly low thermal conductivity [33]. Scientists have also been able to use 3D printing to create highly textured, bulk thermoelectric materials, and incredibly mechanically strong alloys such as BiSbTe, which has great thermoelectric properties and incredible mechanical properties as well [34]. Although specialized materials are often employed for the 3D printing of bulk textured materials, the filler material is often not of great consequence. Scientists explored the use of PLA for its thermoelectric and mechanical properties as an "out-of-the-box" solution for thermoelectric cooling and heating [35]. In addition to PLA, scientists have been using materials such as porous bismuth antimony telluride as it has an incredibly large electrical conductivity of 40,300 S/m, which is also lower than the recorded maximum of 73,600 S/m for bulk BST [36]. Overall, the use of 3D printing is incredibly diverse in all realms of product fabrication and development. The ability to print everything from large-volume fabricated parts to pharmaceuticals, nano-materials, and thermoelectric materials demonstrates the great potential of FDM printing in the fabrication realm.

5.3 Properties of FDM

5.3.1 Mechanical properties

As described in the previous section, FDM printing consists of many layers of material being built up to achieve a 3D object. While this process is excellent for form factor prototyping, it poses limitations as 3D printing moves toward functional component manufacturing. To understand these limitations, we need to analyze the mechanical properties of FDM components. While many mechanical properties can be measured, this section will analyze strength, stiffness, and thermal deformation of FDM prints. The first structure to be considered in 3D printing is the "layer direction." This is the structure of the print created by print layers being added on top of each other. Another is the "raster," which is the print head's direction within a print layer. "In fill" is a user-defined structure that determines how strong the honeycomb structure at the center of a part will be. The "wall thickness," which is also a user preset, determines how thick the outer layer of the print will be. Lastly, layer thickness is determined by the diameter of the nozzle on the printer [37]. The strength of an FDM print is directly related to a force's application in relation to the print's layer direction, raster direction, and layer thickness in addition to if the force is tensile or compressive [2]. In the case of forces applied in parallel to the layer direction, your material selection will play a larger role due to the stresses applied along a single filament strand. For a material such as ABS or PLA tensile strengths that average 32 and 27 MPa, respectively, for a cross hatched 100% fill print [37]. When these same materials are subjected to compressive loading, the strength values become dramatically higher. This due print layers are being squeezed together, which relies on the strength of the filament material versus the adhesion of the individual filament strands to each other. With the strong correlation between material strength and print direction, estimating ultimate tensile strength based on degree offset from parallel to print directions is even possible as a print will always fail along a print layer [38].

Another property that is heavily influenced by both print structure and material selection is material stiffness. Unlike the print's strength, its stiffness is highest in the print plane but perpendicular to the raster direction [39]. This is attributed to the individual strands of filament not being round but actually a rectangular shape giving them better strength across their width. A second factor to be considered that is unique to FDM prints is that air gaps exist at the corners of each raster, which creates voids and allows the material to flex more [39]. Due to these fundamental limitations in the FDM process, researchers have been developing advanced filaments that are incredibly strong and lightweight to provide more stiffness. One example is carbon fiber–infused nylon used in Markforged printers that improve the Flexural Strength of Nylon from 81 to 540 MPa [40]. Unlike the strength and stiffness of FDM components, their thermal deflection properties are entirely reliant on the filament they are printed with. As FDM printing requires a thermoplastic filament, temperature plays a great role in how the material will behave post process and while in use. As is noted in Section 5.2.1, PLA has a melting temperature of 180°C, while PC has a melting point of 310°C, the closer you are to this temperature, the more plastic behavior you will observe [10,16]. One example case is nylon, which has a melting of 230–300°C; however, its thermal deflection temperature is 41°C [12,40]. This again is a case where additives such as carbon fiber, fiberglass, and kevlar are used to provide increased resistance to deflection in FDM prints. Research is being conducted to determine the best orientations for 3D prints based on their end uses to combat these limitations. Things like offset raster directions angled printing and specialized FEA are being used. In studies using ABS plastics, a raster offset of 0° produced parts with shear characteristics comparable to that of injection molding [41]. Additionally, many print companies suggest that flat surfaces be printed at a slight incline [42]. Likely, the largest improvement to understanding the mechanical properties of FDM is ability to perform FEA analysis of 3D prints. Using Ansys 2020, studies have been performed with customized material properties allowing researchers to create models with varying strengths depending on orientation [43]. This allows the software to calculate deflection and failure based on the structure of the print. Utilizing this technique, the software can simulate the mechanical properties of FDM printing with 7.2% accuracy [43]. As with all things FDM, this simulation will also indicate that the print is strongest when the force is parallel to the raster direction. After you have analyzed the end use strength requirements, designed your part, and analyzed the best print orientation but still need better strength characteristics, you will have to rely on fiber-reinforced filaments to make your prints more robust. Companies like Markfordge sell many reinforced filaments that allow for robust end use parts in combination with selective print orientation. This addition can result in a nearly 12-fold improvement in material, giving it a flexural strength greater than that of aluminum. Ultimately, when working within the limitations of FDM prints, the user must have a much clearer understanding of the print's end use than one needs for traditionally manufactured components. Understanding the loading direction and loading intensity at the print's end use will allow you to orientate your print to make the best use of the filament's material properties, which provides the best strength. One last consideration to make with FDM prints is the consistency of these mechanical properties. Unfortunately, FDM prints tend to have a high deviation from their rated strengths due to print inconsistencies, and thus, these have the potential to fail prematurely [39].

5.3.2 Tribological properties

Tribological properties can have an enormous effect on the mechanical structure and quality of a printed product. Recall FDM is a 3D-printed process that extrudes heated thermoplastic onto a platform building a structure layer by layer. When FDM was first designed, though it was truly a simple and amazing feat to construct products, there were however several disadvantages. For example, building a product, layer by layer, can often leave behind a rough surface. In daily life, we can compare this to a gravel floor as opposed to a waxed hardwood floor. When this rough, unfinished surface comes into contact with another, the friction forces can cause detrimental wear and material deformation on the 3D-printed products [44]. Six major factors were found that either increased or decreased wear [44]. As discussed previously, layer thickness has a significant role in overall surface quality. With recent developments and new ways of extruding thermoplastics, the thinner the layer, the smoother the surface and less wear to the part. It is also seen that decreasing the layer thickness will lead to a smoother surface [29]. In addition, adding an aluminum matrix composite (AMC) aided to create a smoother surface and add to the overall material toughness and resistance to wear [45]. Utilizing a scanning electron microscope, it was also seen that there is a significant amount of porosity in FDM products, causing air gaps and voids, which can lead to numerous mechanical and wear problems [44]. Fig. 5.5 shows how air gaps affect the overall surface and structural quality of the product. It can lead to increased porosity as mentioned earlier, as well as, weaken the internal structure of the 3D material. Imagine a layer of sheath with a similar air gap like the image with a raster angle of *0* degrees in Fig. 5.5. If subjected to a friction force perpendicular to the raster angle, the overall wear would be greater due to the large air gap present, creating a much rougher surface. Use multiple layers, and one can visually imagine what such a structure would look like. If shear stress is placed on the product in the x-direction, the *0* degree axis, the product would simply shear in half. It is similar to how a wood saw cuts with the grain in wood. Now shift the shear by *90* degrees, along the y-direction, you left with a product with a better resistance to shear. The air gap, raster angle are interrelated and have a huge effect on the wear rate of the product. When the raster angle saw an increase, it was followed by a decrease in the wear on the material [44]. Fig. 5.5 below is an excellent representation of the different angles used for this experiment. As

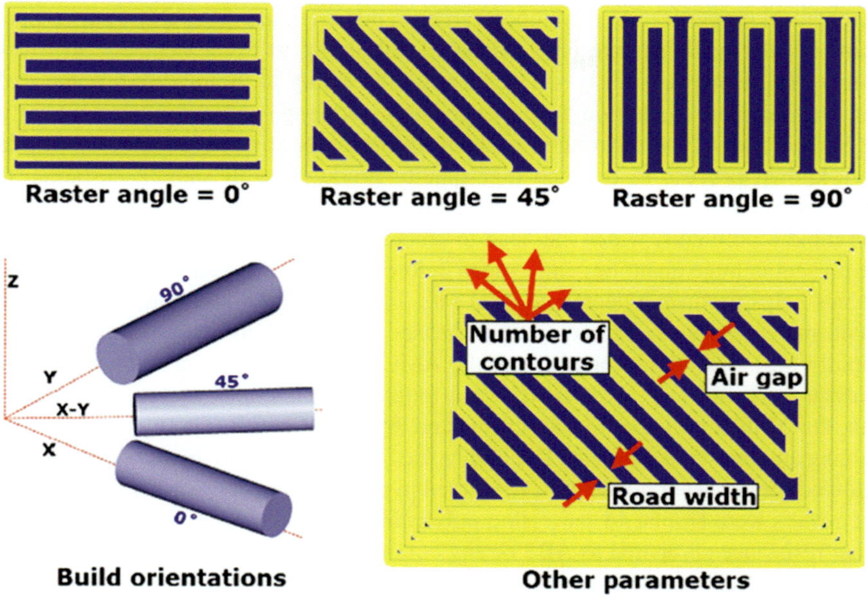

Figure 5.5 This is a detailed representation of what is meant by the different raster angles [44].

seen in Fig. 5.5, the raster angle is simply the route in which the part is made. It is clearly visible how the orientation of any friction force can affect the part. Also, it was observed that whenever the raster angle is parallel with the direction of friction, the amount of wear is less than if the direction is perpendicular.

Moreover, the decrease in build orientation also saw a decrease in the wear of printed parts [44]. How a printed material is set up to be built and the pathway to create it can save time and money. If a part is a pyramid, for example, it is best to start from the base to the top rather than the top to the base. To add, how fast a part slides against another, sliding speed, along with the building orientation, made a huge impact on the surface finish, dimensional accuracy, and porosity [44]. The overall wear rate is lower when the build orientation is below 45 degrees. In addition, another study showed post treatment of a BS part with dimethyl ketone also improved the surface quality [46]. The road width and number of contours also play a role, and it was found that the low and high end of each were the best in preventing wear on the products [44]. Even though the road width is smaller, the increased number of contours allow for a smoother surface. It is directly opposite when the road width is large. The best way to maintain a decreased wear rate is fewer contours. It affected the surface quality and overall mechanical structure of the printed part. To add, another study showed that changing the individual extrusion tips also made a significant difference in the contours and quality of the surface [47].

Even though originally made to print parts based on thermoplastics, recent developments have aided in printing parts containing metal powders and other polymers that have opened the door to even stronger parts with less wear and increased mechanical properties. As mentioned earlier in this section, using AMCs saw a significant impact in the surface finish, quality, and ability to resist wear due to friction forces [45]. However, this was dependent on the ability to thoroughly mix the AMC into the polymer. In addition, fusion deposition modeling—assisted investment casting opens the doorway to a stronger mechanical matrix for any printed part. Casting is a well-known process to produce high-quality products, with smooth surfaces. The stronger composite showed better mechanical qualities. The main disadvantage is the mold's problems, which can lead to cracks in part, causing failure. Air gaps can be present in huge quantities and as seen before in Fig. 5.4, anything below the middle values helps produce a decreased wear rate. In addition, the long duration to make a mold increases the overall cost and decreases the overall ability to produce the part rapidly [45]. The higher presence of the AMC saw an increase of resistance against surface wear and increased strength and toughness. This goes back to the ability to mix the AMC throughout the composite and, by increasing the density, leads to better mixing and resistance to wear. The higher friction coefficient in the denser AMCs also allowed for less wear on the parts [31] It led to others trying to use other materials, and the results were similar. A part using polyamide—iron composite resulted in superior wear resistance compared to ABS or any combination of thermoplastics [27]. The increase in overall density in both combinations, AMC or polyamide-iron, yielded a better part. The wear was minimal, decreasing the air gap and porosity in part, enabling it to carry more mechanical load compared against thermoplastic-based parts.

5.3.3 Corrosion and tribocorrosion properties

In the previous section, tribological properties seen via the FDM process were examined. It is now time to explore how this can affect a parts ability to resist corrosion. If we recall, there are six main factors that affect tribological behavior in 3D-printed parts. They are layer thickness, air gap, raster angle, build orientation, road width, and number of contours. In this section, the main focus will be on layer thickness, air gap, and include extrusion temperature. It may be hard to comprehend how these properties can affect corrosion at first. However, recall that when common thermoplastics are used to form composites with metals, the environment the composite is left in will ultimately determine whether or not you want heavy porosity and air gap in contact in the composite. The type of metal or alloy used is a heavy factor in this. Layer thickness is critical in a part's mechanical strength and wear resistance. However, it is fair to mention that the adhesion between the layers, if programmed wrong, will yield a weak part, which is where continuous fiber reinforcement comes into play [48]. Imagine if a part is weathered down to expose these air gaps, it proves very detrimental to the integrity of the part in terms of utilizing metal alloys in combination with thermoplastics. It yields a composite with higher strength than plastic and is slightly lighter than whole metal parts. One of the promising studies today is investigating metal powder in a nylon or ABS filament [49]. However, the alloy or element used, if exposed to the right environment, will begin to corrode. Gong et al. [50] investigated on Austenitic 316L stainless steel using a novel ultrafuse filament, called Ultrafuse 316LX. They 3D-printed SS 316L using an FDM-based material extrusion printer. Stainless steel has wonderful corrosion resistance and is mechanically suitable for general usage. The much finer grains in the austenitic structure help increase the overall yield strength of this classification of steel alloy. The study compares the SS316L developed through FDM, with SLM, and regular steel properties, as shown in Table 5.2 [50]. As seen with the AMC's and iron mixed composite in the last section [23,24], the heavy presence of metal in the plastic aided in its overall strength and wear resistance. However, no study was conducted to see how this influenced overall resistance to corrosion. In Table 5.2 below, we can see the results of the different parts and see that the cast part obviously produced the best results [50].

Although FDM ranked last in tensile properties, recall comparatively to casting and SLM method, it is much cheaper to produce [50]. In addition, the use of a good corrosion-resistant alloy such as stainless steel or the AMC seen in the last section opens a new field of uses and products capable of being used in everyday life, replacing old metal components. Although this study does not test corrosion resistance, it is clear that a plastic−metal composite using aluminum or stainless steel would not corrode nearly as badly as cast iron or bronze. It is similar to the concept of placing a plastic film or layer over a metal wire; it prevents moisture or other harmful chemicals from touching the metal. Although a composite cannot be used in high-temperature situations, it can be used in a heavy moisture environment, where regular metal parts are not suitable. Parts produced by the FDM process require some fine tuning, as seen in the past section; however, once it is fine-tuned, the part is suitable for most tasks. In our current world, health and environmental safety have been placed at the top of current issues. Introducing parts yielding good corrosion resistance, low environmental toxicity, and the ability to be recycled will turn the heads of every major corporation and consumer. One major application is its use as a medical scaffold [36]. Recall that in the FDM process, there is a presence of small air gaps or pores; this has to do with the layer thickness, the temperature of the extruder when extruded onto the part, and the overall microstructure of the filament. However, the results of this study show only the beginning of a new innovation. With the right mixture of PLA resin, and proper settings, one can produce a medical scaffold capable of housing living cells [51]. Another polymer, polycaprolactone, PCL, when synthesized and mixed in proportion with hexamethylene diisocyanate, has also shown very promising results in being used as a medical scaffold [52]. This opens the gates to replaceable organs, bones, joint replacements, and ligament repair. The main issue with using any FDM process is the variety of porosity seen in the extrusion. Moreover, if the pores are not all

Table 5.2 Tensile properties of SS 316L alloy [50].

Alloy *Properties*	**FDM SS 316L**	**SLM SS 316L**	**AISI SS 316L**
Yield strength (MPa)	167	541	205
UTS (MPa)	465	648	515
Elongation (%)	31	30	60
Young's modulus (Gpa)	152	320	193

consistently the same size, it can be difficult to house any living cells or hope to see cell growth in the scaffold itself [51]. The PLA undergoes a chemical reaction with an epoxy chain to produce foam. When extruded from the FDM, this new foam shows better uniformity of pores and overall better structure for living cells [51].

In another study, it was found that PLA is one of four potential polymers capable of allowing for cell growth and regeneration of ligaments, tissues, and bones. Another highly porous polymer poly(D,L-lactide-co-glycolide mixed with collagen also shows promise in future medical scaffolds [53]. The other two are PCL, polyethylene oxide terephthalate (PEOT), and polybutylene terephthalate (PBT) [54]. Each undergoes month-long testing. The tests essentially introduce loads to test for friction and investigate the deformation that occurs to each scaffold and how this affects cell growth in each of them [54]. The reason they test for friction is that in our everyday life. Our bones, tissues, ligaments, and tendons are all under constant stress and loads. Each accompanied by friction. As seen before in Section 3.2, the friction coefficient plays a huge role in its tribological properties and overall lifespan. Repairing such important tissues and bones is critical for many in this modern age and also plagued with limited success. The good joint function requires a low friction coefficient and the absorption and transmission of loads throughout the joint. To put it into perspective, our ankles and feet, while running, absorb and transmit sometimes three times our body weight with every step depending on the medium one is running on. This wear and tear on our joints is why a low friction coefficient is needed and critical to joint and bone health. Cartilage repair is another complicated affair. Cartilage is the "padding" between bones and joints that prevent bone-on-bone contact, which is very painful. Although two common treatments are available, they require multiple surgeries and are limited in their overall success [54]. The main reason for their ineffectiveness is the immunological rejection of not quite complete donor matches, and the availability of grafting it from another location in a person's body. In addition, in current methods, the graft itself is unable to really expand and grow in the affected area. These problems are all considered during these tests. Its results showed that all four polymers were capable of housing living cells. However, the best was the use of PEOT/PBT in a fluid matrix. Overall, the combination showed the lowest friction coefficient and was most efficient at handling loads transmitted through the scaffold [54]. After almost a month, cells were not only still present but slowly expanding and growing. These four polymers all have the potential as medical scaffolds because of their unique tribological and minimal corrosive properties. In the near future, as more tests are conducted, tissue and ligament repair will most likely see unparalleled success.

5.4 Applications of FDM

5.4.1 Commercialization

FDM has a wide variety of commercial applications. Due to its superior speed and the low cost of use, it is the ideal method for manufacturing for many different industries. One of the most common applications for FDM printers is for industries to produce prototypes [55]. It is ideal for creating prototypes due to its accuracy, high speed, and low cost. By developing a prototype using FDM, the produced model will have properties that reflect the final product in terms of strength, heat resistance, UV stability, and water resistance [56]. Since the prototype has similar properties to the final product, a variety of tests can be performed to determine if the end product will withstand its intended use. By using an FDM prototype, companies can save money on materials and time by rapidly producing prototypes until the desired product and its desired functions are found. By being able to quickly test the concept, manufacturers are able to bring quality prototypes to investors while minimizing cost and maximizing performance. Another application for FDM is the creation of mechanical parts. Many companies use FDM to produce mechanical parts either to sell or for repairs [56]. FDM is ideal for making mechanical parts since it can make complex shapes out of a wide variety of materials. Depending on the use for the mechanical part will influence the material selection. Industries, including automotive and aerospace, use FDM to create custom parts of complex shapes. Another recent application for FDM is the production of biomedical devices [56]. FDM has been used to custom-build prosthetics, delivering the perfect fit for each individual person instead of the generic counterparts. It is also commonly used for creating dental parts. Due to many materials being food safe, FDM is also used for food storage and drug delivery in the pharmaceutical industry [55]. Other applications for FDM are the personalization of products due to the flexibility and ability to produce complex shapes; it is also commonly used for several energy storage devices [55].

Several common industries that utilize FDM to produce mechanical parts are the automotive industry and the aerospace industry. The automotive industry has been using FDM to develop prototypes and low-volume end-use parts. FDM has been used as a cost-effective solution in the automotive industry, which improves measurements, functional testing, vehicle customization, optimized design, and rapid tooling [57]. An example of the use of FDM used in the automotive industry is Bentley Motors Ltd., which first prints almost every detail of a planned future vehicle in a miniature form to perform tests on the vehicles form and tests several practical functions saving money and time to create their final design [57]. Another example of FDM for commercial applications in the automotive industry is at Jaguar Land Rover, which produced a fascia air vent, which used rigid material for the housing and air-deflection blades and an elastic material for the control knobs and air seal [57]. By using FDM, they built the air vent in a single process, enabling them to prove that the hinges on the blades worked and the control knob was what they desired [57]. FDM has also been playing a large role in producing mechanical parts for the aerospace industry. It is commonly used in the aerospace industry for designing, testing, tooling, and production that is used in ground support systems and repair [58]. Several common internal applications include air filter boxes, bezels, brackets, custom cosmetic interior components, display shrouds, environmental control system ducting, fuel tanks, housings and enclosures, oil tanks, clips and clamps, knobs and buttons, and windshield defogger duct nozzles [59]. FDM also has external applications, which include battery compartments, camera mount and gimbal, clips and clamps, component connectors, electrical housings, payload enclosures, plenums, shrouds and closeouts, foreign object debris

(FOD) covers, propulsion components, rocket motors, fuel injectors, thrusters, and combustion chambers [59]. In the aerospace and automotive industries, polymer gears have some advantages over metal gears [60]. An FDM-printed gear is low cost, low weight, high efficiency, and quiet during operation, and it can function without lubrication. A Nylon 66 gear was able to surpass 2.4 million cycles at a load of 5 Nm without any wear, a metal gear was tested with a load factor of 1.2 Nm and failed at 5.94 million cycles [61], showing that using FDM for mechanical applications is a viable way to reduce weight and cut down on costs.

Due to the versatility of FDM materials, a more recent commercial application is in medical industries. FDM has been used to develop custom sized prosthetics, it has printed custom foot orthosis, ankle-foot orthosis, and prosthetic socket in a much faster time compared to traditional prosthetic making methods [62]. Another application for FDM has been explored with the use of oral drug delivery systems [63]. A capsular device was produced to administer the drug orally through a combination of PLA, hydroxypropyl cellulose, polyethylene glycol, acetaminophen, blue-dye, and water sensitive paper-yellow, which can influence the commercial applications for FDM [63]. FDM is becoming more common for medical applications; due to the cost and availability, it is only being used more often. A flexible filament called Bio flex by Filo alfa was tested to determine if it was viable for medical equipment, devices, and long-term internal implants [64]. Spectroscopic, thermal, thermomechanical, and rheological analyses showed that manufacturing the parts did not change anything in the filament structure, also there was no degradation or changes of the printed parts [64]. The study showed that FDM can be applied to construct bone or cartilage, expanding the applications even further than previously thought [64]. Ezigbo et al. [65] performed the structural analysis and testing for several different FDM prosthetics. The test was to determine if a prosthetic was strong enough to withstand impact from a fall by the user [65]. Each load-bearing part received a force of 250 N, and it showed that each part was able to withstand it [65]. Since each part was able to take the force, FDM is a viable cost-effective manufacturing method for different medical applications. FDM has an increasing amount of commercial applications due to its cost, versatility, and its ability to produce complex parts. It is an applicable manufacturing method for industries that want to focus on the concept, engineering models, functional testing, consumer products, high-heat applications, and initial prototypes [66]. With an increase in testing, FDM is increasing its amount of commercial applications proportionally and can be used by businesses to earn a profit due to the low cost of materials, speed of production, and versatile materials to fit almost every requirement.

5.4.2 Rapid manufacturing

FDM has been a game changer in rapid manufacturing and prototyping. Using traditional manufacturing processes, a designer would have to wait weeks between design revisions due to the lead times at machine shops. FDM allows designers to print design revisions at their desk and trial a new design revision daily. This also provides a more environmentally friendly way of producing trial parts than the traditional CNC cutting steel or aluminum parts [67]. As a case study, let us consider FDM's applications throughout electric vehicles' design, manufacturing, and end use. Starting at the concept stage, a designer will need to turn their idea from a digital CAD concept into something tangible. For an automaker like Toyota, this requires approximately 30,000 individual parts [68]. Assuming that 75% of the parts are standard nuts and bolts or are reused from other models that would still leave 7500 custom components that need to be designed and produced.

Let's first consider the traditional method of producing these parts. Parts are designed in CAD, turned into a 2D drawing, provided to manufacturers for quotes, be produced by a manufacturer often with a more than 5-week lead time, trialed, and then revised. This process is then repeated for every design revision of every single part until a prototype car is produced. Now consider this process using FDM. The designer creates a part in CAD, converts it to .stl file, prints it overnight, trials it, and revises it. No 2D drawings, no quotes, and no lead times. This cuts in half the time required by the designer to prepare the CAD for production and removes months of lead times from the design process. In a fiercely competitive segment like automotive and in particular EVs, this advantage can be make or break for a company.

These machines are also made up of thousands of components that need to go through the design and revision process, the same as your prototype car parts. Additionally, some processes just cannot be completed by a machine and require a manual process. This manual process introduces human error into the production process, which creates a whole host of quality problems. The manufacturing team implements a jig to help production associates perform their task more accurately and efficiently to reduce these variances. Again FDM comes into play, allowing the manufacturing team to implement a jig for the manual process literally overnight. This eliminates production hold ups and limits the number of parts that are wasted due to quality concerns. Finally, we will have a successful finished product to send to consumers. While FDM typically will not make it into end-use products due to robustness, surface finish, and production volume, it played an intricate role in bringing final parts into production in a timely fashion.

Taking a deeper dive into FDM in rapid manufacturing, let us compare injection molding and FDM. Injection molding has been and still is the gold standard for high volume parts production. Once a mold has been made, it can be used repeatedly to make hundreds of thousands of identical parts for pennies per part. The drawback to this is that traditionally, making a mold can range from several thousand dollars to hundreds of thousands of dollars [69]. This startup cost is where FDM finds its niche. For individuals, the startup cost for FDM is a printer, which typically ranges from a few hundred dollars to several thousand dollars. After this cost is limited to the filament and powering the printer, depending on your printer and filament selection cost can range from ~\$9.00/lb for PLA to ~\$22.00/lb for carbon fiber infused filament. This means that your first batch of parts can be made for just a few hundred dollars and even better, unlike a mold, you can simply print a different part when you move onto a new project. The same can be said for design revisions, when you make a change, you simply print the revised design, where FDM loses its financial viability is in high volume production. With injection molding, after you have your final design and have had your mold made parts are produced repeatedly with very little maintenance for pennies per part. Additionally, the 3D printer can be used on your next project, meaning your next project's only cost is the cost of injection molding without revisions. This pricing breakpoint is the reason that FDM is best used in the research and development phase of a product and not great as a finished product production process.

5.4.3 Open source

3D printers are available in both closed-source and open-source varieties. Commercially manufactured machines often provide more accuracy and part strength, while open-source 3D printers are much more versatile and can be modified to be able to use a wider variety of print materials and can also be modified to satisfy other print process parameters. As technology improves, the gap in quality between a commercially manufactured 3D printer and an open-source 3D printer is getting smaller. Due to the nature of open-sourced printers, improvements can always be made by individuals and groups of individuals that desire a printer that is capable of making higher quality parts. The term *open source* is typically used in the realm of software and denotes that a software's original source code is freely available to the public and may be modified and redistributed freely. In the case of FDM, open source can refer to both the hardware and the software involved with 3D printers. An open-source 3D printer is one where the code that drives the printer and the information regarding the parts is openly available to the public and may be built, modified, and improved by anyone [70]. This is especially helpful in the advancement of FDM technology because it puts more minds to the tasks and keeps companies in constant competition with open-source models, speeding up innovation in the field. The trend of using and creating an open-source machine is relatively unique to 3D printers when compared to most other machines used for manufacturing. It is not typical to find a classification of machines where open-source versions of those machines are realistically competitive to the proprietary versions of those machines. Most machines used for manufacturing are locked to use only the proprietary materials specified by the manufacturer [71]. If a machine is broken, inadequate, or inefficient, little can be done by the user or even the collective of users to legally or realistically improve the function of the machine without the consent and/or aid of the manufacturer of said machine. Additionally, most closed-source machines eventually become obsolete and unsupported, either due to newer models being pushed by the parent company or the company that makes the machine goes out of business, thus requiring the machine to be decommissioned. For many 3D printers, however, this is not the case. Information regarding the dimensions, function, accessibility, and alternatives to its parts as well as information and source code relevant to the operation of the machine is available to access and modify. This slows the rate at which a 3D printer would become obsolete, and updates can still be made for a printer even when the manufacturer ceases production and development. This opens up a world of opportunity to both professional and recreational operators of 3D printers alike.

Because FDM is very often used for rapid prototyping, a machine that can be modified to serve several different functions and manufacture prototypes of all shapes and sizes is highly desirable. It would be inefficient and not cost-effective to purchase a new machine when the current model in use cannot make a specific prototype. An open source 3D printer allows for modification, and as such, it is much easier to rapidly prototype a part that might not be manufacturable with the current configuration of the machine. This is also important to the users because open-source models tend to be less expensive than the proprietary alternatives [72]. In addition, hobbyists often prefer open-source machines over proprietary ones because they are often interested in the technology itself. Building and modifying their own 3D printer are extremely valuable when trying to learn about the technology, especially when they do not work full time for a company that researches/manufactures FDM printers. A study detailed in Ref. [72] claims that open-source 3D printers can make the same parts to a nearly similar quality as proprietary printers. The open-source printer did suffer some minor deficiencies in thermal warping and surface roughness, but the article also noted that some of the differences could be attributed to limitations in the specific 3D printer that the researchers chose to study. Having an open-source option for FDM accelerates advancements in technology. This is due to both less need for direct aid from the manufacturer of the machines to improve the product and an increase in accessibility to minds and people seeking to optimize the product. People can generate new parts and codes unaffiliated with the original manufacturer that improve the printing processes and part finishes. Having open-source 3D printers on the market also incentivizes producers of proprietary printers to lower prices and research more efficient methods of FDM in order to stay competitive with the cheaper, more customizable models of open-source 3D printers on the market.

Open-source 3D printing has applications overlapping with commercial manufacturing and rapid manufacturing. Essentially, any FDM application that does not require cutting-edge precision or mass manufacturing is an application for open-source 3D printers. Any application where versatility is required in size, material, shape, size, material, and shape is a good application for an open-source 3D printer. Some closed-source printers are extremely good at printing in one material but cannot be used to print any other materials, whereas an open-source printer would be able to use different materials to print with after a few modifications. Also, because of the nature of open-source 3D printers, they can be modified to be large or small, meaning if the printer is currently configured to print small parts and a large part is required, the printer can be modified to accommodate larger parts, whereas a closed-source printer would not be able to do this and a new printer may be needed to print that part. Open-source 3D printers are more effective for prototyping various parts because one cannot tell what parts will be needed for future projects so versatility in a printer is highly desirable.

5.5 Conclusions

AM has proven to be an incredibly valuable process for fast-track prototyping, as well as a possibility for large scale manufacturing. It involves the use of a device resembling a 3D printer whereby there are 3-axes that are computer-controlled with a print head on the z-axis. The x- and y-axis move to specific locations called for in the G-code, while the extruder head on the z-axis translates vertically and heats material to be deposited on the bed. The bed lies at the bottom of the z-axis within the envelope of the x- and y-axis. In conventional manufacturing processes, prototypes requiring a high degree of precision require that they be machined on a mill, lathe, or wire EDM machine. In addition to the machine cost and setup cost, the material cost played a significant role in the cost of prototypes. This limited design engineers to fewer prototypes at a much higher expense. With the advent of 3D printing, the cost of prototyping dropped significantly as models could be made relatively cheaply, while still retaining their strength through the use of high-strength plastics. FDM is an excellent example of this, as the prototypes can be made cheaply and with fantastic precision closely resembling the production parts that can be produced using the same process. By adjusting the FDM printer setup characteristics, wear resistance can be added to parts while using the original material. These modifications include adjustment of the print angle, which can reduce the amount of air within the part, thus increasing wear resistance. FDM printing is an excellent solution for fast-track prototyping of accurate models but can also be used in a production environment. Along with the prints' speed and accuracy, FDM printing can also be used for parts that need to be food safe by simply changing the material used. A common food-safe material is ABS-M30i. Therefore, FDM is useful in pharmaceuticals, biomedical, and implants applications. Overall, FDM manufacturing has been changing the way industrial parts, life-saving drugs can be made and how rapid prototypes can be created.

References

[1] K.V. Wong, A. Hernandez, A Review of Additive Manufacturing, International Scholarly Research Network Mechanical Engineering 2012 (2012), 208760, https://doi.org/10.5402/2012/208760. https://downloads.hindawi.com/archive/2012/208760.pdf. (Accessed 2 April 2022).

[2] M. Attaran, Additive manufacturing: the most promising technology to alter the supply chain and logistics, Journal of Service Science and Management 10 (189–205) (2017) [Online]. Available at: researchgate.net/publication/313904937_Additive_Manufacturing_The_Most_Promising_Technology_to_Alter_the_Supply_Chain_and_L logistics. (Accessed 1 September 2020).

[3] 5 Places AM Is Unstoppable, 2020. https://www.stratasysdirect.com/manufacturing-services/3d-printing/unstoppable-industries-using-additive-manufacturing. (Accessed 1 September 2020).

[4] E. Palermo, FDM: Most Common 3D Printing Method, September 2013. https://www.livescience.com/39810-fused-deposition-modeling.html. (Accessed 1 September 2020).

[5] The Complete Guide to 3D Printing Plastic Parts with FDM, 2020. https://www.stratasysdirect.com/technologies/fused-deposition-modeling. (Accessed 1 September 2020).
[6] B. Varroosis, Introduction to FDM 3D Printing, 2020. https://www.3dhubs.com/knowledge-base/introduction-fdm-3d-printing/#pros-cons. (Accessed 1 September 2020).
[7] M.C. Bhushan, An Overview of AM (3D Printing) for Microfabrication, March 2017 [Online]. Available: https://link-springer-com.unr.idm.oclc.org/article/10.1007/s00542-017-3342-8. (Accessed 1 September 2020).
[8] Ultimate 3D Printing Materials Guide, 2020. https://www.simplify3d.com/support/materials-guide/. (Accessed 14 September 2020).
[9] 3D Printing ABS Filament, 2019. https://gizmodorks.com/abs-3d-printer-filament/#:~:text=ABS%20(Acrylonitrile%20Butadiene%20Styrene)%20plastic,other%20plastics%20-known%20as%20thermoplastics.&text=ABS%20filament%20is%20a%20hard,when%20-compared%20to%20PLA%20filament. (Accessed 14 September 2020).
[10] V. Carlota, All You Need to Know About PLA for 3D Printing, August 2019 [Online], https://www.3dnatives.com/en/pla-3d-printing-guide-190820194/#!. (Accessed 14 September 2020).
[11] Everything You Need to Know about Nylon 3D Printing, 2020 [Online]. Available: https://www.makerbot.com/stories/design/nylon-3d-printing/. (Accessed 14 September 2020).
[12] Plastic Material Melt and Mould Temperatures, 2020 [Online]. Available: https://www.plastikcity.co.uk/useful-stuff/material-melt-mould-temperatures. (Accessed 14 September 2020).
[13] ASA, 2020. https://www.simplify3d.com/support/materials-guide/asa/#:~:text=ASA%2C%20also%20known%20as%20Acrylic,with%20properties%20similar%20to%20ABS.&text=ASA%20is%20known%20for%20high,resistance%2C%20and%20increased%20printing%20difficulty. (Accessed 14 September 2020).
[14] HIPS, 2020. https://www.simplify3d.com/support/materials-guide/hips/#:~:text=HIPS%2C%20or%20High%20Impact%20Polystyrene,markings%20caused%20by%20support%20removal. (Accessed 14 September 2020).
[15] PETG, 2020. https://www.simplify3d.com/support/materials-guide/petg/. (Accessed 14 September 2020).
[16] Polycarbonate, 2020. https://www.simplify3d.com/support/materials-guide/polycarbonate/. (Accessed 14 September 2020).
[17] Flexible, 2020. https://www.simplify3d.com/support/materials-guide/flexible/. (Accessed 14 September 2020).
[18] PVA, 2020. https://www.simplify3d.com/support/materials-guide/pva/. (Accessed 14 September 2020).
[19] Polypropylene, 2020. https://www.simplify3d.com/support/materials-guide/polypropylene/#:~:text=Overview,it%20challenging%20to%203D%20print. (Accessed 14 September 2020).
[20] Composite 3D Printing: An Emerging Technology With a Bright Future, February 2020 [Online], https://amfg.ai/2020/02/25/composite-3d-printing-an-emerging-technology-with-a-bright-future/#:~:text=Most%203D%20printers%20capable%20of,object%20layer%2-Dby%2Dlayer. (Accessed 15 September 2020).
[21] The Complete Guide to 3D Printing Plastic Parts With FDM, 2020. https://www.stratasysdirect.com/technologies/fused-deposition-modeling#:~:text=FDM%20utilizes%20-strong%2C%20engineering%2Dgrade,strength%2Dto%2Dweight%20ratios. (Accessed 15 September 2020).

[22] S. Wickramasinghe, T. Do, P. Tran, FDM-based 3D printing of polymer and associated composite: a review on mechanical properties, defects and treatments, Polymers 12 (2020) 1529.
[23] Z. Liu, Q. Lei, S. Xing, Mechanical characteristics of wood, ceramic, metal and carbon fiber-based PLA composites fabricated by FDM, Journal of Materials Research and Technology 8 (5) (October 2019) [Online]. Available: https://www.sciencedirect.com/science/article/pii/S2238785419301905. (Accessed 25 September 2020).
[24] FDM 3D Printing Materials Compared, 2020 [Online]. Available: https://www.3dhubs.com/knowledge-base/fdm-3d-printing-materials-compared/. (Accessed 26 September 2020).
[25] M.N. Belgacem, A. Gandini, Chapter 21: Polylactic acid: synthesis, properties and applications, in: Monomers, Polymers and Composites from Renewables Resources, Elsevier, Oxford, 2008, pp. 436–437.
[26] An Introduction to Single Screw Extrusion, March 7, 2019 [Online]. Available: https://www.azom.com/article.aspx?ArticleID=13566. (Accessed 16 November 2020).
[27] B. Giemza, M. Domański, M. Deliś, D. Kapica, Tribological properties of 3D printed composites, Journal of KONBIN 48 (1) (2018) 447–463, https://doi.org/10.2478/jok-2018-0066. https://doaj.org/article/bd79e29a010e4001a1163ece543e62f9. (Accessed 16 September 2020).
[28] D.O. Kazmer, A. Colon, Injection printing: additive molding via shell material extrusion and filling, Additive Manufacturing 36 (2020). https://www.sciencedirect.com/science/article/pii/S2214860420308411?via%3Dihub. (Accessed 17 September 2020).
[29] F. Akasheh, H. Aglan, Fracture toughness enhancement of carbon fiber–reinforced polymer composites utilizing AM fabrication, Journal of Elastomers & Plastics 51 (7–8) (2018) 698–711, https://doi.org/10.1177/0095244318817867.
[30] T.C. Okwuosa, B.C. Pereira, B. Arafat, M. Cieszynska, A. Isreb, M.A. Alhnan, Fabricating a shell-core delayed release tablet using dual fdm 3D printing for patient-centred therapy, Pharmaceutical Research 34 (2) (2017) 427–437, https://doi.org/10.1007/s11095-016-2073-3.
[31] G. Kollamaram, D.M. Croker, G.M. Walker, A. Goyanes, A.W. Basit, S. Gaisford, Low temperature fused deposition modeling (FDM) 3D printing of thermolabile drugs, International Journal of Pharmaceutics 545 (1–2) (2018) 144–152, https://doi.org/10.1016/j.ijpharm.2018.04.055.
[32] B.J. Park, et al., Pharmaceutical applications of 3D printing technology: current understanding and future perspectives, Journal of Pharmaceutical Investigation 49 (6) (2019) 575–585, https://doi.org/10.1007/s40005-018-00414-y.
[33] M. He, Y. Zhao, B. Wang, Q. Xi, J. Zhou, Z. Liang, 3D Printing: 3D printing fabrication of amorphous thermoelectric materials with ultralow thermal conductivity, Small 11 (44) (2015) 5888, https://doi.org/10.1002/smll.201570266.
[34] J. Qiu, et al., 3D Printing of highly textured bulk thermoelectric materials: mechanically robust BiSbTe alloys with superior performance, Energy & Environmental Science 12 (10) (2019) 3106–3117, https://doi.org/10.1039/C9EE02044F.
[35] J. Wang, H. Li, R. Liu, L. Li, Y.-H. Lin, C.-W. Nan, Thermoelectric and mechanical properties of PLA/Bi0·5Sb1·5Te3 composite wires used for 3D printing, Composites Science and Technology 157 (2018) 1–9, https://doi.org/10.1016/j.compscitech.2018.01.013.
[36] J. Shi, H. Chen, S. Jia, W. Wang, 3D printing fabrication of porous bismuth antimony telluride and study of the thermoelectric properties, Journal of Manufacturing Processes 37 (2019) 370–375, https://doi.org/10.1016/j.jmapro.2018.11.001.

[37] C.O. Balderrama-Armendariz, et al., Torsion analysis of the anisotropic behavior of FDM technology, The International Journal of Advanced Manufacturing Technology 96 (1–4) (2018) 307–317, https://doi.org/10.1007/s00170-018-1602-0.
[38] B. Banjanin, G. Vladic, M. Pál, S. Balos, M. Dramicanin, M. Rackov, I. Knezevic, Consistency analysis of mechanical properties of elements produced by FDM AM technology, Matéria (Rio de Janeiro) 23 (4) (2018), https://doi.org/10.1590/S1517-707620180004.0584. https://www.scielo.br/scielo.php?pid=S1517-70762018000400441&script=sci_arttext. (Accessed 17 September 2020).
[39] S. Sheth, R.M. Taylor, Numerical Investigation of Stiffness Properties of FDM Parts as a Function of Raster Orientation, January 1, 2017. http://utw10945.utweb.utexas.edu/sites/default/files/2017/Manuscripts/NumericalInvestigationofStiffnessPropertiesof.pdf. (Accessed 17 September 2020).
[40] Continuous Carbon Fiber - High Strength 3D Printing Material. [Online]. Available at: https://markforged.com/materials/continuous-fibers/continuous-carbon-fiber?mfa=sga-na-mof-compositematerial&adg=80188521729&kw= carbon fiber 3d printer&device=c&gclid=CjwKCAjwkoz7BRBPEiwAeKw3qxoJ1IWO3Amc00WzH5XdA9ZAN_7eL9kNMi0yeQ7KLxc1u-bswAleORoC8LMQAvD_BwE. [Accessed 17 September 2020].
[41] Support.formlabs.com. 2020. Customer_V2. [online] Available at: https://support.formlabs.com/s/article/Model-Orientation?language=en_US#: ~ :text=of%20a%20print.-,Tilting%20-a%20flat%20surface,print%20has%20with%20the%20tank. [Accessed 1 October 2020].
[42] M. Domingo-Espin, et al., Mechanical property characterization and simulation of FDM Polycarbonate parts, Materials and Design 83 (2015) 670–677, https://doi.org/10.1016/j.matdes.2015.06.074.
[43] Markforged, Continuous Carbon Fiber - High Strength 3D Printing Material, 2020. https://markforged.com/materials/continuous-fibers/continuous-carbon-fiber. (Accessed 1 October 2020).
[44] O. Ahmed Mohameda, S.H. Masooda, J.L. Bhowmikb, A.E. Somersc, Investigation on the tribological behavior and wear mechanism of parts processed by fused deposition, Journal of Manufacturing Processes (2017) 149–159. https://www-sciencedirect-com.unr.idm.oclc.org/science/article/pii/S1526612517301834. (Accessed 16 September 2020).
[45] S. Singh, R. Singh, Study on tribological properties of Al-Al_2O_3 composites prepared through FDMAIC route using reinforced sacrificial patterns, Journal of Manufacturing Engineering 138 (2015). https://asmedigitalcollection-asme-org.unr.idm.oclc.org/manufacturingscience/article/138/2/021009/376049/Study-on-Tribological-Properties-of-Al-Al2O3. (Accessed 16 September 2020).
[46] S.H.R. Sanei, D. Popescu, 3D-printed carbon fiber reinforced polymer composites: a systematic review, Journal of Composites Science 4 (3) (2020) 98.
[47] A. Armillotta, Assessment of surface quality on textured FDM prototypes, Rapid Prototyping Journal 12 (1) (2006) 35–41, https://doi.org/10.1108/13552540610637255. (Accessed 15 November 2020).
[48] H.K. Garg, R. Singh, Modelling the peak elongation of nylon 6 and FE powder based composite wire for FDM feedstock filament, Journal of the Institution of Engineers (India) (2016), https://doi.org/10.1007/s40032-016-0250-0. https://link-springer-com.unr.idm.oclc.org/article/. (Accessed 15 November 2020).
[49] B. Akhoundi, A.H. Behravesh, A. Bagheri Saed, Improving mechanical properties of continuous fiber-reinforced thermoplastics composites produced by FDM 3D printer, Sage Journals 38 (3) (2018) 99–116, https://doi.org/10.1177/0731684418807300. https://journals-sagepub-com.unr.idm.oclc.org/. (Accessed 15 November 2020).

[50] H. Gong, S. Dean, K. Kardel, C. Andres, Comparison of stainless steel 316L Parts made by FDM and SLM AM processes, JOM: The Journal of the Minerals, Metals & Materials Society 71 (2019). (Accessed 13 October 2020).

[51] W.J. Choi, K.S. Hwang, H.J. Kwon, C. Lee, C.H. Kim, T.H. Kim, S.W. Heo, J.-H. Kim, J.-Y. Lee, Rapid development of dual porous poly(lactic acid) foam FDM (FDM) 3D printing for medical scaffold application, Materials Science and Engineering: C 110 (110693) (2020) 1−9, https://doi.org/10.1016/j.msec.2020.110693.

[52] A. Haryńska, J. Kucinska-Lipka, A. Sulowska, I. Gubanska, M. Kostrzewa, H. Janik, Medical-Grade PCL based polyurethane system for FDM 3D printing- characterization and fabrication, Materials 12(6) (887) (2019) 1−18, https://doi.org/10.3390/ma12060887. https://doaj.org/article/de61cbbb709147188caf7f194c43b7da.

[53] H.-J Yen, C. Tseng, S.-H Hsu, C.-L Tsai, Evaluation of chondrocyte growth in the highly porous scaffolds made by fused deposition manufacturing (FDM) filled with type II collagen, Biomedical Microdevices 11 (2009) 615−624, https://doi.org/10.1007/s10544-008-9271-7. https://link.springer.com/article/10.1007/s10544-008-9271-7. (Accessed 15 November 2020).

[54] W.J. Hendrikson, X. Zeng, J. Rouwkema, C.A. van Blitterswijk, E. van der Heide, L. Moroni, Biological and tribological assessment of poly(Ethylene Oxide Terephthalate)/Poly(Butylene Terephthalate), polycaprolactone, and poly (L/DL) lactic acid plotted scaffolds for skeletal tissue regeneration, Advanced Healthcare Material 5 (2015) 232−243, https://doi.org/10.1002/adhm.201500067.

[55] F.M. Mwema, E.T. Akinlabi, Basics of FDM (FDM), 2020. https://www.ncbi.nlm.nih.gov/pmc/articles/PMC7257444/#CR19. (Accessed 13 October 2020).

[56] FDM, 2020. https://www.prototypeprojects.com/technologies/fdm-prototyping/#:~:text=FDM%20parts%20can%20be%20used,heat%20resistance%2C%20durability%20and%20strength. (Accessed 13 October 2020).

[57] B. Conlin, More than Prototypes: A Look at the 3D Printing Industry, October 2018. https://www.businessnewsdaily.com/9297-3d-printing-for-business.html. (Accessed 13 October 2020).

[58] Five Ways 3D Printing Is Transforming the Automotive Industry, 2020. https://purpleplatypus.com/wp-content/uploads/2018/05/White-Paper-5-Ways-3D-Printing-is-Transforming-the-Automotive-Industry-EN-A4.pdf. (Accessed 13 October 2020).

[59] 3D Printing in the Aerospace Industry: The Ultimate Guide, 2020 [Online]. Available: https://www.stratasysdirect.com/industries/aerospace.

[60] Y. Zhang, C. Purssell, K. Mao, S. Leigh, A physical investigation of wear and thermal characteristics of 3D printed nylon spur gears, Tribology International 141 (2020). https://www.sciencedirect.com/science/article/pii/S0301679X19304724.

[61] D. Zorko, S. Kulovec, J. Duhovnik, J. Tavcar, Durability and design parameters of a Steel/PEEK gear pair, Mechanism and Machine Theory 140 (2019). https://www.sciencedirect.com/science/article/pii/S0094114X19308341.

[62] Y. Jin, J. Plot, R. Chen, J. Wensman, A. Shih, AM of custom orthoses and prostheses − a review, Procedia CIRP 36 (2015). https://www.sciencedirect.com/science/article/pii/S2212827115004370S0094114X19308341.

[63] A. Melocchi, F. Parietti, G. Loreti, A. Maroni, A. Gazzaniga, L. Zema, 3D printing by FDM (FDM) of a swellable/erodible capsular device for oral pulsatile release of drugs, Journal of Drug Delivery Science and Technology 30 (2015). https://www.sciencedirect.com/science/article/pii/S177322471500132X.

[64] A. Haryriska, I. Carayon, P. Kosmela, K. Szeliski, M. Lapinski, M. Pokrywczynska, J. Kucinska-Lipka, H. Janik, A comprehensive evaluation of flexible FDM/FFF 3D printing

filament as a potential material in medical application, European Polymer Journal 138 (2020). https://www.sciencedirect.com/science/article/pii/S0014305720316724. (Accessed 25 October 2020).
[65] P. Ezigbo, F. Kelechi, N. Opara, N. Chukwuchekwa, Development of 3D printable prosthetic arm for amputees using computer aided design and fused deposition modelling, International Journal of Mechatronics, Electrical and Computer Technology 10 (2020).
[66] FDM, 2020. https://www.3dsystems.com/on-demand-manufacturing/fused-deposition-modeling/applications. (Accessed 25 October 2020).
[67] Comparing environmental impacts of AM vs traditional machining via life-cycle assessment, Emerald Insight (2020), https://doi.org/10.1108/RPJ-07-2013-0067/full/html. https://www.emerald.com/insight/content/.
[68] T. Corporation, How Many Parts Is Each Car Made Of, 2020. Available: https://www.toyota.co.jp/en/kids/faq/d/01/04/. (Accessed 15 October 2020).
[69] C.K. Matthew Franchetti, An economic analysis comparing the cost feasibility of replacing injection molding processes with emerging AM techniques, The International Journal of Advanced Manufacturing Technology 88 (2016).
[70] J. Flynt, Best Open Source 3D Printers of 2018, April 22, 2018. Available: 3dinsider.com/open-source-3d-printers/. (Accessed 15 October 2020).
[71] A. Ferreira, et al., Retrofitment, open-sourcing, and characterisation of a legacy fused deposition modelling system, The International Journal of Advanced Manufacturing Technology 90 (9−12) (2017) 3357−3367, https://doi.org/10.1007/s00170-016-9665-2.
[72] W.M. Johnson, et al., Comparative evaluation of an open-source FDM system, Rapid Prototyping Journal 20 (3) (2014) 205−214, https://doi.org/10.1108/RPJ-06-2012-0058.

Additive manufacturing: process and microstructure

6

Leslie T. Mushongera [1] *and Pankaj Kumar* [2]
[1]Department of Chemical & Materials Engineering, University of Nevada, Reno, NV, United States; [2]Department of Mechanical Engineering, University of New Mexico, Albuquerque, NM, United States

6.1 Introduction

Additive manufacturing (AM) is considered a new paradigm to design and manufacture components for various critical applications ranging from aerospace and automobile to biomedical applications [1–3]. The unique capability of building components by progressive addition of thin layer allows the process to customize components according to the individual application [4]. AM is fundamentally different from the conventional manufacturing technique. It is a bottom-up manufacturing approach that allows freedom to manufacture geometrically complex components with precise control at a minimum raw materials waste [5,6]. Due to rapid manufacturing and design flexibility, the fusion-based AM attracted wide attention from the scientific community and the manufacturing industries [7–10]. The basic principle of this technique involves local melting of raw materials in the form of either powder or wire using a high-energy beam followed by rapid solidification—the individual metal particles melt along with neighbor particles locally and are welded together after the solidification [11]. Broadly, the basic principle of AM can be thought of as local fusion welding [12]. However, these two techniques are significantly different in terms of the volume of molten metal and their respective solidification rates [11,12]. A completely different physics is involved in AM due to the small volume and high solidification rates. We understand, intuitively, the amount of molten metal for a given time is related directly to the energy per unit area/volume (energy density) supplied for the melting—the higher the energy density, the higher should be the volume of molten metal. This understanding leads us to recognize the critical variable in fusion-based AM. In powder-based AM, depending on the energy density supplied to melt the metal powders, the quality of the AM varies accordingly. The energy beam either uses a laser or electron to melt the particles. In either case, the beam diameter is typically in the range of micrometers [1]. To achieve the desired energy density to melt the metal powder, a very high-power laser or electron is used for a certain time. Because of the beam diameter, the local melting of powder is restricted to the micron area. However, the local melt volume (meltpool size) depends on the energy (laser or electron) beam power and scanning time of these beams on the powder bed [1]. The scanning time decides the time the laser beam is focused at the local metal powder bed. The slower the scan speed, the higher the laser will spend time at a powder bed location. This leads to a

Tribology of Additively Manufactured Materials. https://doi.org/10.1016/B978-0-12-821328-5.00006-8

relatively larger molten metal volume and, therefore, the performance of the local metal welding. Thus, an optimum scanning speed for the respective energy beam power to melt the respective metals is needed to manufacturing high-quality AM products. The fundamental variables can be controlled in fusion-based AM beam power and the beam speed at which it moves from one point to another, that is, the scanning speed. The scan time dictates the time of production of AM components, while higher power leads to a high energy cost. Both of these negatively affect the production cost of AM components [13]. Also, the higher melt volume may be beneficial in welding, but in AM, the larger local volume (because of higher melting depth) may create the defects [11,12]. Due to large local energy, the local phase transformation from liquid to gas can take place [14,15]. Due to the high solidification rate, the gases can entrap in the structure, creating unwanted porosity [16,17]. This phenomenon is also known as the "keyhole effect." Also, an overlap of the local melt pool is needed to create a successful layer that in turn helps to create the 3D components [18]. A quantity defines this as hatch distance in the AM processes. Hatch distance has a significant impact on the quality of the product [19,20]. The too-high value of hatch distance will not allow joining of the individual metal particles, thus, the 3D product may not form. A smaller hatch distance may lead to significant overlapping of the beam spot leading to the large melt pool size. Depending on meltpool size, the solidification rate changes. The larger the meltpool size, the solidification rate will be relatively lower. Depending on the solidification rate, the microstructure evolution in the materials will be different. In addition, the time required to build 3D components with a smaller hatch distance will be comparatively more than with the larger hatch distance, therefore reducing overall productivity. Therefore, developing a fundamental understanding of three key parameters, energy density, scan speed, and hatch distance, is essential in achieving the desired quality and productivity of AM components.

Also, the microstructure of AM components is the key factor in achieving the required properties for the application. The thermal evolution, temperature gradient (G), and the solidification growth rate in AM processes are the major deciding factors for the final microstructure [21]. However, more than 130 AM process parameters affect the characteristics of the microstructure [22,23]. These parameters are the complex function of the following principal parameters: energy density, scan speed, hatch distance, and layer thickness [23,24]. Therefore, developing an understanding of the effect of parameters on microstructure evolution is not a trivial task. The experimental approach alone is not sufficient to optimize the processing condition for optimal microstructure. Computational tools, therefore, can be a great help to identify the optimum AM manufacturing parameters. In the literature, a wide range of computational tools ranging from 2D cellular automata to cellular automata-finite element to phase field are present. A multiscale modeling approach has also been conducted in the literature for the microstructure evaluation in AM components. These models, however, restricted to very local microstructure evolution.

In this chapter, we review the impact of processing parameters on the quality of the AM components. A phase-field model is also conducted in this study to evaluate the microstructure development and develop an understanding of the microstructure evolution in the AM components utilizing the laser powder bed fusion (LPBF) techniques.

6.2 Effect of processing parameters on porosity development

Undesirable porosity is a major problem in fusion-based AM techniques [25,26]. The presence of porosity in the microstructure is known to reduce the mechanical performance of the materials [27–29]. In AM, various sizes and shapes of porosity can be formed [25,30–32]. These sizes and shapes largely depend on the processing parameters of AM components [13,33,34]. For example, in LPBF, porosity due to lack of fusion occurs when sufficient energy density is not available, and the scan speeds are relatively higher [13]. In this case, due to the low energy, a shallow and thin meltpool and lack of overlapping pattern cause the lack of fusion defect to be crescent-shaped. If the energy density is much higher, it vaporizes the metal while the gas may be trapped, creating small spherical pores in the microstructure [35]. Extensive studies on the impact of processing parameters have been reported in the literature [25,36–40]. Although limited material systems has been studied, a similar relationship of the porosity and the AM parameters has been observed in all the materials studied [1,41–43] (Fig. 6.1).

Kumar et al. [13] studied the effect of energy density and laser scan speed in the LPBF technique on the porosity development in Inconel 718 alloy. In their study, they demonstrated the individual effect of energy density and scan speed on the porosity fraction in the microstructure. In that study, it is shown that as the energy density increased from ~ 1 J/mm^2 to ~ 2.5 J/mm^2, the porosity reduced significantly from $\sim 100\%$ to 25% with a laser power of 75 W (Fig. 6.1). A close to theoretical density has been achieved when the energy density increased to ~ 5 J/mm^2 using 165 W. With

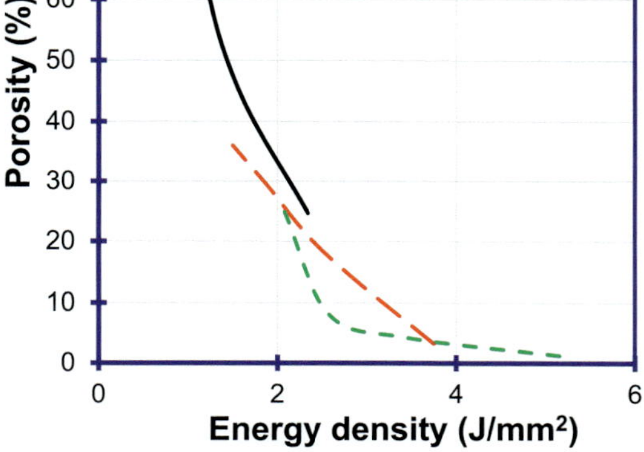

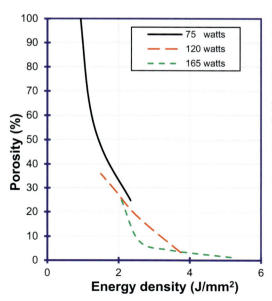

Figure 6.1 Porosity and energy density curve at different laser powers at the same scan speed.
Reproduced from Ref. P. Kumar, J. Farah, J. Akram, C. Teng, J. Ginn, M. Misra, Influence of laser processing parameters on porosity in Inconel 718 during additive manufacturing, The International Journal of Advanced Manufacturing Technology. 103 (2019) 1497–1507. https://doi.org/10.1007/s00170-019-03655-9.

75 and 120 W, an energy density of ~5 J mm^2 could be achieved for achieving very low porosity. Also, the behavior of porosity development in the microstructure varies as the laser power value changes. Also, the scan speed significantly changes the porosity generation in the microstructure. Higher the scan speed for a given power, the higher the porosity (Fig. 6.2). This demonstrates that at a higher scan speed, not enough time is available to melt the local metal powder to weld the powder leading to the lack of fusion porosity. It could be observed that with lower laser power, desired energy density and optimum scan spend could not be achieved for near theoretical density. For every laser power, the porosity formation behaviors are as shown in Figs. 6.1 and 6.2. The difference in behavior could be attributed to a change in the mechanisms by which the porosity forms. It seems for 75 and 120 W, the energy may not be sufficient to form the desire meltpool and leave the lack of fusion porosity of the laser scanning. At 165W, in both Figs. 6.1 and 6.2, there is a clear change in slope, which could indicate that two different porosity formation mechanisms are operating in two different energy density levels. It is possible that below ~2.5 J/mm^2, the lack-of-fusion mechanism is operative, and pore sizes are relatively larger, while above ~2.5 J/mm^2, the mode of the porosity formation may be changed from lack-of-fusion to keyhole mechanism to form the smaller pores in the microstructure. Depending on the process parameters, the meltpool geometry changes leading to a change in the melting characteristics from the conduction to keyhole mode, thereby the final geometry of the pores in the microstructure. However, there is no sharp boundary of meltpool geometry defined for conduction to keyhole mode. A study reports an analytical solution for the transition of melting from conduction to keyhole mode as a function of material and the laser power and scan speed [35]. The energy required for the transition of melting mode from conduction to keyhole mode is directly proportional to the

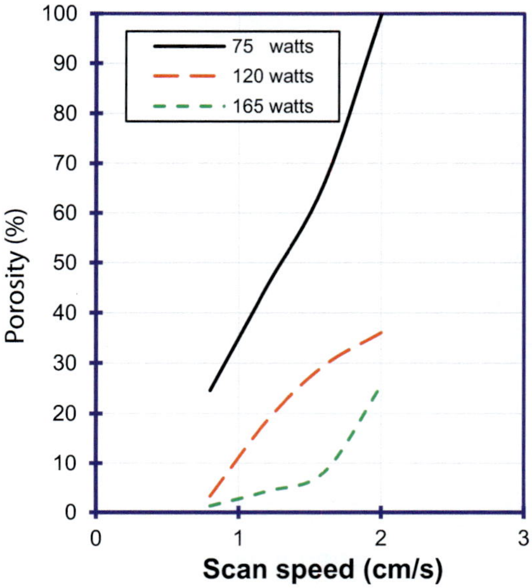

Figure 6.2 Porosity and scan speed curve at different laser powers at the same laser power.
Reproduced from Ref. P. Kumar, J. Farah, J. Akram, C. Teng, J. Ginn, M. Misra, Influence of laser processing parameters on porosity in Inconel 718 during additive manufacturing, The International Journal of Advanced Manufacturing Technology. 103 (2019) 1497–1507. https://doi.org/10.1007/s00170-019-03655-9.

power while inversely proportional to the square root of scan speed. Therefore, to avoid keyhole formation, either reduce the laser power or increase the scan speed.

As discussed earlier, due to the lack of complete overlapping of meltpool due to incomplete melting of powders, lack-of-fusion pores are formed. At the low laser power and high scanning speed, lack-of-fusion pores are likely to form. Therefore, it is possible to determine the optimal condition if we know the meltpoool geometry that develops as a function of laser power. In the literature, the lack of fusion index, which is defined as the ratio of melt pool depth and layer thickness, is utilized to quantify powders' complete melting [11]. The analytical solution for the fluid flow is also conducted for the meltpool to understand the melting of the powder in order to optimize the processing parameters to reduce the pore formation [1]. Many studies use a single laser scan to identify the meltpool geometry to identify the optimum laser power and scan speed [38]. For the lack-of-fusion pores, the hatch distance and the layer thickness are critical. The layer thickness and hatch distance are to be chosen in such a way that for the given power and scan speed, there should be complete interlayer melting to achieve the highest possible density of the components Fig. 6.3.

Also, at very high energy density and scan speed, the molten pool destabilizes, elongates, and separates into small spherical balls known as the "balling effect or

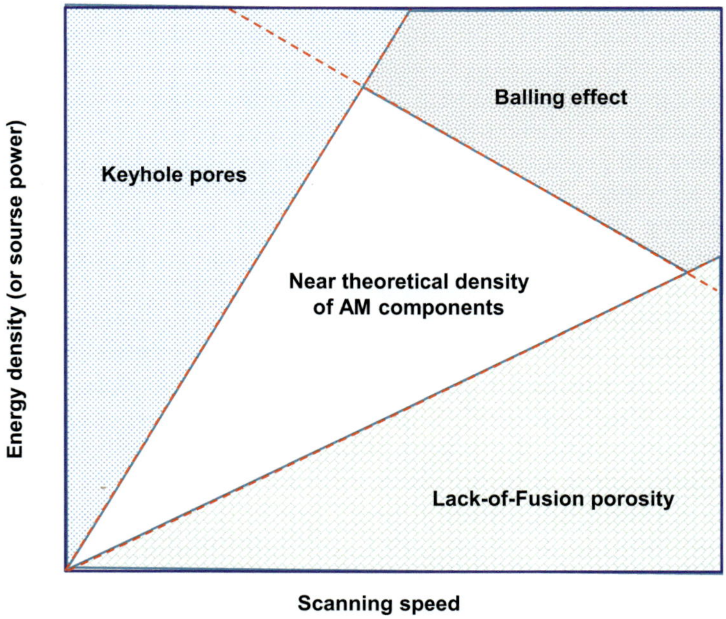

Figure 6.3 Schematic showing the effect of additive manufacturing (AM) processing parameters on the pore formation in AM components.
Reproduced from Ref. J.J.S. Dilip, S. Zhang, C. Teng, K. Zeng, C. Robinson, D. Pal, B. Stucker, Influence of processing parameters on the evolution of melt pool, porosity, and microstructures in Ti-6Al-4V alloy parts fabricated by selective laser melting, Progress in Additive Manufacturing. 2 (2017) 157−167. https://doi.org/10.1007/s40964-017-0030-2.

Plateau-Rayleigh instability" in LPBF [1,44]. A similar destabilization process can occur but form a half-cylinder at high scan spend in the directed energy deposition (DED) [45,46]. Due to the balling effect, the interlayer bonding may be restricted, and interball porosity in irregular shapes forms preferentially at the interlayer locations [44,47]. Therefore, each of the principal process parameters, including laser power, scan speed, hatch distance, and layer thickness, required to optimize to achieve the desired density of the AM components. The influence of processing parameters on the porosity is illustrated in Fig. 6.3.

6.3 Effect of processing parameters on the surface roughness

Surface roughness of as-manufactured AM component is the major barrier to use in applications [48,49]. The surface roughness has a significant impact on the structural performance, especially under fatigue loading conditions [50–52]. Typically, high-end structure components require a surface roughness of <1 μm [53]. Several AM processing factors affect the surface roughness of the AM components [54–56]. One of them is the "stair effect," which is a stepped approximation by the inclined surface and the layers of curves [57–59]. This effect can be observed in all AM techniques. The stair-step effect can be reduced by reducing the layer thickness to obtain a good-quality surface finish [60–63]. A relationship of average surface roughness parameter (R_a) to the layer thickness has been established in the literature [64]. Although the strain case approximation can give an initial idea of the surface roughness, it fails to predict the experimentally observed trends [65]. A new approach has been proposed to overcome the shortcoming of the "stair effect" approximation to predict the surface roughness Fig. 6.4 [65]. Nevertheless, the geometrical consideration of AM processing of powders is critical in achieving the desired surface quality.

Also, the surface roughness can be controlled by the energy input parameters such as energy density, power, and scanning speed [1,54,66]. Many studies focused on evaluating the surface roughness as the function of AM processing parameters. Cherry et al. [67] and Wang et al. [68] independently studied the effect of laser parameters on the surface roughness, R_a, of stainless steel (SS) in LPBF and observed that the R_a is significantly impacted by the laser power and scan speed. Fig. 6.4 shows the variation in R_a as a function of volumetric energy density (J/mm^3). In both studies, the R_a decreases as the energy density increased initially and reaches to minimum R_a but increased when the energy density increased to very high values. This behavior demonstrates that the different mechanisms are operative for the surface roughness at low energy density and very high energy density, respectively. At low energy, the input heat is not enough to completely melt the powders leading to the sticking of the powder particles at the interlayer, and the surface of the components causes a relatively large R_a. As the energy increases, the heat input also increases, which allows forming the stable meltpool to weld particles together. In this condition, the surface roughness decreases. With an increase in the energy density, it is expected that the surface

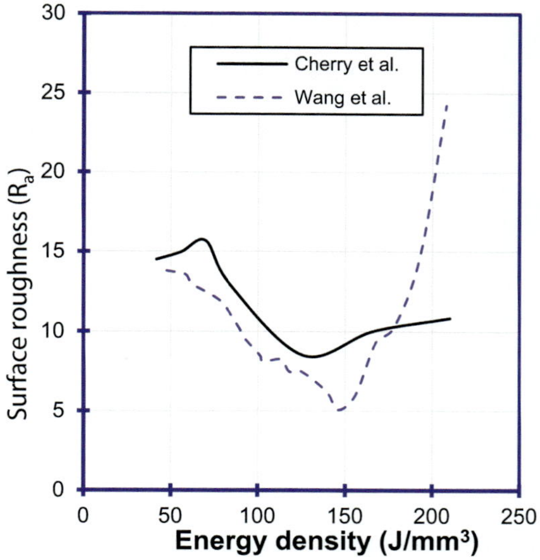

Figure 6.4 Surface roughness, R_a, variation as a function of energy density (J/mm^3).
Reproduced from Refs. J.A. Cherry, H.M. Davies, S. Mehmood, N.P. Lavery, S.G.R. Brown, J. Sienz, Investigation into the effect of process parameters on microstructural and physical properties of 316L stainless steel parts by selective laser melting, The International Journal of Advanced Manufacturing Technology. 76 (2015) 869–879. https://doi.org/10.1007/s00170-014-6297-2, D. Wang, Y. Liu, Y. Yang, D. Xiao, Theoretical and experimental study on surface roughness of 316L stainless steel metal parts obtained through selective laser melting, Rapid Prototyping Journal. (2016).

roughness decreases. The surface roughness also depends on how the high energy density has been achieved. If large scanning speeds are involved, the balling phenomena can cause an increased surface roughness [65,69]. Therefore, a stable meltpool is needed, which can be achieved by high laser power and low scan speed, to achieve the low surface roughness [70,71]. From Fig. 6.4, it can be observed that at a critical energy density, the R_a increases and reaches the maximum at a very high energy density. The increased R_a behavior has been attributed to the balling phenomena. Spherical beads are often observed at very high energy density on the surface, which confirms the balling phenomena that leads to the high R_a values of AM components. However, the critical values of energy density that lead to minimum surface R_a will depend on the materials, powder size, and shapes. From this discussion, it can be understood that the various input AM processing parameters are required to optimize, from the powder shape and size to power and scan speed, to achieve the lowest surface roughness. A similar condition related to the surface roughness may be desired in other AM techniques but at different meltpool sizes and energy requirements.

6.4 Microstructure evolution

The limited material systems have been studied for AM components [4,72–78]. All AM components show typical rapid solidification microstructure irrespective of the material systems. The columnar and equiaxed morphologies are the typical grain structure observed in the AM components (Fig. 6.5A–C) [79–84].

Due to the extremely large temperature gradient, the columnar grains readily form, which causes the directional properties of the AM components martin [2,85]. The

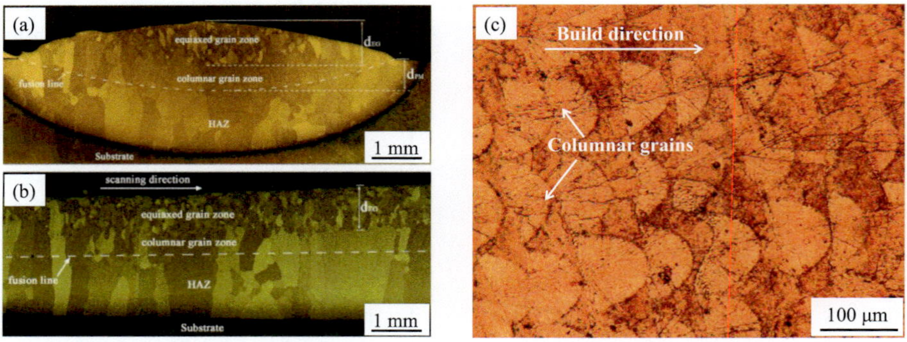

Figure 6.5 Representative microstructure of (A) transverse and (B) longitudinal plane of single layer deposited by laser additive-manufactured (AM) Ti–6.5Al–3.5Mo–1.5Zr–0.3Si ref [88]. (C) Representative microstructure of LPBF manufactured Inconel 718 along the build direction (ref. [13]).

isotropic properties are desired, which can be achieved by the equiaxed microstructure in the AM component [86,87]. Although columnar grains dominate in the AM microstructure, the equiaxed grains also form near the solidification front due to constitutional undercooling [88,89]. If constitutional undercooling is promoted during the solidification, the probability of equiaxed grain formation will be higher. Partially melted particle and grain refiner elements such as refractory elements support the constitutional undercooling and provide the nucleation site for the equiaxed grains in AM components [90,91]. The solidification microstructures are dictated by the temperature gradient, G, solidification rate, R, and undercooling ΔT [92–95]. Many studies constructed the G-R maps for different materials to relate the solidification microstructure [96]. Depending on the G/R ratio, the planar, columnar, cellular, and equiaxed microstructures can form in AM components. The low G/R ratio promotes the equiaxed microstructure, while high G/R leads to the planar microstructure [92–95]. The desired microstructure can be achieved by controlling the G/R ratio, which is directly related to the AM processing conditions. Changing the solidification growth front by rapidly changing the scan strategy can change the G/R ratio and control the microstructure. The detailed analysis of solidification parameters impacting the microstructure evolution in AM is given in Sections 5,6,7 and 8.

Also, the grain size observed in the AM components is related to many factors, but the cooling rate dominates over the other factors [97–99]. Larger the cooling rate, smaller the grain size [73,100]. Due to the nature of AM, the cooling rate decreases as the layer increases in the built direction. As the height increases, the previously solidified layer acting as the preheated substrate for the melting layer causes the reduction in the cooling rate. Therefore, a gradient in grain size has often been observed [79,101]. At the bottom, the finer grain were present, while coarser grains are observed as the height increases. The larger grain sizes were observed at the upper layers of the AM components [1,102].

The nickel-based superalloys have been widely studied for the AM [25,102–105]. The heterogeneous microstructure has been observed in these alloys due to the elemental segregation near the interdendritic regions both in LEBF and DED techniques [102,104,106–108]. The intermetallic Laves phases and delta phase are typically observed due to the segregation [109,110]. For example, niobium (Nb) segregation in the interdendritic promotes the formation of Laves phase [111–113]. Therefore, the microstructure is affected by the distribution of Nb-rich molten metal in the dendritic regions. This intermetallic phase formation is mainly affected by the solidification trend. Since these phases form in the final stage of solidification, these are observed as a long chain along the dendrites [110,114]. Larger the dendritic arm, larger will be the chain of the Lavas phase. Therefore, large columnar dendrites show a large chain of the Lavas phase along the interdendritic region. These phases are deleterious for mechanical performance [25,115,116]. The size of precipitated intermetallic phases depends on the cooling rate. Also, by controlling the G/R ratio and the cooling rate by AM processing parameters, the continuous length of the intermetallic phases in the microstructure of nickel-based AM components can be controlled.

The AM microstructure of austenitic SS typically consists of the cell or the columnar structure due to the high thermal gradient [117,118]. The equiaxed grains have rarely been observed in AM-fabricated SS [119–121]. Austenitic SS is the most common type of steel subjected to AM, consisting of a full austenite phase [122,123]. In AM, however, a small fraction of ferrite has also been observed [122,123]. Due to segregation, the intercellular or intercolumnar region becomes chromium- and molybdenum-rich molten liquid, which are ferrite stabilizers, leading to the formation of ferrite in the microstructure [118]. The ferrite content, however, depends on the cooling rate during AM [124].

Similarly, the rapid solidification microstructures have been observed in two-phase (α and β) Ti-6Al-4V alloy. Dendritic columnar grains are often observed in AM-manufactured Ti-6Al-4V alloy [125–127]. The primary α transforms to β when the temperature reaches the eutectoid temperature. The β transformed back either to α or martensitic α' depending on the cooling rate in AM [128]. A very high cooling rate causes β to transfer in α'. The α' in the microstructure, although increasing the strength, significantly reduces the alloy's ductility. Postheat treatment is often considered to transform α' phase to ($\alpha + \beta$) to achieve the desired mechanical properties. In AM, in situ phase transformation of α' to ($\alpha + \beta$) may occur due to repeated heating and cooling of layers in AM. Also, the columnar spacing of the ($\alpha + \beta$) in the microstructure is significantly controlled by the cooling rate [1,127].

In summary, the characteristics microstructure of AM alloys corresponds to the solidification microstructures. The grain structure and its characteristics are controlled by the AM processing parameters and the chemical compositions, which significantly impact the final properties of the materials. It is, therefore, of great practical importance to establish a fundamental understanding of AM parameters on the microstructure evolution and develop a relationship between them. To understand microstructure evolution in AM metal and alloys, it is essential to conduct the fundamental studies using theoretical modeling and computation. In the following section (section 5), a phase-field approach is presented to develop a fundamental understanding of microstructure evolution in rapid solidification in AM alloys.

6.5 Phase-field modeling of rapid solidification

In the last few years or so, phase-field modeling has become an important tool for the quantitative study of the solidification microstructure during the laser-based AM. The major advantage of this approach is given by its great flexibility of modeling, which allows addressing even complex systems involving several different physical mechanisms at the same time. The general idea behind phase-field modeling is to include phase-field order parameter $\phi(x,t)$ that is a function of space and time to denote the phases in a given system. Taking, for example, a two-phase melt-substrate system, the order parameter has a constant value in each bulk phase, for example, $\phi = 0$ in the melt and $\phi = 1$ in the forming solid substrate. Then, the interface between different phases is represented by a smooth transition region where the order parameter varies smoothly from 0 to 1. Thus, the interface width is smeared over a finite width. Microstructural changes are reflected by the spatio-temporal changes in the phase field variable. In phase-field modeling, free moving interfaces between different phases are not treated as geometric boundaries, that is, boundary conditions do not have to be applied explicitly at the interfaces. Rather, all the information about the motion and exact location of the phase boundaries is implicitly contained in the phase field, which obeys a partial differential equation that is solved within the whole computational domain.

The formulation of a phase-field model typically starts from a relevant thermodynamic potential such as a free energy functional that includes the necessary material descriptions. To model the melt to substrate transition in rapid solidification, the free energy functional should at least account for the interfacial energy and thermodynamics. Depending on the system, as many as possible contributions can be incorporated into the free energy function. For solidification, the free energy functional can be written as

$$F[\phi(x,t),c] = \int_V (g_{int} + g_{chem})dV, \# \tag{6.1}$$

where, g_{int} is the interfacial free energy density, which accounts for the energetic costs for the presence of interfaces and g_{chem} is the thermodynamic contribution that accounts for the redistribution of solutes within the system. To address the solidification in binary, ternary, and other high-order alloys during laser-based AM, the thermodynamics of materials have to be included in the phase field model very precisely. The thermodynamics is included by constructing chemical free energy densities for the individual phases, defined in a parabolic form determined by the equilibrium concentrations and the thermodynamic factors. Two approaches are used to determine the parameters for the free energies. The first approach involves directly coupling the phase-field model to CALPHAD databases to retrieve the temperature-dependent concentrations and the thermodynamic factors. The benefit of using such a formulation is that it directly employs thermodynamic parameters of materials and considers the derived thermodynamic factors as constants within the solidification interval, which improves the computational efficiency and precision as compared to the model with the driving force written in terms of entropy change. The second approach involves fitting the free energies for the individual phases based on the equilibrium data calculated CALPHAD databases. The benefit of this approach lies in its simplicity in implementation. The driving force for solidification is determined from the chemical-free energy density as [#], [**]:

$$\Delta g_{ch}(\phi, c) = \frac{m_l \Delta S(c - c^{eq})}{\phi + (1-\phi)k_{ls}} \cdot \# \tag{6.2}$$

The mean equilibrium concentration c^{eq} is defined as:

$$c^{eq} = \phi c_s^{eq} + (1-\phi)c_l^{eq} \cdot \# \tag{6.3}$$

where c_s^{eq} and c_l^{eq} are the equilibrium concentrations in the solid and liquid phases at a certain temperature T, respectively. The equilibrium concentrations are calculated using the expression:

$$c_{s/l}^{eq} = c_{s/l}^{eq}(T_L) + \frac{T - T_L}{m_{s/l}}, \# \tag{6.4}$$

where T_L is the liquidus temperature. The parameters m_s and m_l are the solidus and liquidus slopes obtained from the phase diagram, respectively, and $k_{ls} = m_l/m_s$ is the isothermal partition coefficient. The entropy change between the phases can be estimated by $\Delta S = \sigma V_m / \Gamma$, where σ, V_m, and Γ are the interface energy, molar volume, and Gibbs–Thomson coefficient, respectively. The temporal and spatial evolution of ϕ and c represent the microstructure evolution and are governed by the phase equation [129,130]:,

$$\frac{\tau_0}{W^2} a(\theta)^2 \frac{\partial \phi}{\partial t} = \overrightarrow{\nabla}\left[a(\theta)^2 \overrightarrow{\nabla}\phi\right] + \partial_x \left[\left|\overrightarrow{\nabla}\phi\right|^2 a(\theta) \frac{\partial a(\theta)}{\partial(\partial_x \phi)}\right] + \partial_z \left[\left|\overrightarrow{\nabla}\phi\right|^2 a(\theta) \frac{\partial a(\theta)}{\partial(\partial_z \phi)}\right]$$
$$- \frac{1}{W^2}\frac{\partial h(\phi)}{\partial \phi} + \frac{a_1}{W\sigma}\frac{\partial g(\phi)}{\partial \phi}\Delta g_{ch}, \tag{6.5}$$

and the redistribution of solutes is accounted for by a diffusion equation formulated on the basis of the continuity equation and Ficks first law of diffusion [129,130]:

$$\frac{\partial c}{\partial t} = \overrightarrow{\nabla}\left[(D_s + D_l(1-\phi))\overrightarrow{\nabla}\frac{c - c^{eq}}{\phi + (1-\phi)k_{ls}} + J_{at} \cdot \#\right. \tag{6.6}$$

The parameters τ_0 and W are the relaxation time and interface width, respectively. An efficient phase-field model should correctly reproduce the experimental energies of the interfaces. When the size of the domain is much larger than the interfacial width, the microstructural evolution is barely controlled by the interfacial width. If one is not

Figure 6.6 Double potential showing a two-phase equilibrium state. Reproduced from ref. L.T. Mushongera, M. Fleck, J. Kundin, Y. Wang, H. Emmerich, Effect of Re on directional γ'-coarsening in commercial single crystal Ni-base superalloys: a phase field study, Acta Materialia. 93 (2015) 60–72. https://doi.org/10.1016/j.actamat.2015.03.048.

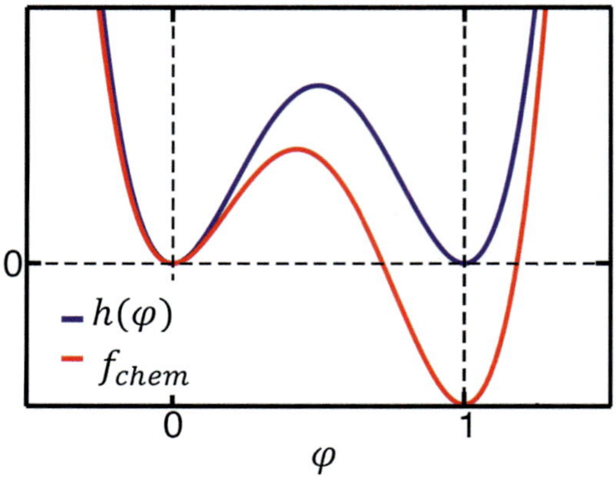

interested in the initial stages of microstructural evolution, then reproducing experimental interfacial energies in the model is unnecessary. Furthermore, for diffusion-limited transformations, the interface width should be much larger than the mesh size Δx. An optimal choice for the interface width should be around $W = 4 \sim 10\Delta x$. The mesh size Δx for phase-field simulations should be small enough to resolve interfaces. It is worth noting that W is a numerical parameter rather than a physical entity whose value is chosen purely for numerical reasons. The double well function $h(\phi) = \phi^2(1-\phi)^2$ shown in Fig. 6.6 [131] guarantees that the free energy functional has two equilibrium states, at $\phi = 0$ and $\phi = 1$, corresponding to the bulk phases [132]. This potential represents a two-phase equilibrium scenario whereby the two bulk phases corresponding to the equally deep wells have the same free energy densities. However, this scenario is rather unphysical in material systems. A rather more physical system is that which represents two phases with different thermodynamic free energy densities. This representation can be achieved by the inclusion of an interpolation function $g(\phi)$ to couple the different free energy densities. The addition of interpolated free energy densities to the double-well potential tilts it by an amount that is proportional to the local driving force for phase transformation. In this case, the solid phase is favored to grow at the expense of the melt. A typical interpolation function used in phase-field models is of form $g(\phi) = \phi^3(10 - 15\phi + 6\phi^2)$ since this is the minimal polynomial expression satisfying the necessary interpolation condition, $g(0) = 1$, $g(1) = 1$, and having also vanishing slope at $\phi = 0$ and $\phi = 1$, in order to not shift the bulk states. The term $a(\theta) = 1 + \varepsilon\cos(4\theta)$ describes the fourfold surface energy anisotropy, where ε is the anisotropy strength and $\theta = \arctan(\partial_z\phi / \partial_x\phi)$ is the angle between the interface normal and the x-axis. The parameters D_s and D_l in Eq. (6.6) are the solute diffusion coefficients in solid and liquid, respectively. The term J_{at} is the antitrapping current expressed as [130] **]:

$$J_{at} = \frac{W}{\sqrt{2}}\left(c_s^{eq} - c_l^{eq}\right)\frac{\vec{\nabla}\phi}{|\vec{\nabla}\phi|}\frac{\partial\phi}{\partial t}\cdot\# \tag{6.7}$$

The antitrapping current is used to eliminate spurious solute-trapping effects at the interface, which emanates because diffusivity in the liquid is several orders of magnitude faster than in the forming solid. The relaxation time τ_0 is linked to the physical quantities by,

$$\tau_0 = \frac{a_1 a_2 W^3 \left| m_l \left(c_s^{eq} - c_l^{eq} \right) \right| \Delta S}{\sigma D_l}, \# \quad (6.8)$$

Where a_1 and a_2 are numerical constants, given the length-scale of the columnar grains in rapid solidification, it is important for the model to employ a shifting frame along the temperature gradient direction such that the undercooling is always finite. This will also allow getting the steady-state solutions in rapid solidification. In a shifting frame, the temperature field becomes [130]:

$$T(x, z, t) = T_L - \Delta T + G\left(z - \frac{3}{4} Z \right) - V_c t + G n_s l_s, \# \quad (6.9)$$

where, n_s is the number of shifts and l_s is the shift length. The undercooling, ΔT, is linked to the cooling rate and temperature gradient G by the following relations:

$$\Delta T = \Delta T_d \Delta T_f, \# \quad (6.10)$$

$$\Delta T_d = \left(\frac{\pi^4 R_d}{820\varepsilon} \right)^{1/5.4}, \# \quad (6.11)$$

where, ΔT_d is the dimensionless undercooling and ΔT_f is the solidification interval. $R_d = R d_0 / D_l$ is the dimensionless solidification rate. $R = V_c / G$ is the actual solidification rate and,

$$d_0 = \frac{\sigma}{\left| m_l \left(c_s^{eq} - c_l^{eq} \right) \right| \Delta S}. \# \quad (6.12)$$

is the chemical capillary length. This phase-field model is applied to study the influence of variation in local chemistry on site-specific microstructural evolution in rapid solidification of Ti−45Al (at.%) alloys. The objective is to understand the role of nonequilibrium partitioning of the Al solute on interfacial instabilities and primary dendritic arm spacing (PDAS).

6.6 Melt thermodynamics and interfacial instabilities

The microstructural morphologies of alloys in solidification depend on thermal and compositional features that can induce local instabilities at the liquid/solid interface during growth, which makes the planar form, typical of pure metals, transform into cellular and dendritic morphologies. In the solidification of pure materials, the interface between the growth front is typically planar. In supercooled pure materials, solidification is purely driven by the conduction of latent heat into the liquid. This does not favor the growth of any instabilities or perturbations at the growth front. However, in the solidification of alloys, the diffusion of solutes into the liquid, which in a sense is similar to the conduction of latent heat into the liquid, completely modifies the dynamics at the growth front. In the transient stages of alloy solidification, conduction of latent heat into the liquid is a dominant mechanism, thus the growth front looks planar. As the interface moves and solutes are rejected in the liquid, the growth front deviates from being planar. Now let is look at how thermodynamic conditions in the melt pool influence instabilities at the solidification front. A growing phase will always reject the species (solutes) that lower the melting point. Thus, the local melting (freezing) temperature of the liquid (even just ahead of the interface) is lower than that of the initial liquid. If the temperature gradient ahead of an initially planar interface is gradually reduced below a critical temperature T_{crit}, the first stage in the breakdown of the interface is the formation of a cellular structure. As the composition of the liquid decreases with distance ahead of the interface, the local melting temperature increases. This causes the so-called constitutional undercooling Fig. 6.7.

Assume for a moment that the temperature at all points along the liquid–solid interface is equal to the equilibrium liquidus temperature. Any "bump" or protuberance extending from the solid into the liquid "tests" the local temperature and determines if the bump's temperature (ahead of the flat interface) is above or below the local liquidus temperature. If it finds the liquid is supercooled, the bump continues to grow. If it finds the liquid is superheated, the bump melts back. It is important to

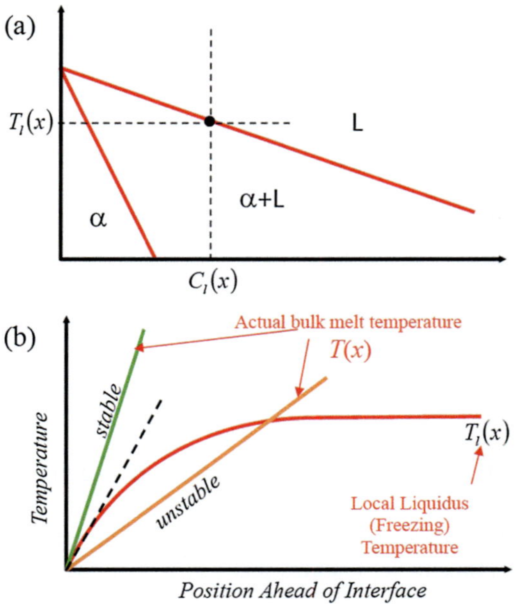

Figure 6.7 (A) Determination from a binary phase diagram, of the local liquidus temperature, $T_l(x)$ for a given local liquid composition ahead in the melt just ahead of the growth front, $C_l(x)$. (B) The stability of the liquid-solid growth front is determined by comparing the actual melt temperature ahead of the interface, $T(x)$ with the local liquidus temperature $T_l(x)$.

note two important aspects. The actual melt temperature field is not uniform, rather it is locally different.

Furthermore, the composition of the melt is not uniform, and therefore, the local liquidus temperature varies with position. The "testing" of the undercooling/superheating state is carried out with the aid of a phase diagram. If we assume steady-state, diffusion-limited transport in the liquid, then we can write the solute composition in the liquid at steady state as:

$$\frac{C_l(x)}{C_0} = 1 + \left(\frac{1 - k_{ls}}{k_{ls}}\right) e^{-\frac{vX}{D_l}} \tag{6.13}$$

where X is the distance ahead of the moving interface (in moving coordinates), $k_{ls} = C_l / C_s$ is the partition coefficient, C_0 is the initial (average) alloy composition, and D_l is the diffusivity in the liquid. Knowing the local liquid composition ahead in the melt just ahead of the growth front, $C_l(x)$, the phase diagram (Fig. 6.7A) can be used to determine the local equilibrium solidification temperature, that is, the local liquidus temperature, $T_l(x)$. To identify the stability of the liquid—solid growth front, the actual melt temperature ahead of the interface, $T(x)$, is compared with the local liquidus temperature $T_l(x)$. Fig. 6.7B shows a comparison of $T(x)$ and $T_l(x)$. If the melt near the growth front is colder than the liquidus, the interface is "unstable," and a perturbation will grow. Thus, either cellular or dendritic growth occurs depending on the degree of undercooling. This will be looked at in-depth in the following. If the melt temperature $T(x)$ is just below local liquidus temperature $T_l(x)$, cellular growth as shown in Fig. 6.8A is observed. If the melt is considerably warmer than the liquidus, that is, the actual melt temperature is above T_{cr}, the interface is "stable" and the perturbation will shrink, and the interface reverts to being planar.

Local microstructural pattern selection in laser-based AM resulting in either cellular or dendritic growth is usually determined by the degree of constitutional undercooling of the melt. As mentioned earlier, a growing phase will always reject the species that lowers the melting point. Thus, the local melting (freezing) temperature of the alloy melts with solutes, and other impurities (even just ahead of the interface) are always at a temperature lower than that of the initial liquid. At relatively low growth rates, as the melt temperature is gradually reduced below the critical temperature, the interface breaks down to form a protrusion. The formation of the first protrusion causes solute to be rejected laterally and pile up at the root of the protrusion. This further lowers the local melt temperature at the root, causing recesses to form, which triggers further instability of the plane front since any protuberance forming on the interface would find itself in supercooled liquid and, therefore, would be stable. Eventually, protrusions develop into long cells growing parallel to the direction of heat flow, as typically observed in laser-based 3D printing. At high growth rates, a solute-rich boundary layer builds up in front of the interface, which lowers the temperature *T(x)* below the liquidus, Tl(x), giving rise to constitutional undercooling (Fig. 6.8B), which triggers the formation of dendritic structures. The effectiveness of different solutes on constitutional undercooling varies widely. For solutes with a very small partition coefficient, cellular or dendritic growth can be caused by the addition of a very small fraction of a solute.

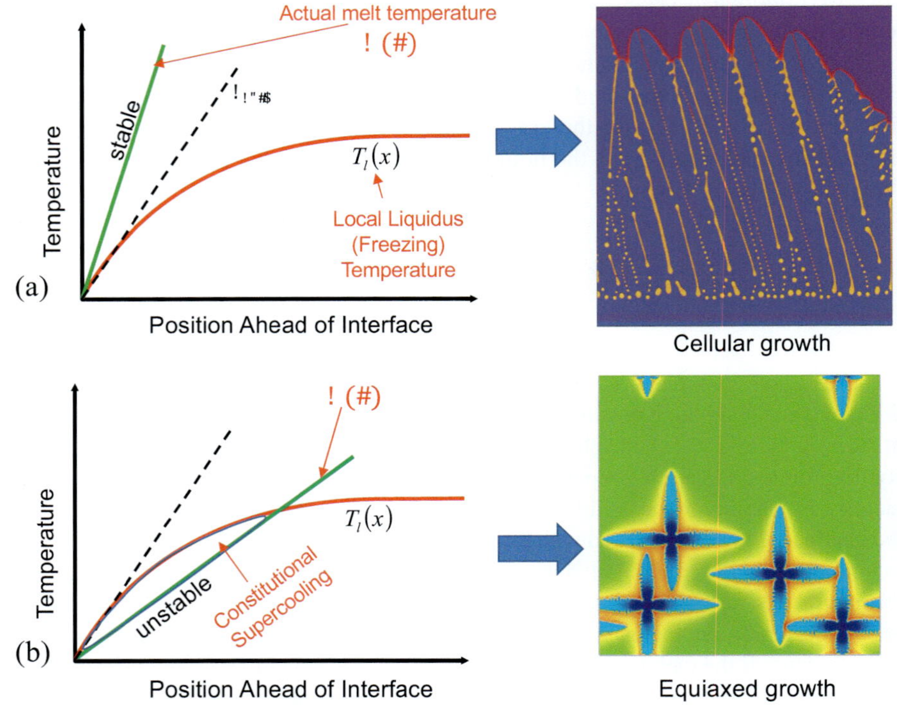

Figure 6.8 (A) If the melt temperature $T(x)$ is above the local liquidus temperature $T_l(x)$, cellular growth is observed. (B) If a solute-rich boundary layer builds up in front of the interface, the $T(x)$ lowers below $T_l(x)$ giving rise to constitutional undercooling, which triggers equiaxed growth.

6.7 Microsegregation in rapid solidification

In a convectional solidification process like casting, where the solidification front advances at a much slower velocity (order of μm/s) compared to the interface diffusion rate V_D (0.1−1 m/s), the complete solute redistribution occurs, and the interface can reach a local equilibrium concentration following the phase diagram. However, for the rapid solidification, the solidification front advances with at a faster velocity of as compared to the V_D, leading to insufficient time for solute redistribution or reaching the local equilibrium state. At the beginning of alloy solidification, with the growth of initial nuclei, the solute is rejected to the liquid and starts to accumulate at the solid−liquid interface because of different solute diffusivities in solid and liquid phases. As a result, the liquidus temperature varies at the interface with the solute concentration. On condition that the thermal parameters satisfy the criterion:

$$\frac{G}{R} = \frac{\Delta T_f}{D_l}, \tag{6.14}$$

The variation of liquidus temperature leads to constitutional undercooling. This induces local perturbations at the growth front, which later on are amplified and evolve into protrusions along the temperature gradient direction, as shown in Fig. 6.10, during rapid solidification of Ti–45Al. During this intermediate stage, the initial cellular structure adjusts itself via a series of dynamic events, including cessation, merging, and splitting of some cells to reduce interfacial energy and achieve stable cell tips. As the microstructure evolves, the Al solutes is rejected by the growing solid into the liquid. As can be seen in Fig. 6.9, most of the solutes are rejected predominantly to the intercellular regions. Additionally, circular droplets pinched off from the bottom of the intercell regions can be observed in Fig. 6.10 at 0.25 ms, which are formed to

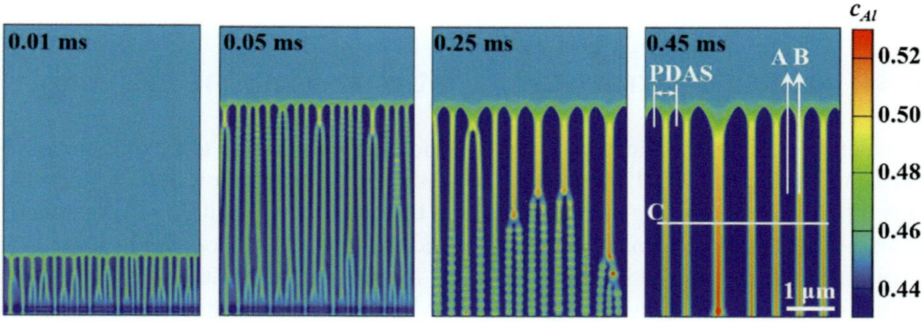

Figure 6.9 Evolution of cellular structure at different solidification times, represented by solute concentration during rapid solidification of Ti–45Al.
Reproduced from Ref. X. Zhang, B. Mao, L. Mushongera, J. Kundin, Y. Liao, Laser powder bed fusion of titanium aluminides: an investigation on site-specific microstructure evolution mechanism, Materials & Design. 201 (2021) 109501. https://doi.org/10.1016/j.matdes.2021.109501.

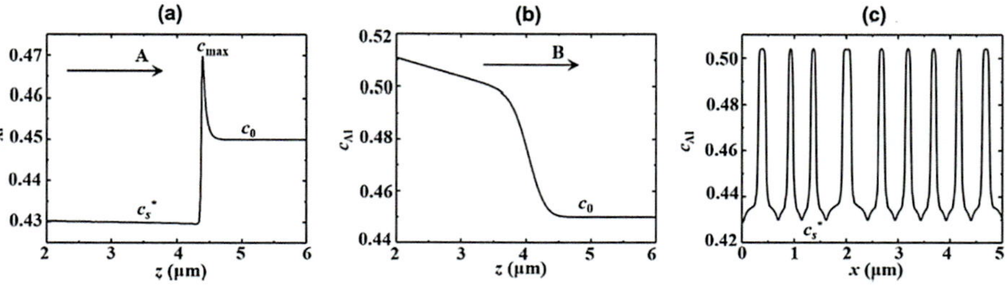

Figure 6.10 Al concentration variations along line (c) A, (d) B, and (e) C shown in Fig. 6.9.
Reproduced from Ref. X. Zhang, B. Mao, L. Mushongera, J. Kundin, Y. Liao, Laser powder bed fusion of titanium aluminides: an investigation on site-specific microstructure evolution mechanism, Materials & Design. 201 (2021) 109501. https://doi.org/10.1016/j.matdes.2021.109501.

maintain the intercell spacing. Despite that these droplets are eliminated in the final stages of solidification via the aforementioned dynamic events, they may survive at certain thermal conditions and become highly solute enriched. Eventually, secondary solid phases may precipitate directly from the liquid phase in these solute-enriched intercell regions. In the latter stages, the number and geometrical features of primary cells remain the same as solidification proceeds, achieving the final steady cellular structure. Note that interface instabilities are not observed at the sides of cells in this case, which could be caused by the significantly reduced constitutional undercooling at the intercell regions due to the small area of these regions and the employed large G/R. By decreasing the value of G/R, it is possible to initiate the instabilities at the cell sides and obtain cellular-dendritic structure (or even equiaxed dendrites).

6.8 Local microstructural variations within the melt pool

Powder particles in the vicinity of the melt pool are heated due to factors including proximity to the melt pool, thermal contact resistance, powder morphology, and thermal diffusivity. Depending on the conditions, the powder will melt into the melt pool, remain solid, or partially melt and become solid again. The melt pool temperature is of great importance to deposition quality in laser metal deposition processes. As mentioned above, the temperature fields vary spatially within the melt pool due to the constitutional undercooling effect. Fig. 6.11A shows a typical temperature distribution during the single-track selective laser melting. The nonuniform temperature distribution implies that the cooling rate V_c and temperature gradient G also varies spatially in each region of the melt pool. Consequently, a PDAS map can be obtained by substituting the various thermal variables, that is, V_c and G at different locations into the phase-field model. Fig. 6.11B shows the PDAS map obtained in selective laser processing of titanium alloys. The PDAS range in the figure shows that a significant microstructural change can take place within a single melt pool. To understand such microstructural change as affected by the thermal parameters, the Kurz–Fisher model can be used.

$$PDAS = 4.3e(1 - \Delta T_d)^{0.5}\left(\frac{\Gamma \Delta T_f D_l k_{ls}}{G V_c}\right)^{0.25} \cdot \# \tag{6.15}$$

In most publications, the term $(1 - \Delta T_d)^{0.5}$ is ignored by assuming the cell/dendrite tip temperature is close to the liquidus temperature (ΔT_d tends to 0); therefore, temperature gradient and cooling rate have the same impact on the PDAS by: PDAS $\sim G^{-0.25} V_c^{-0.25}$. However, during rapid solidification, the tip temperature may quickly decrease to the solidus temperature, resulting in a high undercooling (ΔT_d tends to 1) that is also a function of G and V_c. This will weaken the influence of temperature gradient on the PDAS. For instance, it has been found that the PDAS of selective laser-built manganese steel was related to the thermal parameters by: PDAS $\sim G^{0.09} V_c^{-0.6}$. The value of PDAS during rapid solidification is highly

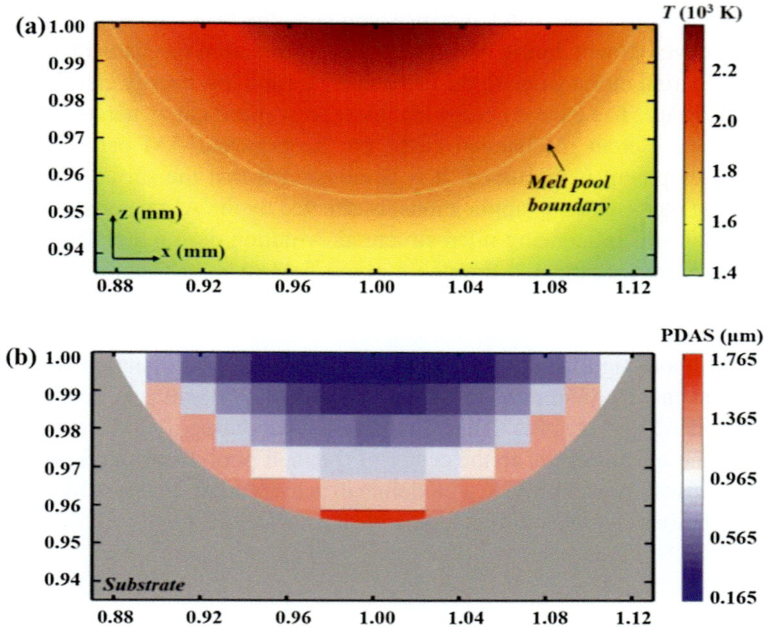

Figure 6.11 (A) Simulated temperature distribution across the transverse section of the melt pool during selective laser melting. (B) Variation in primary dendritic arm spacing (PDAS) across the melt pool.

Reproduced from Ref. X. Zhang, B. Mao, L. Mushongera, J. Kundin, Y. Liao, Laser powder bed fusion of titanium aluminides: an investigation on site-specific microstructure evolution mechanism, Materials & Design. 201 (2021) 109501. https://doi.org/10.1016/j.matdes.2021.109501.

dependent on the cooling rate rather than the temperature gradient. On the other hand, increasing the cooling rate reduces the concentration of rejected solutes in the lateral direction because of less diffusion time and thus requires a smaller PDAS to avoid constitutional undercooling. As a result, the PDAS increases when moving from the center to the boundary of the melt pool due to the change in the cooling rate. Nevertheless, the PDAS may vary within a smaller range for a real scenario because the cellular grains can continuously grow out of the small regions. Additionally, it is found that the microstructural change along the x-direction is less pronounced compared with that along the z-direction. This phenomenon could be attributed to the less thermal variations in the x-direction.

6.9 Summary

The chapter identifies the fundamental processing parameters that affect the quality and microstructure in AM components. Each of the principal parameters, such as energy density, scan speed, hatch distance, and layer thickness, effects have an impact on porosity development during AM. The porosity formation mechanism has been discussed. The partial melting due to the large scan rate and low energy can develop lack-of-fusion porosity. The excess energy and slow scan speed can generate keyhole-type porosity. The different mechanism based on the principal AM parameters for the porosity development is discussed. It is demonstrated that the surface factor, which is the limiting factor in many applications, is affected by the AM parameters. The governing mechanisms influencing microstructural evolution during laser-based AM of a binary alloy were investigated through a phase-field method. Also, it is shown that laser-based AM of alloys leads to the site-specific cellular structure across the melt-pool. It is found that site-specific microstructure variations along the build (vertical) direction are more significant than those along the transverse direction since thermal variations are more pronounced in the build direction. The microstructure evolution in laser-based AM is sensitive to constitutional undercooling. As a result, finer structures are found at sites closer to the center of the melt pool due to increased microsegregation. This study will provide a guideline to achieve high-quality AM components in terms of porosity, surface roughness, and microstructure.

References

[1] T. DebRoy, H.L. Wei, J.S. Zuback, T. Mukherjee, J.W. Elmer, J.O. Milewski, A.M. Beese, A. Wilson-Heid, A. De, W. Zhang, Additive manufacturing of metallic components – process, structure and properties, Progress in Materials Science 92 (2018) 112–224, https://doi.org/10.1016/j.pmatsci.2017.10.001.

[2] J.H. Martin, B.D. Yahata, J.M. Hundley, J.A. Mayer, T.A. Schaedler, T.M. Pollock, 3D printing of high-strength aluminium alloys, Nature 549 (2017) 365–369.

[3] J. Giannatsis, V. Dedoussis, Additive fabrication technologies applied to medicine and health care: a review, The International Journal of Advanced Manufacturing Technology 40 (2009) 116–127.

[4] A. Bandyopadhyay, Y. Zhang, S. Bose, Recent developments in metal additive manufacturing, Current Opinion in Chemical Engineering 28 (2020) 96–104, https://doi.org/10.1016/j.coche.2020.03.001.

[5] S. Singh, S. Ramakrishna, R. Singh, Material issues in additive manufacturing: a review, Journal of Manufacturing Processes 25 (2017) 185–200, https://doi.org/10.1016/j.jmapro.2016.11.006.

[6] P. Promoppatum, S.-C. Yao, Influence of scanning length and energy input on residual stress reduction in metal additive manufacturing: numerical and experimental studies, Journal of Manufacturing Processes 49 (2020) 247–259, https://doi.org/10.1016/j.jmapro.2019.11.020.

[7] S. Cooke, K. Ahmadi, S. Willerth, R. Herring, Metal additive manufacturing: technology, metallurgy and modelling, Journal of Manufacturing Processes 57 (2020) 978–1003.

[8] J.V. Gordon, S.P. Narra, R.W. Cunningham, H. Liu, H. Chen, R.M. Suter, J.L. Beuth, A.D. Rollett, Defect structure process maps for laser powder bed fusion additive manufacturing, Additive Manufacturing 36 (2020) 101552.

[9] N. Putra, M. Mirzaali, I. Apachitei, J. Zhou, A. Zadpoor, Multi-material additive manufacturing technologies for Ti-, Mg-, and Fe-based biomaterials for bone substitution, Acta Biomaterialia 109 (2020) 1−20.

[10] B. Ahuja, M. Karg, M. Schmidt, Additive manufacturing in production: challenges and opportunities, in: Laser 3D Manufacturing II, International Society for Optics and Photonics, 2015, p. 935304, https://doi.org/10.1117/12.2082521.

[11] J.P. Oliveira, T.G. Santos, R.M. Miranda, Revisiting fundamental welding concepts to improve additive manufacturing: from theory to practice, Progress in Materials Science 107 (2020) 100590, https://doi.org/10.1016/j.pmatsci.2019.100590.

[12] E. Karayel, Y. Bozkurt, Additive manufacturing method and different welding applications, Journal of Materials Research and Technology 9 (2020) 11424−11438, https://doi.org/10.1016/j.jmrt.2020.08.039.

[13] P. Kumar, J. Farah, J. Akram, C. Teng, J. Ginn, M. Misra, Influence of laser processing parameters on porosity in Inconel 718 during additive manufacturing, The International Journal of Advanced Manufacturing Technology 103 (2019) 1497−1507, https://doi.org/10.1007/s00170-019-03655-9.

[14] S.A. Khairallah, A.T. Anderson, A. Rubenchik, W.E. King, Laser powder-bed fusion additive manufacturing: physics of complex melt flow and formation mechanisms of pores, spatter, and denudation zones, Acta Materialia 108 (2016) 36−45.

[15] Z. Gan, O.L. Kafka, N. Parab, C. Zhao, L. Fang, O. Heinonen, T. Sun, W.K. Liu, Universal scaling laws of keyhole stability and porosity in 3D printing of metals, Nature Communications 12 (2021) 2379, https://doi.org/10.1038/s41467-021-22704-0.

[16] P. Tan, R. Kiran, K. Zhou, Effects of sub-atmospheric pressure on keyhole dynamics and porosity in products fabricated by selective laser melting, Journal of Manufacturing Processes 64 (2021) 816−827.

[17] W. Ge, J.Y. Fuh, S.J. Na, Numerical modelling of keyhole formation in selective laser melting of Ti_6Al_4V, Journal of Manufacturing Processes 62 (2021) 646−654.

[18] J.-N. Zhu, E. Borisov, X. Liang, E. Farber, M.J.M. Hermans, V.A. Popovich, Predictive analytical modelling and experimental validation of processing maps in additive manufacturing of nitinol alloys, Additive Manufacturing 38 (2021) 101802.

[19] J. Huang, M. Li, J. Wang, Z. Pei, P. Mclntyre, C. Ma, Selective laser melting of tungsten: effects of hatch distance and point distance on pore formation, Journal of Manufacturing Processes 61 (2021) 296−302.

[20] Y.H. Zhou, W.P. Li, L. Zhang, S.Y. Zhou, X. Jia, D.W. Wang, M. Yan, Selective laser melting of Ti−22Al−25Nb intermetallic: significant effects of hatch distance on microstructural features and mechanical properties, Journal of Materials Processing Technology 276 (2020) 116398.

[21] L. Du, D. Gu, D. Dai, Q. Shi, C. Ma, M. Xia, Relation of thermal behavior and microstructure evolution during multi-track laser melting deposition of Ni-based material, Optics & Laser Technology 108 (2018) 207−217, https://doi.org/10.1016/j.optlastec.2018.06.042.

[22] M. Averyanova, P. Bertrand, B. Verquin, Studying the influence of initial powder characteristics on the properties of final parts manufactured by the selective laser melting technology, Virtual and Physical Prototyping 6 (2011) 215−223, https://doi.org/10.1080/17452759.2011.594645.

[23] O. Zinovieva, A. Zinoviev, V. Ploshikhin, Three-dimensional modeling of the microstructure evolution during metal additive manufacturing, Computational Materials Science 141 (2018) 207−220, https://doi.org/10.1016/j.commatsci.2017.09.018.

[24] I. Yadroitsev, I. Smurov, Surface morphology in selective laser melting of metal powders, Physics Procedia 12 (2011) 264−270, https://doi.org/10.1016/j.phpro.2011.03.034.

[25] K. Moussaoui, W. Rubio, M. Mousseigne, T. Sultan, F. Rezai, Effects of Selective Laser Melting additive manufacturing parameters of Inconel 718 on porosity, microstructure and mechanical properties, Materials Science and Engineering: A 735 (2018) 182–190, https://doi.org/10.1016/j.msea.2018.08.037.
[26] A. du Plessis, I. Yadroitsava, I. Yadroitsev, Effects of defects on mechanical properties in metal additive manufacturing: a review focusing on X-ray tomography insights, Materials & Design 187 (2020) 108385.
[27] T. Ronneberg, C.M. Davies, P.A. Hooper, Revealing relationships between porosity, microstructure and mechanical properties of laser powder bed fusion 316L stainless steel through heat treatment, Materials & Design 189 (2020) 108481.
[28] P. Kumar, K.S. Ravi Chandran, F. Cao, M. Koopman, Z.Z. Fang, The nature of tensile ductility as controlled by extreme-sized pores in powder metallurgy Ti-6Al-4V alloy, Metallurgical and Materials Transactions A (2016), https://doi.org/10.1007/s11661-016-3419-5.
[29] P. Kumar, K.S.R. Chandran, Strength–Ductility property maps of powder metallurgy (PM) Ti-6Al-4V alloy: a critical review of processing-structure-property relationships, Metallurgical and Materials Transactions A 48 (2017) 2301–2319, https://doi.org/10.1007/s11661-017-4009-x.
[30] B. Zhang, S. Liu, Y.C. Shin, In-Process monitoring of porosity during laser additive manufacturing process, Additive Manufacturing 28 (2019) 497–505.
[31] W. Ren, J. Mazumder, In-situ porosity recognition for laser additive manufacturing of 7075-Al alloy using plasma emission spectroscopy, Scientific Reports 10 (2020) 1–11.
[32] H. Taheri, M.R.B.M. Shoaib, L.W. Koester, T.A. Bigelow, P.C. Collins, L.J. Bond, Powder-based additive manufacturing-a review of types of defects, generation mechanisms, detection, property evaluation and metrology, International Journal of Additive and Subtractive Materials Manufacturing 1 (2017) 172–209.
[33] R. Cunningham, S.P. Narra, T. Ozturk, J. Beuth, A.D. Rollett, Evaluating the effect of processing parameters on porosity in electron beam melted Ti-6Al-4V via synchrotron X-ray microtomography, JOM: The Journal of the Minerals, Metals & Materials Society 68 (2016) 765–771.
[34] R. Cunningham, S.P. Narra, C. Montgomery, J. Beuth, A.D. Rollett, Synchrotron-based X-ray microtomography characterization of the effect of processing variables on porosity formation in laser power-bed additive manufacturing of Ti-6Al-4V, JOM: The Journal of the Minerals, Metals & Materials Society 69 (2017) 479–484.
[35] W.E. King, H.D. Barth, V.M. Castillo, G.F. Gallegos, J.W. Gibbs, D.E. Hahn, C. Kamath, A.M. Rubenchik, Observation of keyhole-mode laser melting in laser powder-bed fusion additive manufacturing, Journal of Materials Processing Technology 214 (2014) 2915–2925, https://doi.org/10.1016/j.jmatprotec.2014.06.005.
[36] P. Ferro, R. Meneghello, S.M.J. Razavi, F. Berto, G. Savio, Porosity inducing process parameters in selective laser melted AlSi10Mg aluminium alloy, Physical Mesomechanics 23 (2020) 256–262.
[37] G. Kasperovich, J. Haubrich, J. Gussone, G. Requena, Correlation between porosity and processing parameters in $TiAl_6V_4$ produced by selective laser melting, Materials & Design 105 (2016) 160–170, https://doi.org/10.1016/j.matdes.2016.05.070.
[38] J.J.S. Dilip, S. Zhang, C. Teng, K. Zeng, C. Robinson, D. Pal, B. Stucker, Influence of processing parameters on the evolution of melt pool, porosity, and microstructures in Ti-6Al-4V alloy parts fabricated by selective laser melting, Progress in Additive Manufacturing 2 (2017) 157–167, https://doi.org/10.1007/s40964-017-0030-2.

[39] B. Verlee, T. Dormal, J. Lecomte-Beckers, Density and porosity control of sintered 316L stainless steel parts produced by additive manufacturing, Powder Metallurgy 55 (2012) 260−267.
[40] W. Stopyra, K. Gruber, I. Smolina, T. Kurzynowski, B. Kuźnicka, Laser powder bed fusion of AA7075 alloy: influence of process parameters on porosity and hot cracking, Additive Manufacturing 35 (2020) 101270.
[41] H. Bikas, P. Stavropoulos, G. Chryssolouris, Additive manufacturing methods and modelling approaches: a critical review, The International Journal of Advanced Manufacturing Technology 83 (2016) 389−405.
[42] T.D. Ngo, A. Kashani, G. Imbalzano, K.T. Nguyen, D. Hui, Additive manufacturing (3D printing): a review of materials, methods, applications and challenges, Composites Part B: Engineering 143 (2018) 172−196.
[43] N. Li, S. Huang, G. Zhang, R. Qin, W. Liu, H. Xiong, G. Shi, J. Blackburn, Progress in additive manufacturing on new materials: a review, Journal of Materials Science & Technology 35 (2019) 242−269.
[44] A. Sola, A. Nouri, Microstructural porosity in additive manufacturing: the formation and detection of pores in metal parts fabricated by powder bed fusion, Journal of Advanced Manufacturing and Processing 1 (2019) e10021, https://doi.org/10.1002/amp2.10021.
[45] P. Kiani, A.D. Dupuy, K. Ma, J.M. Schoenung, Directed energy deposition of AlSi10Mg: single track non scalability and bulk properties, Materials & Design 194 (2020) 108847, https://doi.org/10.1016/j.matdes.2020.108847.
[46] H. Kyogoku, T.-T. Ikeshoji, A review of metal additive manufacturing technologies: mechanism of defects formation and simulation of melting and solidification phenomena in laser powder bed fusion process, Mechanical Engineering Reviews 7 (2020) 19−00182.
[47] D. Gu, Y.-C. Hagedorn, W. Meiners, G. Meng, R.J.S. Batista, K. Wissenbach, R. Poprawe, Densification behavior, microstructure evolution, and wear performance of selective laser melting processed commercially pure titanium, Acta Materialia 60 (2012) 3849−3860, https://doi.org/10.1016/j.actamat.2012.04.006.
[48] A. Yadollahi, N. Shamsaei, Additive manufacturing of fatigue resistant materials: challenges and opportunities, International Journal of Fatigue 98 (2017) 14−31, https://doi.org/10.1016/j.ijfatigue.2017.01.001.
[49] R. Shrestha, J. Simsiriwong, N. Shamsaei, Fatigue behavior of additive manufactured 316L stainless steel parts: effects of layer orientation and surface roughness, Additive Manufacturing 28 (2019) 23−38.
[50] H. Masuo, Y. Tanaka, S. Morokoshi, H. Yagura, T. Uchida, Y. Yamamoto, Y. Murakami, Influence of defects, surface roughness and HIP on the fatigue strength of Ti-6Al-4V manufactured by additive manufacturing, International Journal of Fatigue 117 (2018) 163−179.
[51] B. Vayssette, N. Saintier, C. Brugger, M. El May, E. Pessard, Numerical modelling of surface roughness effect on the fatigue behavior of Ti-6Al-4V obtained by additive manufacturing, International Journal of Fatigue 123 (2019) 180−195.
[52] D. Kotzem, P. Dumke, P. Sepehri, J. Tenkamp, F. Walther, Effect of miniaturization and surface roughness on the mechanical properties of the electron beam melted superalloy Inconel® 718, Progress in Additive Manufacturing (2019) 1−10.
[53] K. Mumtaz, N. Hopkinson, Top surface and side roughness of Inconel 625 parts processed using selective laser melting, Rapid Prototyping Journal 15 (2009) 96−103, https://doi.org/10.1108/13552540910943397.

[54] J. Gockel, L. Sheridan, B. Koerper, B. Whip, The influence of additive manufacturing processing parameters on surface roughness and fatigue life, International Journal of Fatigue 124 (2019) 380–388, https://doi.org/10.1016/j.ijfatigue.2019.03.025.
[55] B. Whip, L. Sheridan, J. Gockel, The effect of primary processing parameters on surface roughness in laser powder bed additive manufacturing, The International Journal of Advanced Manufacturing Technology 103 (2019) 4411–4422.
[56] A.M. Ralls, P. Kumar, P.L. Menezes, Tribological properties of additive manufactured materials for energy applications: a review, Processes 9 (2021) 31, https://doi.org/10.3390/pr9010031.
[57] G. Moroni, W.P. Syam, S. Petro, Towards early estimation of part accuracy in additive manufacturing, Procedia CIRP 21 (2014) 300–305.
[58] N. Kumbhar, A. Mulay, Post processing methods used to improve surface finish of products which are manufactured by additive manufacturing technologies: a review, Journal of The Institution of Engineers (India): Series C. 99 (2018) 481–487.
[59] F. Kaji, A. Barari, Evaluation of the surface roughness of additive manufacturing parts based on the modelling of cusp geometry, IFAC-PapersOnLine 48 (2015) 658–663.
[60] P. Das, R. Chandran, R. Samant, S. Anand, Optimum part build orientation in additive manufacturing for minimizing part errors and support structures, Procedia Manufacturing 1 (2015) 343–354.
[61] S. Pereira, A.I.F. Vaz, L.N. Vicente, On the optimal object orientation in additive manufacturing, The International Journal of Advanced Manufacturing Technology 98 (2018) 1685–1694.
[62] M.A. Matos, A.M.A. Rocha, A.I. Pereira, Improving additive manufacturing performance by build orientation optimization, The International Journal of Advanced Manufacturing Technology (2020) 1–13.
[63] F. Cabanettes, A. Joubert, G. Chardon, V. Dumas, J. Rech, C. Grosjean, Z. Dimkovski, Topography of as built surfaces generated in metal additive manufacturing: a multi scale analysis from form to roughness, Precision Engineering 52 (2018) 249–265.
[64] S. Rahmati, E. Vahabli, Evaluation of analytical modeling for improvement of surface roughness of FDM test part using measurement results, The International Journal of Advanced Manufacturing Technology 79 (2015) 823–829.
[65] G. Strano, L. Hao, R.M. Everson, K.E. Evans, Surface roughness analysis, modelling and prediction in selective laser melting, Journal of Materials Processing Technology 213 (2013) 589–597.
[66] J.C. Fox, S.P. Moylan, B.M. Lane, Effect of process parameters on the surface roughness of overhanging structures in laser powder bed fusion additive manufacturing, Procedia CIRP 45 (2016) 131–134, https://doi.org/10.1016/j.procir.2016.02.347.
[67] J.A. Cherry, H.M. Davies, S. Mehmood, N.P. Lavery, S.G.R. Brown, J. Sienz, Investigation into the effect of process parameters on microstructural and physical properties of 316L stainless steel parts by selective laser melting, The International Journal of Advanced Manufacturing Technology 76 (2015) 869–879, https://doi.org/10.1007/s00170-014-6297-2.
[68] D. Wang, Y. Liu, Y. Yang, D. Xiao, Theoretical and experimental study on surface roughness of 316L stainless steel metal parts obtained through selective laser melting, Rapid Prototyping Journal (2016).
[69] L.-X. Lu, N. Sridhar, Y.-W. Zhang, Phase field simulation of powder bed-based additive manufacturing, Acta Materialia 144 (2018) 801–809.
[70] L.E. Criales, Y.M. Arısoy, B. Lane, S. Moylan, A. Donmez, T. Özel, Laser powder bed fusion of nickel alloy 625: experimental investigations of effects of process parameters on melt pool size and shape with spatter analysis, International Journal of Machine Tools and Manufacture 121 (2017) 22–36, https://doi.org/10.1016/j.ijmachtools.2017.03.004.

[71] V. Gunenthiram, P. Peyre, M. Schneider, M. Dal, F. Coste, I. Koutiri, R. Fabbro, Experimental analysis of spatter generation and melt-pool behavior during the powder bed laser beam melting process, Journal of Materials Processing Technology 251 (2018) 376–386.
[72] W.E. Frazier, Metal additive manufacturing: a review, Journal of Materials Engineering and Performance 23 (2014), https://doi.org/10.1007/s11665-014-0958-z.
[73] J.J. Lewandowski, M. Seifi, Metal additive manufacturing: a review of mechanical properties, Annual Review of Materials Research 46 (2016) 151–186, https://doi.org/10.1146/annurev-matsci-070115-032024.
[74] W.J. Sames, F. List, S. Pannala, R.R. Dehoff, S.S. Babu, The metallurgy and processing science of metal additive manufacturing, International Materials Reviews 61 (2016) 315–360.
[75] V. Bhavar, P. Kattire, V. Patil, S. Khot, K. Gujar, R. Singh, A review on powder bed fusion technology of metal additive manufacturing, in: Additive Manufacturing Handbook, 2017, pp. 251–253.
[76] S.K. Everton, M. Hirsch, P. Stravroulakis, R.K. Leach, A.T. Clare, Review of in-situ process monitoring and in-situ metrology for metal additive manufacturing, Materials & Design 95 (2016) 431–445.
[77] A. Bandyopadhyay, K.D. Traxel, Invited review article: metal-additive manufacturing—modeling strategies for application-optimized designs, Additive Manufacturing 22 (2018) 758–774.
[78] A. Gisario, M. Kazarian, F. Martina, M. Mehrpouya, Metal additive manufacturing in the commercial aviation industry: a review, Journal of Manufacturing Systems 53 (2019) 124–149.
[79] J.H.K. Tan, S.L. Sing, W.Y. Yeong, Microstructure modelling for metallic additive manufacturing: a review, Virtual and Physical Prototyping 15 (2020) 87–105, https://doi.org/10.1080/17452759.2019.1677345.
[80] C. Körner, M. Markl, J.A. Koepf, Modeling and simulation of microstructure evolution for additive manufacturing of metals: a critical review, Metallurgical and Materials Transactions A 51 (2020) 4970–4983.
[81] A. Basak, S. Das, Epitaxy and microstructure evolution in metal additive manufacturing, Annual Review of Materials Research 46 (2016) 125–149.
[82] S. Gorsse, C. Hutchinson, M. Gouné, R. Banerjee, Additive manufacturing of metals: a brief review of the characteristic microstructures and properties of steels, Ti-6Al-4V and high-entropy alloys, Science and Technology of Advanced Materials 18 (2017) 584–610, https://doi.org/10.1080/14686996.2017.1361305.
[83] S. Kelly, S. Kampe, Microstructural evolution in laser-deposited multilayer Ti-6Al-4V builds: Part I. Microstructural characterization, Metallurgical and Materials Transactions A 35 (2004) 1861–1867.
[84] M. Seifi, A. Salem, J. Beuth, O. Harrysson, J.J. Lewandowski, Overview of materials qualification needs for metal additive manufacturing, JOM: The Journal of the Minerals, Metals & Materials Society 68 (2016) 747–764.
[85] T. DebRoy, T. Mukherjee, J.O. Milewski, J.W. Elmer, B. Ribic, J.J. Blecher, W. Zhang, Scientific, technological and economic issues in metal printing and their solutions, Nature Materials 18 (2019) 1026–1032.
[86] T. Gatsos, K.A. Elsayed, Y. Zhai, D.A. Lados, Review on computational modeling of process–microstructure–property relationships in metal additive manufacturing, JOM: The Journal of the Minerals, Metals & Materials Society 72 (2020) 403–419.

[87] A. Saboori, A. Aversa, G. Marchese, S. Biamino, M. Lombardi, P. Fino, Microstructure and mechanical properties of AISI 316L produced by Directed Energy Deposition-based additive manufacturing: a review, Applied Sciences 10 (2020) 3310.
[88] T. Wang, Y.Y. Zhu, S.Q. Zhang, H.B. Tang, H.M. Wang, Grain morphology evolution behavior of titanium alloy components during laser melting deposition additive manufacturing, Journal of Alloys and Compounds 632 (2015) 505–513.
[89] J. Akram, P. Chalavadi, D. Pal, B. Stucker, Understanding grain evolution in additive manufacturing through modeling, Additive Manufacturing 21 (2018) 255–268, https://doi.org/10.1016/j.addma.2018.03.021.
[90] M.J. Bermingham, D.H. StJohn, J. Krynen, S. Tedman-Jones, M.S. Dargusch, Promoting the columnar to equiaxed transition and grain refinement of titanium alloys during additive manufacturing, Acta Materialia 168 (2019) 261–274.
[91] H. Li, Y. Huang, S. Jiang, Y. Lu, X. Gao, X. Lu, Z. Ning, J. Sun, Columnar to equiaxed transition in additively manufactured CoCrFeMnNi high entropy alloy, Materials & Design 197 (2021) 109262, https://doi.org/10.1016/j.matdes.2020.109262.
[92] J.P. Oliveira, A.D. LaLonde, J. Ma, Processing parameters in laser powder bed fusion metal additive manufacturing, Materials & Design 193 (2020) 108762, https://doi.org/10.1016/j.matdes.2020.108762.
[93] C.J. Todaro, M.A. Easton, D. Qiu, D. Zhang, M.J. Bermingham, E.W. Lui, M. Brandt, D.H. StJohn, M. Qian, Grain structure control during metal 3D printing by high-intensity ultrasound, Nat Commun 11 (2020) 142, https://doi.org/10.1038/s41467-019-13874-z.
[94] W. Kurz, B. Giovanola, R. Trivedi, Theory of microstructural development during rapid solidification, Acta Metallurgica 34 (1986) 823–830, https://doi.org/10.1016/0001-6160(86)90056-8.
[95] J.J. Blecher, T.A. Palmer, T. DebRoy, Solidification map of a nickel-base alloy, Metallurgical and Materials Transactions A 45 (2014) 2142–2151, https://doi.org/10.1007/s11661-013-2149-1.
[96] P.A. Kobryn, S. Semiatin, Microstructure and texture evolution during solidification processing of Ti–6Al–4V, Journal of Materials Processing Technology 135 (2003) 330–339.
[97] H. Attar, M. Bermingham, S. Ehtemam-Haghighi, A. Dehghan-Manshadi, D. Kent, M. Dargusch, Evaluation of the mechanical and wear properties of titanium produced by three different additive manufacturing methods for biomedical application, Materials Science and Engineering: A 760 (2019) 339–345.
[98] X. Ji, E. Mirkoohi, J. Ning, S.Y. Liang, Analytical modeling of post-printing grain size in metal additive manufacturing, Optics and Lasers in Engineering 124 (2020) 105805.
[99] J. Shao, G. Yu, X. He, S. Li, R. Chen, Y. Zhao, Grain size evolution under different cooling rate in laser additive manufacturing of superalloy, Optics & Laser Technology 119 (2019) 105662.
[100] N. Raghavan, S. Simunovic, R. Dehoff, A. Plotkowski, J. Turner, M. Kirka, S. Babu, Localized melt-scan strategy for site specific control of grain size and primary dendrite arm spacing in electron beam additive manufacturing, Acta Materialia 140 (2017) 375–387.
[101] D. Zhang, A. Prasad, M.J. Bermingham, C.J. Todaro, M.J. Benoit, M.N. Patel, D. Qiu, D.H. StJohn, M. Qian, M.A. Easton, Grain refinement of alloys in fusion-based additive manufacturing processes, Metallurgical and Materials Transactions A 51 (2020) 4341–4359.
[102] H.L. Wei, G.L. Knapp, T. Mukherjee, T. DebRoy, Three-dimensional grain growth during multi-layer printing of a nickel-based alloy Inconel 718, Additive Manufacturing 25 (2019) 448–459.

[103] C. Pleass, S. Jothi, Influence of powder characteristics and additive manufacturing process parameters on the microstructure and mechanical behaviour of Inconel 625 fabricated by selective laser melting, Additive Manufacturing 24 (2018) 419–431.
[104] Y. Tian, D. McAllister, H. Colijn, M. Mills, D. Farson, M. Nordin, S. Babu, Rationalization of microstructure heterogeneity in INCONEL 718 builds made by the direct laser additive manufacturing process, Metallurgical and Materials Transactions A 45 (2014) 4470–4483.
[105] G.L. Knapp, N. Raghavan, A. Plotkowski, T. Debroy, Experiments and simulations on solidification microstructure for Inconel 718 in powder bed fusion electron beam additive manufacturing, Additive Manufacturing 25 (2019) 511–521.
[106] P.D. Nezhadfar, A.S. Johnson, N. Shamsaei, Fatigue behavior and microstructural evolution of additively manufactured Inconel 718 under cyclic loading at elevated temperature, International Journal of Fatigue 136 (2020) 105598.
[107] Y.L. Hu, X. Lin, Y.L. Li, S.Y. Zhang, X.H. Gao, F.G. Liu, X. Li, W.D. Huang, Plastic deformation behavior and dynamic recrystallization of Inconel 625 superalloy fabricated by directed energy deposition, Materials & Design 186 (2020) 108359.
[108] Y. Hu, X. Lin, Y. Li, S. Zhang, Q. Zhang, W. Chen, W. Li, W. Huang, Influence of heat treatments on the microstructure and mechanical properties of Inconel 625 fabricated by directed energy deposition, Materials Science and Engineering: A 817 (2021) 141309.
[109] A.N. Jinoop, C.P. Paul, S.K. Mishra, K.S. Bindra, Laser Additive Manufacturing using directed energy deposition of Inconel-718 wall structures with tailored characteristics, Vacuum 166 (2019) 270–278, https://doi.org/10.1016/j.vacuum.2019.05.027.
[110] H.E. Helmer, C. Körner, R.F. Singer, Additive manufacturing of nickel-based superalloy Inconel 718 by selective electron beam melting: processing window and microstructure, Journal of Materials Research 29 (2014) 1987–1996, https://doi.org/10.1557/jmr.2014.192.
[111] S.G.K. Manikandan, D. Sivakumar, K.P. Rao, M. Kamaraj, Effect of weld cooling rate on Laves phase formation in Inconel 718 fusion zone, Journal of Materials Processing Technology 214 (2014) 358–364.
[112] H. Xiao, P. Xie, M. Cheng, L. Song, Enhancing mechanical properties of quasi-continuous-wave laser additive manufactured Inconel 718 through controlling the niobium-rich precipitates, Additive Manufacturing 34 (2020) 101278.
[113] J.S. Zuback, P. Moradifar, Z. Khayat, N. Alem, T.A. Palmer, Impact of chemical composition on precipitate morphology in an additively manufactured nickel base superalloy, Journal of Alloys and Compounds 798 (2019) 446–457.
[114] Q. Jia, D. Gu, Selective laser melting additive manufacturing of Inconel 718 superalloy parts: densification, microstructure and properties, Journal of Alloys and Compounds 585 (2014) 713–721, https://doi.org/10.1016/j.jallcom.2013.09.171.
[115] B. Farber, K.A. Small, C. Allen, R.J. Causton, A. Nichols, J. Simbolick, M.L. Taheri, Correlation of mechanical properties to microstructure in Inconel 718 fabricated by direct metal laser sintering, Materials Science and Engineering: A 712 (2018) 539–547.
[116] V.A. Popovich, E.V. Borisov, A.A. Popovich, V.S. Sufiiarov, D.V. Masaylo, L. Alzina, Functionally graded Inconel 718 processed by additive manufacturing: crystallographic texture, anisotropy of microstructure and mechanical properties, Materials & Design 114 (2017) 441–449, https://doi.org/10.1016/j.matdes.2016.10.075.
[117] A.S. Wu, D.W. Brown, M. Kumar, G.F. Gallegos, W.E. King, An experimental investigation into additive manufacturing-induced residual stresses in 316L stainless steel, Metallurgical and Materials Transactions A 45 (2014) 6260–6270.
[118] Z. Wang, T.A. Palmer, A.M. Beese, Effect of processing parameters on microstructure and tensile properties of austenitic stainless steel 304L made by directed energy deposition additive manufacturing, Acta Materialia 110 (2016) 226–235.

[119] J. Li, D. Deng, X. Hou, X. Wang, G. Ma, D. Wu, G. Zhang, Microstructure and performance optimisation of stainless steel formed by laser additive manufacturing, Materials Science and Technology 32 (2016) 1223−1230.
[120] B.M. Morrow, T.J. Lienert, C.M. Knapp, J.O. Sutton, M.J. Brand, R.M. Pacheco, V. Livescu, J.S. Carpenter, G.T. Gray, Impact of defects in powder feedstock materials on microstructure of 304L and 316L stainless steel produced by additive manufacturing, Metallurgical and Materials Transactions A 49 (2018) 3637−3650.
[121] R.I. Revilla, M. Van Calster, M. Raes, G. Arroud, F. Andreatta, L. Pyl, P. Guillaume, I. De Graeve, Microstructure and corrosion behavior of 316L stainless steel prepared using different additive manufacturing methods: a comparative study bringing insights into the impact of microstructure on their passivity, Corrosion Science 176 (2020) 108914.
[122] A. Adeyemi, E.T. Akinlabi, R.M. Mahamood, Powder bed based laser additive manufacturing process of stainless steel: a review, Materials Today: Proceedings. 5 (2018) 18510−18517.
[123] P. Bajaj, A. Hariharan, A. Kini, P. Kürnsteiner, D. Raabe, E.A. Jägle, Steels in additive manufacturing: a review of their microstructure and properties, Materials Science and Engineering: A 772 (2020) 138633, https://doi.org/10.1016/j.msea.2019.138633.
[124] B. Zheng, Y. Zhou, J. Smugeresky, J. Schoenung, E. Lavernia, Thermal behavior and microstructure evolution during laser deposition with laser-engineered net shaping: part II, Experimental Investigation and Discussion, Metallurgical and Materials Transactions A 39 (2008) 2237−2245.
[125] B.E. Carroll, T.A. Palmer, A.M. Beese, Anisotropic tensile behavior of Ti−6Al−4V components fabricated with directed energy deposition additive manufacturing, Acta Materialia 87 (2015) 309−320, https://doi.org/10.1016/j.actamat.2014.12.054.
[126] S.S. Al-Bermani, M.L. Blackmore, W. Zhang, I. Todd, The origin of microstructural diversity, texture, and mechanical properties in electron beam melted Ti-6Al-4V, Metallurgical and Materials Transactions A 41 (2010) 3422−3434, https://doi.org/10.1007/s11661-010-0397-x.
[127] W. Xu, M. Brandt, S. Sun, J. Elambasseril, Q. Liu, K. Latham, K. Xia, M. Qian, Additive manufacturing of strong and ductile Ti−6Al−4V by selective laser melting via in situ martensite decomposition, Acta Materialia 85 (2015) 74−84, https://doi.org/10.1016/j.actamat.2014.11.028.
[128] N. Sridharan, A. Chaudhary, P. Nandwana, S.S. Babu, Texture evolution during laser direct metal deposition of Ti-6Al-4V, JOM: The Journal of the Minerals, Metals & Materials Society 68 (2016) 772−777.
[129] J. Kundin, L. Mushongera, H. Emmerich, Phase-field modeling of microstructure formation during rapid solidification in Inconel 718 superalloy, Acta Materialia 95 (2015) 343−356, https://doi.org/10.1016/j.actamat.2015.05.052.
[130] X. Zhang, B. Mao, L. Mushongera, J. Kundin, Y. Liao, Laser powder bed fusion of titanium aluminides: an investigation on site-specific microstructure evolution mechanism, Materials & Design 201 (2021) 109501, https://doi.org/10.1016/j.matdes.2021.109501.
[131] L. Mushongera, Strain-induced γ'-coarsening during Aging of Ni-Base Superalloys under Uniaxial Load: Modeling and Analysis (Ph.D. thesis), University of Bayreuth, 2016.
[132] L.T. Mushongera, M. Fleck, J. Kundin, Y. Wang, H. Emmerich, Effect of Re on directional γ'-coarsening in commercial single crystal Ni-base superalloys: a phase field study, Acta Materialia 93 (2015) 60−72, https://doi.org/10.1016/j.actamat.2015.03.048.

Development of surface roughness from additive manufacturing processing parameters and postprocessing surface modification techniques

7

Alessandro M. Ralls [1], Carlos Flores [1], Thomas Kotowski [1], Cody Lee [1], Pankaj Kumar [2,3] and Pradeep L. Menezes [1]

[1]Department of Mechanical Engineering, University of Nevada, Reno, NV, United States; [2]Department of Mechanical Engineering, University of New Mexico, Albuquerque, NM, United States; [3]Department of Chemical and Materials Engineering, University of Nevada, Reno, NV, United States

7.1 Introduction

As defined by the American Society for Testing and Materials (ASTM), additive manufacturing (AM) is a process in which materials are fabricated through layer-by-layer deposition [1,2]. Derived from the Charles Hull's historical introduction of the stereolithography (SLA) process [3], the application of AM extensively spans from at home 3D printing to the large industrial scale production of specialized parts [4]. Through the large-scale application of AM technologies, manufacturers of all scales tremendously benefit from this technology from both an engineering and business point of view. From the perspective of an engineer, AM reduces the total lead time of geometrically complex products due to the elimination of postprocessing technologies that otherwise would be used [5]. Although the benefits of AM for commercial businesses are not as extensively covered as the general engineering benefits, business sectors such as supply chains (SCs) are able to rapidly adjust to changes in product demand while satisfying the "seven rights of logistics," being to "deliver the right product, in the right quantity and right condition, to the right place at the right time for the right customer at the right price" [6].

Contrast to more traditional subtractive manufacturing techniques, AM yields a series of production-based advantages, which have helped its popularization since its introduction [2,7,8]. From a design aspect, the construction of complex geometries is far easier, faster, and in some cases, not possible with subtractive methods [9]. In addition to this, the amount of material wastage is minimized, which can help preserve the need for tooling maintenance and overall material utilization, especially in the case of metals [2,10]. However, in a general sense, the current status of layer-by-layer

Tribology of Additively Manufactured Materials. https://doi.org/10.1016/B978-0-12-821328-5.00007-X

process of production does impede the absolute speed of manufacturing production [11]. In essence, metal AM consists of utilizing metal in the form of a powder, which in turn will be melted by an energy beam in the form of either a laser or electron based. Through the influence of the beam, the powder is then melted and transformed into a uniform and solidified form. This process is then repeated by a layer-by-layer basis, which replicates a premade three-dimensional design using a computer interface. As the layers are deposited, these layers tend to have a thickness range between 20 μm and 1 mm depending on the processing conditions. In addition to this, it should also be kept in consideration that the type part design, its material composition, and its application will dictate the proper layer thickness needed for an optimized part [12–14].

Within the literature, the most utilized AM powder-based techniques include but are not limited to selective laser melting (SLM), selective laser sintering (SLS), electron beam melting (EBM), and laser metal deposition (LMD). Although the end product is the same between these techniques, the differences lie in the process in which the part is created. For SLS, the powder is first deposited to the working area from which the laser beam will partially melt them, contrast to the full melting induced from SLM. Similarly, EBM operates the same as SLM/SLS; however, an electron beam is utilized rather than a laser beam. Lastly, the LMD process operates on the same laser beam principle as SLM/SLS; however, rather than the powder being distributed on a bed, the powder is supplied though a specialized feeding nozzle [12]. Aside from these processes, alternative forms of metal AM technologies also exist, with the most popular being cold spray (CS). In essence, CS is a process in which constructs components through supersonically projecting metallic particles to a working area, thus inducing metallurgical bonding and the formation of operation components. Understanding this, it should also be taken into consideration that independent manufactures have fabricated similar processes with minor changes; however, for the scope of this chapter, an emphasis will be placed on the most common laser processes, being SLM, direct metal laser sintering (DMLS), EBM, and CS [9,13,15].

Despite AM having many advantages over other traditionally used manufacturing techniques, there is still room for improvement from a processing aspect of AM-based builds. One of the core issues that researchers and engineers have encountered pertains to the finishing surface quality of the as-built substrate. Based on the average surface roughness, also widely known as R_a from ISO 4287:1997 [16], there exists an inverse relationship on the deviation of asperities on the material surface and its functionality. More specifically, if there is a greater R_a value, the part accuracy is diminished thus requiring the need for postprocessing machineries [17,18]. In addition to this, the orientation in which the surface roughness leans toward will also have an impact in contacting-based operations, as surface texture can dictate mechanical and tribological performance [19].

In order to combat this, ultrasonic mechanical surface engineering techniques have been extensively used to reduce the surface roughness of as-built components. These processes include but are not limited: ultrasonic burnishing (UB), ultrasonic impact treatment (UIT), ultrasonic elliptical vibration cutting (UEVC), and electropulsing-assisted ultrasonic rolling process (EP-USRP) [20–24]. Although utilizing these techniques, the plastic deformation induced from these processes has the potential of controlling the surface roughness of the substrate thus mitigating the issues present with surface quality. In this chapter, an analysis of the processing parameters, which dictate surface roughness will be discussed as well as postprocessing methods, which can be used to improve the total surface quality of the substrates.

7.2 Additive manufacturing processes

7.2.1 Selective laser melting

Acting on the principles of traditional powder bed-based processes, SLM acts as a reliable and time-tested technology in metal AM. Building off of a three-dimensional computer-aided design (CAD) design, a metal powder is spread in thin layers and solidified through the heat deposited by the laser beam. Once the layer is complete, the action is repeated continuously until the part is complete. Fig. 7.1 [18] provides an insightful schematic of how the configuration of a traditional SLM device.

Due to the induce heat transfer modes, the thermal gradients induced by laser and molten pool allow for increased heat absorbance of surrounding particles, thus producing a dense part [25]. The cooling rates post melting is also accentuated, thus limiting the kinetic movement of the affected crystals thus anisotropically refining the system grains [26]. As per the Hall-Petch relationship [27], the mechanical properties of a component are inversely proportional to its average grain size, which indicates an increase in mechanical performance. With SLM, the process is quite adaptable to the variety of materials, which can be processed. Such materials that we studied largely span from 316L SS, IN625, IN712, and Ti-6AL-4V [28,29].

However, one major drawback of this process pertains to the limited surface quality. When surface quality is reduced, the processed particles become less dense due to

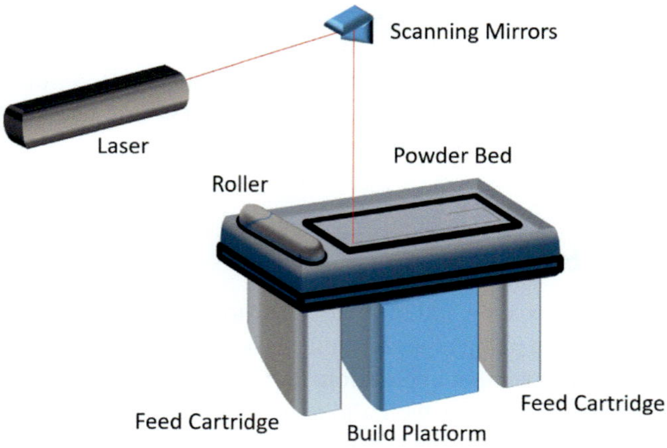

Figure 7.1 A schematic of a selective laser melting device [18].

improper fusion [30]. In cyclical loading applications, a decrease in surface quality indicates increased localized stresses, which can hamper the fatigue life of metal-based components [31]. This is due to the increased likelihood of fatigue cracking, which can result with early part failure [32]. Therefore, by having a lower surface quality, manufactured components may require postsurface treatments, which may result with additional costs and manufacturing time.

In order to adjust the surface finish, the laser processing parameters must be optimized [29]. Specifically, the energy density shows to have the greatest influence on the final surface finish [30], which can be defined as:

$$Energy\ Density = \frac{P}{V * h * d}$$

Where P is the laser power in Watts, V is the scanning velocity in $\frac{mm}{s}$, h is the hatch distance in mm, and d is the layer thickness in mm [15]. By manipulating these variables, the final surface finish will be directly impacted, which can dictate part performance.

7.2.2 Direct metal laser sintering

Direct metal laser sintering is a powder bed fusion process, similar to SLM. Initially, a layer of power is spread of the manufacturing surface where a laser sinters the metal powder together. After this process is complete, the build platform then lowers and a new layer of powder is spread across the part. This process is repeated layer by layer until the part is completely produced.

Compared to SLM, the key difference between DMLS and SLM lies with the melting of the powder. In DMLS, the powder is not reduced to a completely melted state, contrast to SLM [2]. Similar to SLM, DMLS can be used with several different materials, with the most common being titanium and steel alloys [15]. With the incomplete fusion of particles, DMLS tends to result with a higher surface roughness when compared with traditional subtractive manufacturing methods [8]. However, the ability to rapidly produce parts, reduce the amount of waste material, and enhance metallurgical properties provides a greater incentive for application [2]. Similar to SLM, the energy density of the laser beam acts as the largest dictator of surface roughness for DMLS components. By modifying the variables which constitute this value, the surface quality can be improved.

7.2.3 Electron beam melting

Electron beam melting is a method of AM similar to SLM. The process is largely the same, except for the laser being replaced with a beam of electrons, as shown in Fig. 7.2 [18]. This method has common applications in the aerospace and medical industries due to the increased accuracy of the building process when compared to SLM [33]. In EBM, the power bed is preheated, allowing for the melted particles to undergo a phase transformation thus enhancing the part properties. In order to improve the

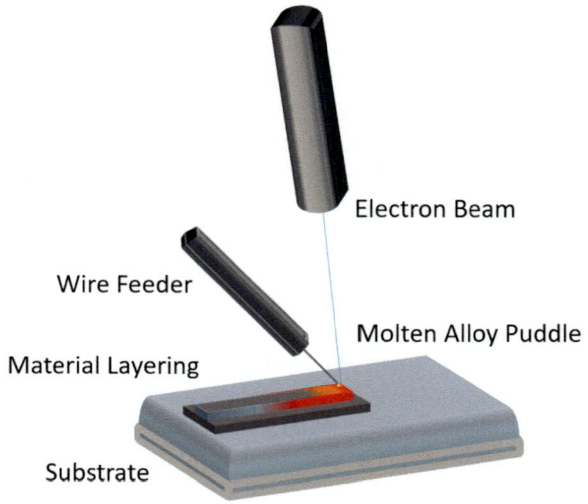

Figure 7.2 A schematic of an electron beam melting device [18].

surface quality, the melt pool size should be increased in order to decrease the temperature gradient of the laser. By having a smaller temperature gradient, there will be fewer surface tensions thus refining the quality of the surface [20].

7.2.4 Cold-spray-based additive manufacturing

A third popular technique that has been increasingly used is cold-spray additive manufacturing (CSAM). By rapidly depositing metallic particles onto a working surface via supersonic gas jet, plastic deformation and metallurgical bonding are induced, as depicted in Fig. 7.3 [34,35]. Typically, the powders sprayed are typically a finely powdered material, typically different from that of the substrate. The powder is

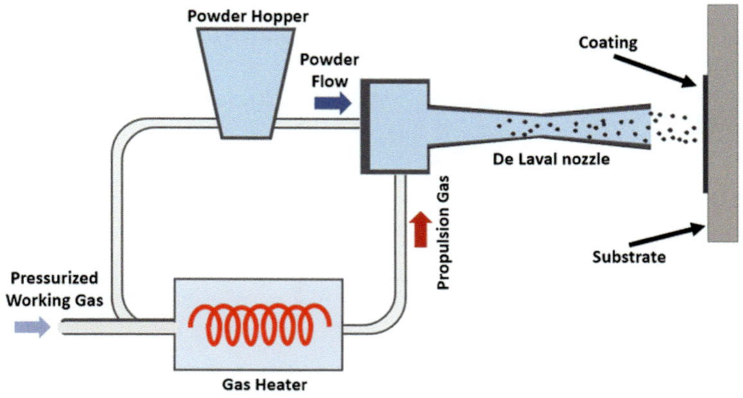

Figure 7.3 A schematic of a carbon spray device [35].

sprayed at a high enough velocity to bond to the substrate surface. The interlocking induced from the process is a byproduct of the high kinetic forces present in area where the particles are contacted with each other [36]. The variables that are used to control the CS process typically are material, particle size, spray velocity, sweeping velocity of spray tip, and distance between the substrate and spray tip.

Cold-spray additive manufacturing is a very commonly used process due to its ability to retain the base material properties by fusing the launched particles under their melting points [36]. In addition to this, oxide-free surfaces can be achieved given that the processing parameters are optimized. Highly dense deposits can also be achieved as a byproduct of this, which can improve corrosion resistance and the lifespan of the build [37]. The compressive forces from the CS process also contribute to the improved metallurgical properties of CS deposits. However, some negative effects of CS typically lie in the cost and surface roughness induced from the process. Typically, if the process is not completely optimized, the course surface finish will indicate a lack of adhesion between particles, thus decreasing the integrity of the deposition.

7.3 Effect of laser processing parameters on surface roughness

7.3.1 Selective laser melting parameters

With surface roughness being one of the key issues for SLM processes [8], many have studied the effects of laser processes for different metallic systems. Wang et al. [30] is among many who have studied the development of surface roughness for 316L SS. In this work, the scanning speed, laser power, and overlap ratio of multiple melt tracks were varied and studied as a function of surface roughness. In this work, surface roughness, R_a, was varied from 4.79 to 24.16 μm, indicating that the laser parameters do have a significant impact on the surface roughness development. As can be seen in Fig. 7.4, increasing the scan speed of the laser resulted with a higher R_a value.

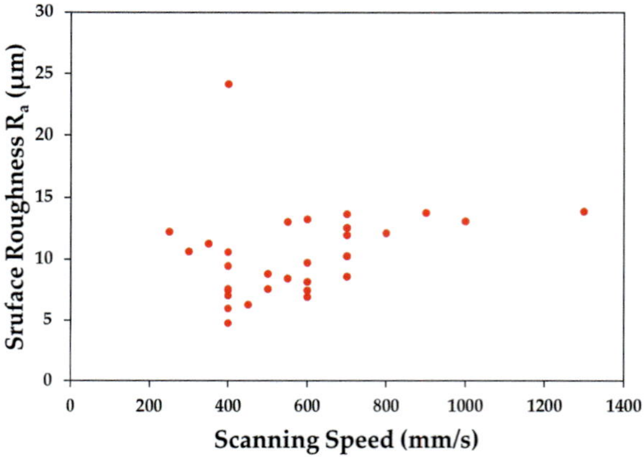

Figure 7.4 The relationship of scanning speed and surface roughness for SLM 316L stainless steel.

Conversely, a lower scanning speed results in a more complete melt of the power bed, as represented with the lower R_a value.

In combination to this, the total influence of volumetric energy density $\left(\frac{J}{mm^3}\right)$ has shown to have the clearest trend of scanning speed and final surface roughness. When laser energy density is below 75 J/mm^3, the surface roughness exists in a range between 10 and 15 μm. This can be attributed to the lack of bonding due to the insufficient amount of volume energy density. Similarly, when increasing past 160 J/mm^3, the surface roughness begins to increase again due to surface over melting defects such as the balling effect. It is proposed that within these two value ranges, specifically at 156.25 $\frac{J}{mm^3}$, the surface roughness is optimized due to complete particle formation. In general, within this range, the combination of high laser power, low scanning speed, and small powder layer thickness produces the most consistently low surface roughness.

Similar to the work by Wang et al. [30], Balbaa et al. [38] also studied the effects of laser parameters to surface roughness for IN625. Contrast to Wang's study, the surface roughness values were significantly lower, spanning from 1.68 to 4.75 μm. Fig. 7.5 displays the effect of laser power compared to surface roughness. As can be seen, increasing the laser power dramatically reduces the surface roughness. This decrease in surface roughness is caused by a larger melt pool. Increasing the size of the melt pool results in a decreased surface roughness due to the increased bonding of the material [26]. However, when comparing the scanning speed with consistent laser power, the surface roughness tends to deviate less. Although there are some deviations, this can be primarily attributed to the amount of localized thermal heating from the laser melting process. However, with the decreasing trend with increasing laser power, this demonstrates that processing parameters such as laser power have the greatest influence on the final surface finish due to its direct influence on the melt pool.

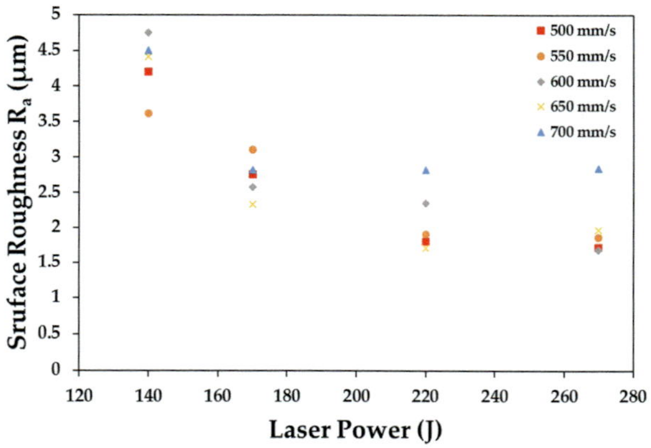

Figure 7.5 The relationship of laser power and surface roughness for SLM IN625 [38].

7.3.2 Direct laser metal sintering parameters

Similar to SLM, surface roughness in DMLS can be controlled by varying laser processing parameters. Scanning speed, laser power, powder thickness, and hatch size are the key parameters that combine to form energy density. Patibandla et al. [15] have worked on Ti-6AL-4V and showed that as energy density increases, the surface roughness decreases to a certain point, whereas it begins to increase after 200 J/mm^3. As depicted in Fig. 7.6, the optimal laser bean density, 55.56 $\frac{J}{mm^3}$, was found to result with the lowest surface roughness, measuring at 7.7 μm. This phenomenon, similarly to SLM-based parameters, is due to either incomplete melting or overheating, which in turn results with surface defects. In relation to this, the porosity associated with different laser intensities also shared a direct relation with surface roughness. In essence, as the surface roughness begins to decrease, the porosity of the surface is also decreased.

For 316L SS, the R_a values range from 10.3 to 13.8 μm. As can be seen in Fig. 7.7, the surface roughness is proportional to the laser energy density due to the average values recorded values gradually increasing. This was found to be due to an increased buildup of the powder on the final layer for this material [15]. However, it was found that there was a higher variance in surface roughness when the energy density is at 200 $\frac{J}{mm^3}$. At this value, the surface roughness varied from 12.2 to 14.4 μm, largely due to the combination of material evaporation and pore generation. The lowest value for the surface roughness was found to be at 10.4 μm when the laser energy density was 66.67 $\frac{J}{mm^3}$, contrast to the surface roughness results found from SLM. In a general sense, it can be insinuated that with lower intensities, the melt pool is sufficient enough for even cooling, whereas with higher intensities, potential defects such as the keyhole porosity can occur thus increasing the surface roughness.

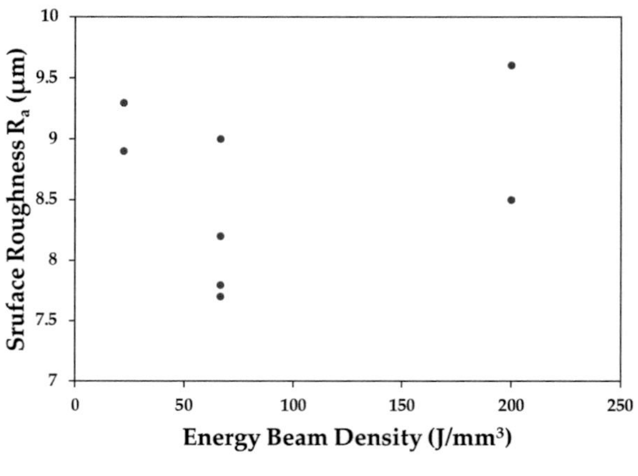

Figure 7.6 The relationship of energy beam density and surface roughness for direct metal laser sintering (DMLS) Ti-6Al-4V.

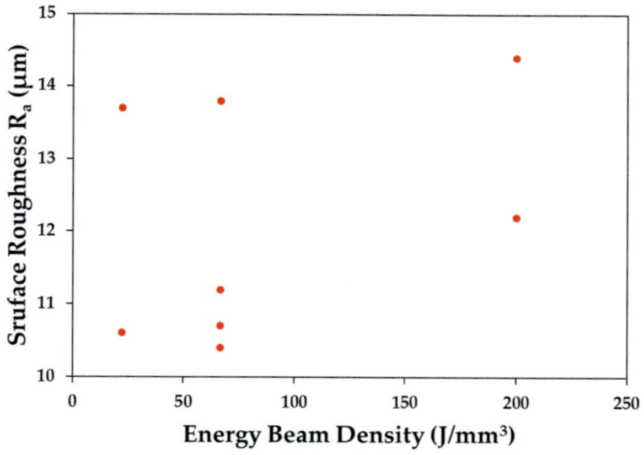

Figure 7.7 The relationship of energy beam density and surface roughness for direct metal laser sintering (DMLS) 316L stainless steel.

7.3.3 Electron beam processing parameters

Electron beam processing uses similar process parameters to that of laser-based AM techniques. These parameters consist of hatch size, scanning speed, layer thickness, and beam current. However, more complex factors such as the sloping angle of how the builds are fabricated also influence the final surface roughness. Galati et al. [39] investigated the effects of sloping angle to surface roughness and found that the build direction of the laser does in fact have an impact on the final surface finish for Ti6Al4V powder. This was accomplished by statistically observing the impacts of upward and downwards surface fabrication through an ANOVA analysis. The processing parameters for the contour scans consisted of a scanning speed of 850 mm/s, focus offset of 6 mA, beam current of 5 mA, and hatch contours at 0.29 mm. The processing parameters for the hatching scans consisted of a speed function of 45, focus offset of 25 mA, beam current of 20 mA, reference length of 45 mm, reference current of 12 mA, and a line offset of 0.2 mm.

When the sloping angle is positive, the surface roughness has been shown to increase linearly [39]. Due to the temperature gradient of the melt pool, the staircasing defect become prevalent, which contribute to the increased roughness values. Based on the predictive model from Minitab, the predictive surface roughness model of increasing sloping angle was found to be:

$$R_a = 6.103 + 0.294 * \text{Sloping Angle } (\mu m)$$

Despite this, downward sloping angles were shown to generally have higher surface roughness values compared to the upward sloping angles due to the larger melting area from the laser [39]. Unlike the upward surfaces, the cooling rates were reduced due to the increased heat accumulation of the process. Due to this, any nonmelted particles will begin to sinter which in turn will increase the surface roughness, as shown in Fig. 7.8 where the downward surfaces near 0° tend to have the greatest change in surface roughness. Based on these results, it is recommended that upward surfaces are used when using EBM processing.

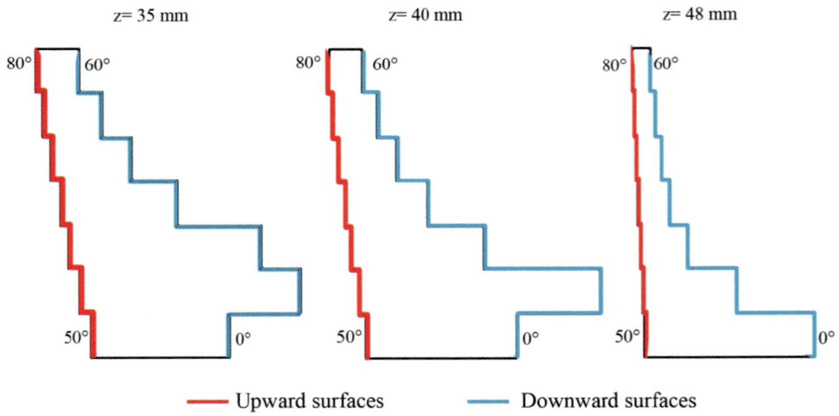

Figure 7.8 The relationship of melting heights compared to upward and downward surfaces [39].

7.3.4 Cold spray processing parameters

Typically, objects that require high polish surfaces do not use CS as a surface processing method due to the high surface roughness from the process [40]. However, by finding the correct process parameters that minimize surface roughness, not only will there be less of a need for postsurface refining techniques, but the integrity of the build will be increased. This could be especially beneficial in applications where corrosion is a major factor due to the increase in surface area with increased surface roughness. In general, the primary parameters which are used to reduce surface roughness include the working gas pressure, working gas temperature, and standoff distance [41]. Working gas pressure refers to the pressure used to force particles out of application nozzle. The temperature is related to what temperature the working gas is brought to before application. Lastly, the standoff distance is the distance between the application nozzle and substrate.

In the work by Goyal et al. [42], the surface roughness was studied and optimized by conducting an ANOVA analysis on the various processing conditions in order to produce oxide free, dense coatings. The substrate materials selected for the study are aluminum (ASTM B 221), brass (ASTM B 36), and nickel (ASTM B 435), all acting as a substrate for copper to be deposited to. These materials were selected because of their common use in electronic components. Material specimens were grit sanded and polished before CS. Each specimen was cold sprayed with the parameters listed in Table 7.1, constituting feed type, substrate material, air pressure (P), air temperature (T), and standoff distance (D). The resulting surface roughness (R_a) is also listed for each test.

The first parameter examined was the working gas pressure. Fig. 7.9 shows the relationship between the increase in gas pressure and the resulting average surface roughness. The data show a lower surface roughness as pressure increases. The average R_a roughly decreases from 10.5 to 7.5 μm, a 2.5 μm change in surface roughness as

Table 7.1 Cold-spraying parameters and their corresponding surface roughness values [42].

Coating material	Substrate material	Feed type	Cold spray parameters			R_a, (μm)
			P, (psi)	**T, (c)**	**D, (mm)**	
Cu	Al	Gravity	104	350	2.5	13.70
		Gravity	112	375	5	10.71
		Gravity	120	400	7.5	7.86
	Brass	Gravity	128	350	5	9.97
		Gravity	136	375	7.5	8.81
		Gravity	144	400	2.5	8.27
	Ni	Gravity	152	375	2.5	11.32
		Gravity	160	400	5	6.91
		Gravity	168	350	7.5	6.96
	Al	Argon	176	400	7.5	8.72
		Argon	184	350	2.5	9.45
		Argon	192	375	5	8.54
	Brass	Argon	200	375	7.5	9.48
		Argon	208	400	2.5	7.72
		Argon	216	350	5	6.51
	Ni	Argon	224	400	5	8.07
		Argon	232	350	7.5	7.72
		Argon	240	375	2.5	7.63

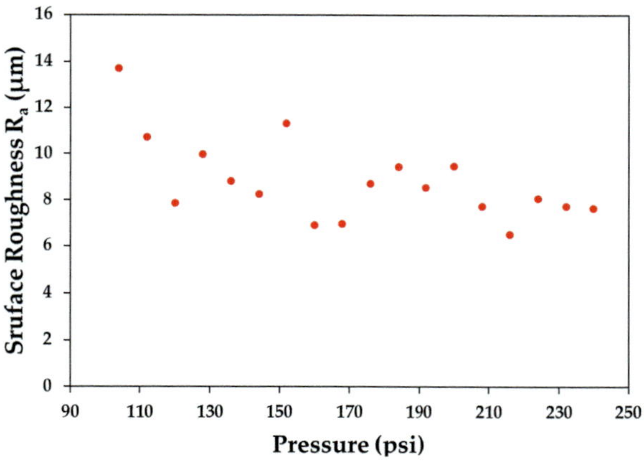

Figure 7.9 A relationship between the gas pressure and surface roughness of cold-sprayed copper [42].

pressure increases by 130 psi, which can be attributed the particle velocity induced from the increased pressure [43]. As the kinetic force of the particles increases, there is greater metallurgical bonding, thus resulting with a finer surface finish [44].

The effect of the working gas temperature on surface roughness is plotted in Fig. 7.10. The graph shows a slight decrease in surface roughness as temperature is

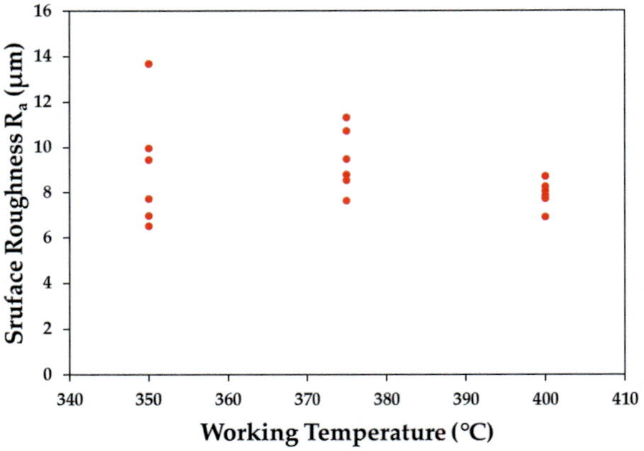

Figure 7.10 A relationship between the working temperature and surface roughness of cold-sprayed copper [42].

increased. The average value of R_a went from 9.5 to 8 μm, a total decrease of 1.5 μm over a 50°C change in temperature. Increase in the working gas temperature has been found to decrease particle velocity [43]. Friction can also be assumed to increase with temperature due to a rise in turbulence. Just like pressure, lower kinetic force will result in smaller deformation and less effect on surface roughness [45,46]. The working temperature seems to have a smaller effect on surface roughness as can be seen by the slope of the fitted line.

Lastly, the effect of the standoff distance is examined against the surface roughness of the CS copper in Fig. 7.11. The surface roughness clearly decreases as the standoff distance increases. The average roughness starts around 9.5 μm and decreases to about 8 μm. This is a change of 1.5 μm over a change of 5 mm in the standoff distance. Short standoff distances expose the substrate to high temperatures that cause a phenomenon called bow-shock that results in lower particle velocities [42]. Particle velocities that are too low can keep the surface from hardening, which can result in higher surface roughness.

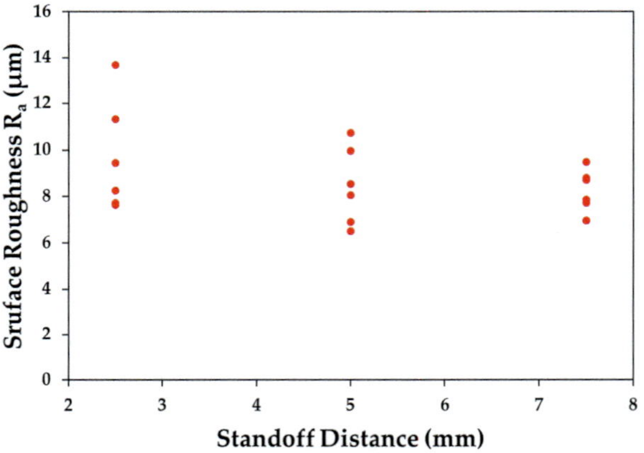

Figure 7.11 A relationship between the standoff distance and surface roughness of cold-sprayed copper [42].

7.4 Postprocessing techniques for surface roughness control

7.4.1 Ultrasonic burnishing

Ultrasonic burnishing is a surface engineering technique, which involves using an ultrasonic generator that delivers a high-frequency electrical signal to a tool, which contains the ultrasonic transducer and amplitude transformer that is used to control the burnishing head. The burnishing head rotates and vibrates up and down simultaneously, while it moves in one direction which is the direction of the feed on the workpiece. UB can be used to refine or improve the properties of metal components. Plastic deformation occurs during the process, and no actual material is removed, instead the irregularities on the surface of the workpiece are removed through this plastic deformation process. The process causes the compressive forces of the tooling to displace the peaks on the surface material to the valleys; UB has the potential to replace other grinding and polishing techniques [47−51].

Many studies in the past have demonstrated that the surface roughness can be controlled using UB techniques [20,47,49]. The work by Jian and Zhanqiang [20] does an exemplary job showing this. In this work, the working piece used for the Rotary UB method consisted of a rectangular workpiece, while the material used was Ti-6Al-4V. The tooling itself is a flat-based burnishing head with four rollers. The burnishing head did not exceed 0.4 mm, which restricts the burnishing depth to 0.4 mm. A total of 18 experimental runs were done with varying factors in order to study the effects these varying factors would have on the surface roughness of the workpiece.

The surface roughness results for the UB operated at varying burnishing parameters of the Ti-6Al-4V specimen are shown in Table 7.2. The different parameters for all of the 18 experimental runs are also shown that produced the different surface roughness values. The specimens all had an initial mean R_a value of 0.228 μm after being machined through a conventional milling technique. The experimental data compare the surface roughness values after UB is carried out to the mean milled surface roughness initial values. The data also show how the varying parameters are changed to produce different R_a results. There were a number of experimental trials that produced UB surface roughness values that were higher than those of the mean-milled surface; however, there are also a number of experimental trials that produced lower surface roughness values through the use of the UB technique.

Fig. 7.12 shows a graphical representation of the surface roughness values attributed to that particular experimental number. The UB surface roughness values are compared to the mean value of the milled surface roughness values. The results show that five data points have lower R_a values than the mean milled surface roughness value. According to the study results, burnishing depth has the largest effect on the surface roughness out of all the different parameters. The experiment showed that surface roughness was much higher for the experimental runs where burnishing depth was at a higher value, most of the higher values for surface roughness were from an experimental run that had the burnishing depth at 0.2 mm or 0.3 mm, which

Table 7.2 Roughness reduction in Ti-6Al-4V alloy by ultrasonic burnishing.

No.	Ra[μm] milled surface (initial)	Ra[μm] after ultrasonic burnishing	Burnish depth h [mm]	Spindle speed S [r/min]	Feed rate F [mm/min]	Frequency of ultrasonic vibration f [kHz]
1	0.228	0.367	0.1	400	40	19
2	0.228	0.37	0.1	600	60	22
3	0.228	0.263	0.1	800	80	25
4	0.228	0.156	0.2	400	60	25
5	0.228	0.152	0.2	600	80	19
6	0.228	0.215	0.2	800	40	22
7	0.228	0.577	0.3	400	80	22
8	0.228	0.662	0.3	600	40	25
9	0.228	0.975	0.3	800	60	19
10	0.228	0.276	0.1	400	60	22
11	0.228	0.208	0.1	600	80	25
12	0.228	0.196	0.1	800	40	19
13	0.228	0.43	0.2	400	40	25
14	0.228	0.39	0.2	600	60	19
15	0.228	1.16	0.2	800	80	22
16	0.228	1.09	0.3	400	80	19
17	0.228	0.748	0.3	600	40	22
18	0.228	1.5	0.3	800	60	25

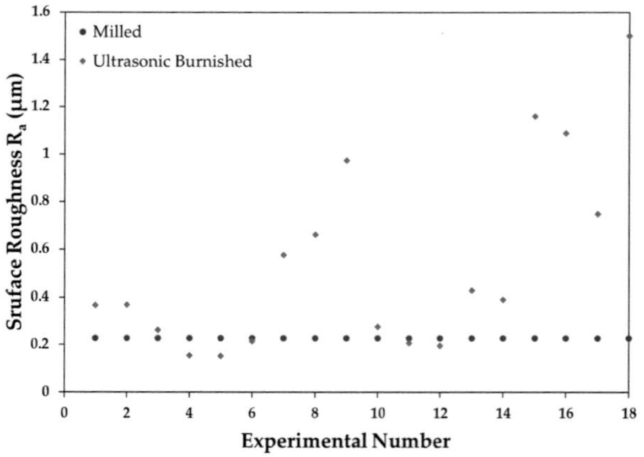

Figure 7.12 The mean surface roughness of the milled substrates compared to the ultrasonic burnished substrates.

is apparent when looking at the data results. Spindle speed also had a large effect on the R_a values and was the second most underlying factor of the different parameters. The data show a trend in higher R_a values when burnishing depth and spindle speed are set to their highest parameters. The lower values for R_a were achieved when the burnishing depth and spindle speed were set to their lowest parameters. In particular, the lowest roughness value of 0.39 μm was recorded at a burnish depth of 0.2 mm and a spindle speed of 600 r/min. According to the study, the feed rate and frequency of ultrasonic vibration had the least significant impact on the values of surface roughness.

For 316L SS, Salmi et al. [52] was able to improve surface roughness through also using UB. For this experimental study, 316L SS was modeled in four hexagonal prisms manufactured using an EOSINT M 280 machine. The steel prisms were processed by utilizing the UB equipment, which was installed into a numerical control machine as shown in Fig. 7.13B. As different parameters were used for this process, the R_a values for this process were typically in the range of 13 ± 5 μm.

The machine used to process the material was a Mazak milling center that had a power of 22 kW, whereas the impact frequency of the tooling was 19 kHz. There were three different parameters that were studied in order to record the data and see how these different parameters affected the final specimens. These parameters include: table feed rate, side shift, and spring compression in the tooling. Fig. 7.14 shows the effects that the varying parameters had on the values of surface roughness. Parameter values for obtaining the lowest surface roughness results for the side shift appear to be 0.025 mm. The feed rate is ideal at values from 300 to 400 mm/min and especially at 1000 mm/min as shown by the curve where it is apparent that surface values begin to increase but suddenly drop when the feed rate is set at that value of 1000 mm/min. While it seems that when the feed rate is set to 5000 mm/min, the surface roughness value drops; this actually caused the tooling to bend thus having the potential to damage the tooling at this particular feed rate. The ideal value for the spring compression in the tooling is at 1 mm where the value for R_a is at its lowest.

Table 7.3 shows the minimum, maximum, and average surface roughness values for the experiment. As expected, the as-built 316L specimen had the highest surface roughness values of all the recorded results. The milled specimen surface roughness results show that the values for this procedure were lower than those of the milled and burnished specimens. While the milled surface roughness values were significantly lower than those of the as-built and milled and burnished specimens, the ultrasonic burnished average surface roughness values for those particular specimens were very close to that of the milled specimens. This difference is so small however that it can be disregarded, and it is evident by the data that these two processes can be used to

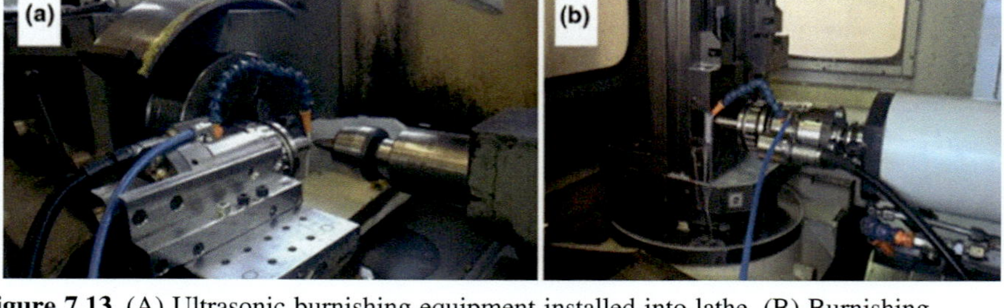

Figure 7.13 (A) Ultrasonic burnishing equipment installed into lathe. (B) Burnishing equipment installed into numerical control machine; this is the method that was utilized for the 316L stainless steel [52].

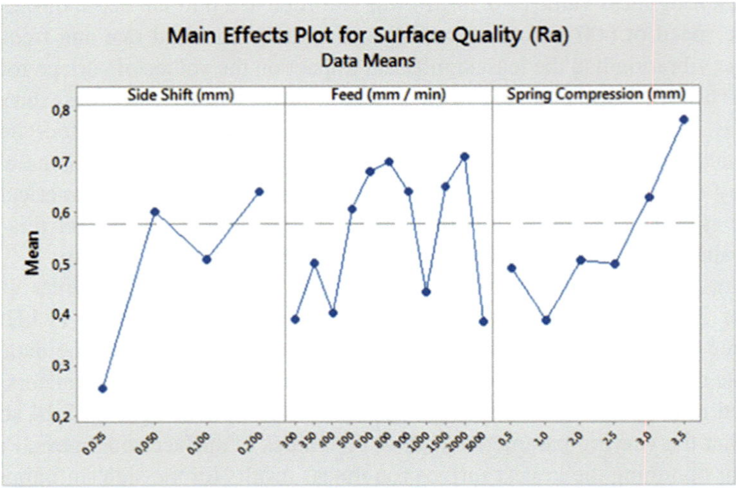

Figure 7.14 The main effects plot of surface roughness with varying parameter factors [52].

Table 7.3 Surface roughness values for 316L stainless steel [37].

	Minimum R_a (μm)	Maximum R_a (μm)	Average R_a (μm)
As-built 316L	5.73	9.27	7.39
Ultrasonic burnished 316L	0.22	1.23	0.55
Milled 316L	0.31	0.66	0.47
Milled and burnished 316L	0.94	1.10	1.00

obtain a similar surface roughness value. UB has the potential to be more cost effective than its milling counterpart, milling often requires that the tooling be changed regularly during the process, whereas with UB the carbide ball that is used in the tooling lasts longer. As mentioned before both methods have the potential to obtain better results for surface roughness but one method has the potential to be more cost effective than the other. There are other parameters for UB that still need to be studied; however, this study shows that UB can be used to produce materials with optimal surface quality [49–51].

7.4.2 Ultrasonic impact treatment

The ultrasonic impact treatment technique consists of the impact head vibrating through the use of an ultrasonic horn at a set frequency controlled by an ultrasonic generator. The impact head impacts the material, which causes changes in the surface structure of the materials being processed. There is a wide range of uses for titanium alloys in a number of industries such as oil and gas, infrastructure, and aerospace. The UIT process can be used to enhance many properties of fatigue as well as in refining the grain properties of titanium alloys through the plastic deformation that takes place during the UIT process. The UIT technique can be used for a wide range of metals such as aluminum, bronze, iron, and titanium. This experiment gives examples and data as to how surface roughness of Ti-6Al-4V alloy is affected during this process.

Dekhtyar et al. [53] studied the impact of UIT on the surface roughness behavior of Ti-6Al-4V alloy. This experiment consisted of preparing Ti-6Al-4V samples using powder metallurgy (PM)-based techniques. The different parameters used during this process are listed in Table 7.4. The ultrasonic generator frequency was set at 18.7 kHz with an amplitude setting of 20 μm, and the power output rating of the UIT device used was 4 kW. The Ti-6Al-4V specimen was placed on a rotating spindle on a lathe machine with a frequency of 0.2 Hz. Afterward, the specimen was then subjected to a number of impacts by a pin and an ultrasonic horn. Similar processes can be found in the work of Mordyuk et al. and Lesyk et al. [54,55].

The results for both of the ultrasonication transfer mechanisms are shown in Table 7.5. These results show that the UIT process produced a sample in which the surface roughness and surface roughness depth were decreased by as much as 4 to 5 times that of the Pristine PM sample. The continuous impact that was created by the UIT process made the sample smoother than the Pristine PM process sample. This is evident by the values of the surface roughness and roughness depth that were recorded and compared to each other. This means that the UIT process can possibly reduce the fatigue life and wear in certain components. Along with benefits to material fatigue life, UIT can also improve corrosion resistance and hardness of the material. This experiment showed that this technique can reduce the surface roughness of material, which can help improve the corrosion resistance and hardness of the material [56,57]. Thus, UIT can be used for AM-fabricated components for manipulating the surface roughness, thus improving the metallurgical properties of the material.

Table 7.4 Ultrasonic impact treatment parameters used for Ti-6Al-4V [53].

Ultrasonic generator frequency, f_{us} (kHz)	18.7
Amplitude, A (μm)	20
Power output, P (kW)	4
Rotating spindle frequency, n (Hz)	0.2
Induced frequency of impacts, f (kHz)	3
Pin surface curvature radius, R (mm)	20
Normal velocity of pin from ultrasonic energy, v_{us} (m/s)	2.35
Shift velocity of specimen surface, v (m/s)	2.51

Table 7.5 Resulting surface roughness of pristine powder metallurgy (PM) and ultrasonic impact treatment (UIT) [54].

Material	Surface roughness, R_a (μm)	Roughness depth, R_z (μm)
Pristine PM (polished)	1.25	4.0
UIT treated	0.32	0.8

7.4.3 *Ultrasonic elliptical vibration cutting*

Ultrasonic elliptical vibration cutting is a technique that consists of an ultrasonic elliptical vibrator attached to the tooling rip during a cutting process. As this process occurs, the tool tips begin to move through some specified angular frequency, which in turn can mitigate the amount of forces present in cutting operations. This in turn creates a finer surface finish while preserving the tool material. Although not widely studied, Strano [58] is among some who have studied the effects of this process for Inconel 718. The study consisted of using a single point diamond cutting tool to study the cutting effects it would have on the Inconel 718 material using the UEVC technique [58]. During the process, the tool tip is vibrated in an elliptical fashion through the use and application of an ultrasonic elliptical vibrator with the goal of reducing surface roughness [59,60].

The different parameters that were used during the experiment are shown in Table 7.6. These parameters consisted of ultrasonic frequency (kHz), feed rate (mm/r), depth of cut (mm), and cutting speed (mm/min). The three levels correspond to the low, medium, and high factors of the experimental study, for instance, level 1 corresponds to the low factor and so on. This part of the experiment was carried out mainly to determine what parameters had the most significant impact on surface finish.

The impact of feed rate, cutting speed, and amplitude ratio to median predicted surface roughness are shown in Fig. 7.15. These main effects plots are based from Minitab 17's Taguchi design. Based on these plots, the feed rate is shown to linearly increase the surface roughness values as the feed rate itself is increased. Cutting speed decreased surface roughness values as the cutting speed was set to increasing speeds. The amplitude ratio also shows a linear increase in surface roughness values as it is

Table 7.6 Ultrasonic elliptical vibration cutting experimental parameters [58].

	Levels		
Parameters	1	2	3
Ultrasonic frequency (kHz)	20	25	30
Feed rate (mm/r)	0.08	0.15	0.2
Depth of cut (mm)	0.05	0.1	0.15
Cutting speed (mm/min)	15	20	25
Nose radius (mm)	0.4	0.8	1.2

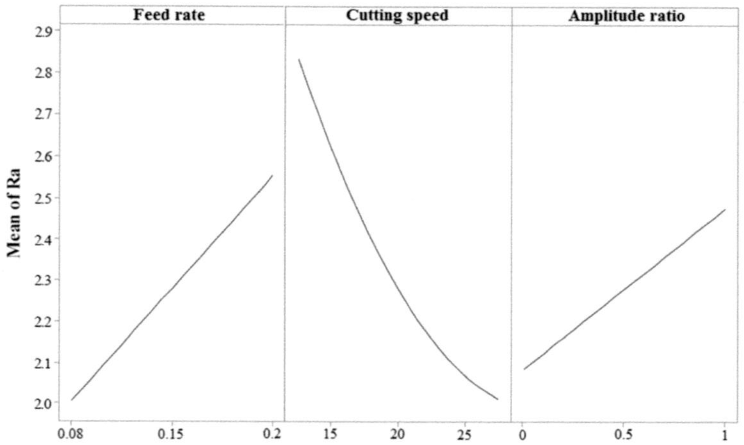

Figure 7.15 Main effects plot for feed rate, cutting speed, and amplitude ratio and their impacts on surface roughness [58].

increased. The minimum, maximum, and averaged R_a value predicted in this work was 0.44, 2.77, and 1.23 μm.

Table 7.7 shows the different parameters for the validation tests, and levels 2, 4, and 5 were used to calculate the predicted and experimental values of surface roughness, which are shown in Table 7.8. Test parameters were increased with each level and the predicted surface roughness values corresponded to the increase in parameter settings. Percentage differences were then calculated using the predicted and actual surface roughness measurements. According to the experimental results, feed rate had the highest effect on the value of surface roughness. As feed rate was increased, the

Table 7.7 Test parameters of ultrasonic elliptical vibration cutting [58].

	Levels		
Parameters	2	4	5
Feed rate (mm)	0.1	0.18	0.2
Cutting speed (mm/s)	17	23	25
Amplitude ratio (x/y)	0.25	0.75	1

Table 7.8 Predicted and actual surface roughness from ultrasonic impact treatment [58].

Levels	**Predicted**	**Actual**	**%Differences**
2	2.287	1.913	16.730
4	2.345	1.833	21.832
5	2.482	2.150	13.357

surface roughness values were subsequently increased; it should also be noted that as the cutting speed along with the ultrasonic frequency is increased the values for surface roughness values decrease.

7.4.4 Ultrasonic surface rolling process

The ultrasonic surface rolling process (USRP) is a technique, which combines a series of principles from deep rolling, low plasticity burnishing, and UIT. This process can be conceptualized as the process of inducing mechanical vibrations through an alternating current force. Through this, plastic deformation is induced, similar to shot peening, where the grains are refined and residual stresses are induced to the surface [61]. What differentiates this process from other traditional processes is the ultrasonic power waves induced from the electrical energy, which creates a unique outcome of reducing surface roughness while limiting the damage done to the surface [62].

Sun et al. [63] studied the different effects on surface quality that USRP exhibited on nickel Inconel 718 material. It also compared how surface roughness was affected when taking this process and adding electropulsing-assisted techniques to USRP. The different surface roughness values were plotted to interpret the data and to see the differences in those values for the three different techniques used in this experimental procedure. The utilized techniques consist of the following: conventional turning, USRP, and electropulsing-assisted ultrasonic rolling (EP-USRP). The differences between USRP and EP-USRP are that with the electropulsing variant, a frequency is induced by applying an electropulsing generator connected to a positive and negative graphite brush attached to each end of the tooling as shown in Fig. 7.16.

The experimental data compared how these processes affected surface roughness. Fig. 7.17 shows how the values for surface roughness differ with the conventional turning process and both of the ultrasonic rolling processes. These surface roughness values were measured and recorded using a surface roughness tester, which was used to measure axial surface roughness. The turning process never reached a surface roughness value as low as the USRP technique even when electropulsing was used, Fig. 7.17B shows that the R_a value for the turning process is 0.47 μm. Looking at Fig. 7.17A, it is apparent that the highest value for surface roughness for the USRP

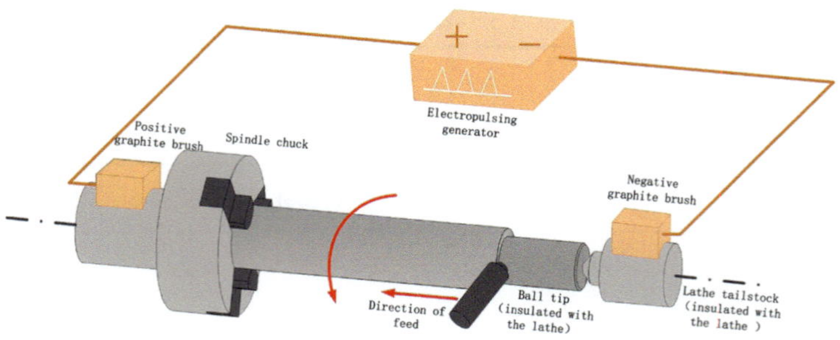

Figure 7.16 Electropulsing ultrasonic surface rolling tooling [63].

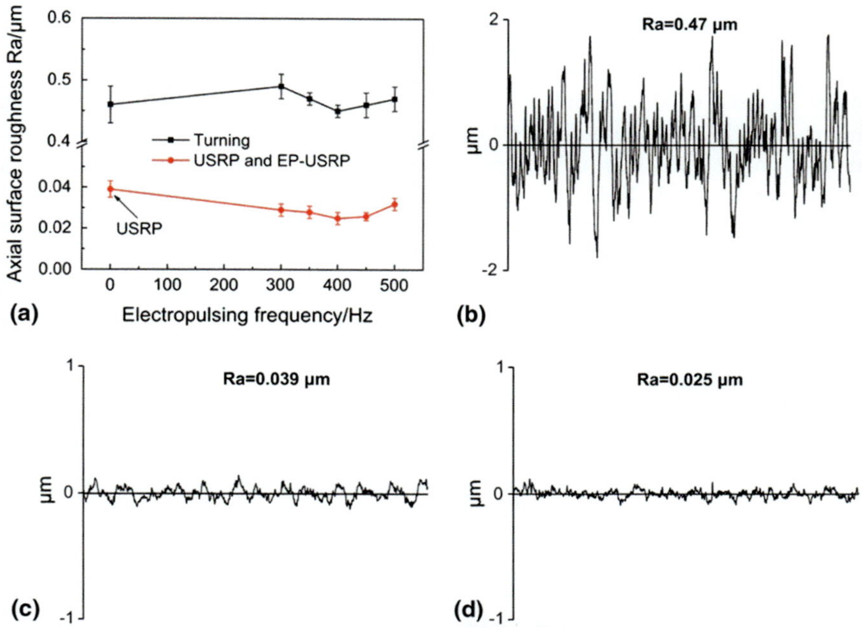

Figure 7.17 Surface roughness values for turning, ultrasonic rolling process, and electropulsing-assisted rolling process. (A) Surface roughness values and curves for the different techniques, (B) turning surface roughness, (C) ultrasonic rolling process surface roughness, and (D) electropulsing-assisted rolling process surface roughness [63].

technique was with no electropulsing assistance. This surface roughness value is shown in Fig. 7.17C and was calculated as 0.039 μm. With the electropulsing-assisted ultrasonic rolling process in Fig. 7.17D, the data show that when the electropulsing frequency was introduced, the surface roughness value was lower than that of the same process with no electropulsing assistance. Fig. 7.17A again shows that surface roughness values dropped for frequencies in the range of 300–500 Hz and was at its lowest at a frequency of 400 Hz. The R_a value at this frequency of 400 Hz was calculated as 0.025 μm.

The obtained results from the experiment show that while the USRP technique can greatly enhance the surface roughness of nickel Inconel 718, the surface roughness can be altered even further when introducing electropulsing assistance to this process. This does however depend on the frequency used during this process, the experimental study shows that at a frequency of 400 Hz, the optimal surface roughness value for this experiment was obtained. It should be noted that with increasing frequency, that surface roughness value begins to increase past the initial surface roughness value without the electropulsing assistance, which means the correct frequency setting should be utilized in order to get the best surface roughness results. This process can be used to alter the characteristics of many other materials as well [64–66].

7.4.5 Laser polishing

Laser polishing is a surface processing method used to polish the rough surfaces of metals. To laser polish, a laser beam is used in order to irradiate the surface of a metal, melting a thin layer of the surface [67,68]. In AM, laser polishing is typically used as a finishing process and is good at working on complex surfaces that other polishing methods cannot reach. Fig. 7.18 depicts the typical set up of a laser polishing experiment. This experiment uses a fiber laser to shoot a laser beam through optics. The test specimens are placed underneath the optics, and the beam is moved across the surface in a zig-zagging trajectory [69]. Table 7.9 holds data points between initial and final

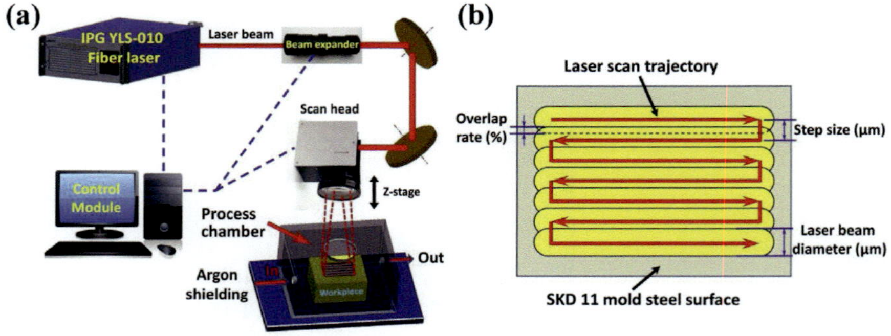

Figure 7.18 This is the experimental setup of a laser polishing experiment where a laser beam removes the top layer of rough metal surface in order to polish it [68].

Table 7.9 A collection of data points from multiple laser polishing experiments [68,70,71].

References	Material	E, (J/mm^3)	Initial R_a, (um)	Final R_a, (um)
[68]	SKD11 tool steel	7.62	3.571	2.437
		9.53	3.571	1.164
		10.81	3.571	0.679
		12.73	3.571	0.332
		19.05	3.571	1.72
[70]	ASP23 steel	5.3E-06	0.22	0.163
		6.9E-06	0.22	0.162
		8.3E-06	0.22	0.158
[71]	AISI H13 steel	15.625	5.33	0.64
		15.625	1.7	0.55
		15.625	2.53	0.46
		15.625	0.49	0.3
[71]	AISI P20 steel	15.625	4.4	0.36
		15.625	1.17	0.32
		15.625	1.52	0.29
		15.625	0.72	0.25

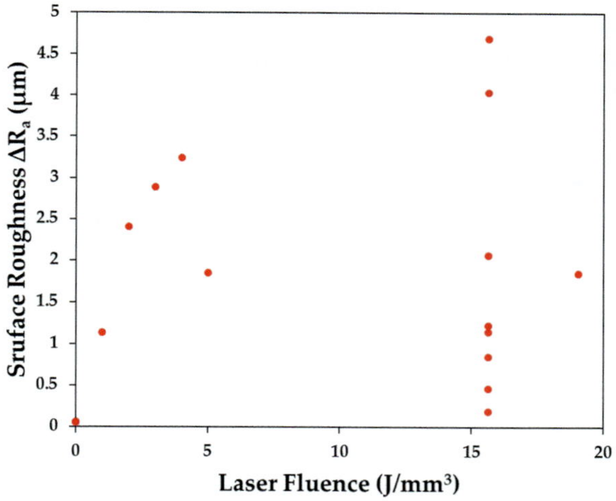

Figure 7.19 The relationship of laser fluence and the change of initial to final surface roughness.

surface roughness of multiple tool steels at different laser fluences. Laser fluence is essentially the energy over an area. This table pulls data from multiple studies on laser polishing. This is done in order to determine the relationship between laser fluence, E, and the change between initial and final surface roughness.

The laser fluence used for laser polishing is examined against the change of the initial and final surface roughness. Fig. 7.19 plots the relationship between energy and the delta R_a. The graph shows that larger removals of surface roughness require higher energy from the laser beam. On average, to remove around 2 μm from the surface, the laser must have a laser fluence around 15 J/mm^2. Further, finer polishing requires an exponentially smaller energy in order to get smaller R_a values. The scanning process of the laser beam leaves behind spatial ripples similar to lines created by machining [68]. These ripples are factored into the surface roughness and can significantly affect smaller values of R_a. Lower energy polishing leaves behind less prominent ripples. This could contribute to why finer polishing needs lower laser fluence.

7.5 Conclusions

AM of metals allows for the rapid production of many different parts and components; however, due to the nature of the process, the surface roughness is relatively high when compared to more traditional manufacturing methods. However, the initial laser parameter settings can have a significant impact on the surface roughness of a component.

For SLM, the scanning speed was found to be the most important laser parameter. The setting of the laser to a slower speed allows for a more complete melt pool to be formed and the powder bed to be more completely bonded. This results in a smoother finish. However, the other laser parameters do have a significant effect. The laser power should be set as high as possible for the material being used.

For DMLS, it was found that by varying the energy density, there appears to be a lack of a relationship of final surface finish. This can be attributed to two different issues. These issues consist of the formation of pores and material vaporization, which can splatter contents of the melt pool around the weld track. In addition to this, the powder particles may have been inconsistent, which can compromise the final surface finish.

For EBM, there was no conclusive parameter, which had the ability to create a specific surface roughness in Ti-6Al-4V. Similar to the work for DMLS materials, the surface roughness was not consistent due to the splattering effect of the electron beam. However, for the most consistent range of values for surface roughness, a 2.0 mA beam current is most desirable. It is suggested that future studies observing other surface parameters are done for EBM work.

For CS, the pressure of the working gas pressure was shown to have the biggest impact on the final surface finish. By continuously increasing the pressure to 250 psi, the average surface finish was steadily decreased, thus indicating a denser formation of the deposition. This can be largely attributed to the mechanical bonding that is induced from the rapid kinetic forces of the particles, contrast the parameters such as the working temperature or standoff distance.

The rotary UB technique is able to obtain better surface roughness results for Ti-6-Al-4V alloys than conventional milling with the correct machine parameters. Burnishing depth was shown to be the most important parameter when using this technique to obtain a better surface finish, while the machine spindle speed was shown to be the second most important parameter. While conventional milling did acquire surface roughness values that were lower than the rotary UB values in some cases, studies have shown that this technique is viable in obtaining better results with the correct parameters in place.

The UB technique can also be applied to 316L stainless steel materials to produce optimal surface roughness results. The results of the study show that the three parameters, table feed rate, side shift, and spring compression in the tooling, affect the surface roughness when manipulated in different ways. Comparing the milled and UB average surface roughness values, it was apparent that the milled surface had a lower value, this value difference can be neglected however due to it being such a small difference in value making it insignificant. This shows that the UB technique can produce very similar surface roughness results in 316L stainless steels when compared to conventional milling.

UIT was able to produce a significant change in surface roughness of Ti-6Al-4V alloy through the continuous impact brought on by the machine tooling. The decrease in surface roughness was shown to reduce by as much as four to five times than that of the pristine PM (polished) specimen. Surface roughness reduction can be achieved by means of this technique along with other beneficial factors such as improving corrosion resistance and improving the hardness of the material.

Surface roughness of Inconel 718 can be reduced using the UEVC technique. It was shown that the main parameter affecting the surface roughness is the feed rate while the cutting speed and nose radius of the tooling were the second most important parameters in determining surface roughness. Nickel Inconel 718 surface roughness can also be reduced using the USRP technique, it was also shown to be further reduced using the EP-USRP technique. The study results showed that USRP was able to produce surface roughness values lower than those of the turning process and could be further reduced by introducing electropulsing, at 400 Hz, the value for surface roughness was at its minimum.

Lastly, for laser polishing, the amount of laser fluence used has a direct impact on the amount of material that is present on the surface. Generally, using a laser fluence of around 15 J/mm^3 will have a removal of 2 μm of material from the surface. However, in order to create a fine polish, the laser fluence must be decreased due to the rippling effect induced from the laser.

Funding

The authors thank NASA RRR CAN, award number: NV-80NSSC19M0172 for the support.

References

[1] ASTM Compass. https://compass.astm.org/EDIT/html_annot.cgi?ISOASTM52900+15 (accessed January 10, 2021).

[2] W.E. Frazier, Metal additive manufacturing: a review, Journal of Materials Engineering and Performance 23 (6) (June 2014) 1917−1928, https://doi.org/10.1007/s11665-014-0958-z.

[3] C.W. Hull, Apparatus for production of three-dimensional objects by stereolithography, March 11, 1986. US4575330A.

[4] A. Ben-Ner, E. Siemsen, Decentralization and localization of production: the organizational and economic consequences of additive manufacturing (3D printing), California Management Review 59 (2) (February 2017) 5−23, https://doi.org/10.1177/0008125617695284.

[5] R. Handal, An implementation framework for additive manufacturing in supply chains, Journal of Operations and Supply Chain Management 10 (2) (December 2017), https://doi.org/10.12660/joscmv10n2p18-31.

[6] Seven 'rights' of logisticsSEVEN 'RIGHTS' OF LOGISTICS, in: P.M. Swamidass (Ed.), Encyclopedia of Production and Manufacturing Management, Springer US, Boston, MA, 2000, p. 684.

[7] R. Hague, S. Mansour, N. Saleh, Material and design considerations for rapid manufacturing, International Journal of Production Research 42 (22) (November 2004) 4691−4708, https://doi.org/10.1080/00207840410001733940.

[8] J. Schmelzle, E.V. Kline, C.J. Dickman, E.W. Reutzel, G. Jones, T.W. Simpson, (Re) Designing for Part Consolidation: understanding the challenges of metal additive manufacturing, Journal of Mechanical Design 137 (111404) (October 2015), https://doi.org/10.1115/1.4031156.

[9] D. Herzog, V. Seyda, E. Wycisk, C. Emmelmann, Additive manufacturing of metals, Acta Materialia 117 (September 2016) 371−392, https://doi.org/10.1016/j.actamat.2016.07.019.

[10] D. Bourell, et al., Materials for additive manufacturing, CIRP Annals 66 (2) (January 2017) 659−681, https://doi.org/10.1016/j.cirp.2017.05.009.

[11] J.R. Tumbleston, et al., Continuous liquid interface production of 3D objects, Science 347 (6228) (March 2015) 1349−1352, https://doi.org/10.1126/science.aaa2397.

[12] N. Guo, M. Leu, Additive manufacturing: technology, applications and research needs, Frontiers of Mechanical Engineering 8 (3) (September 2013) 215–243, https://doi.org/10.1007/s11465-013-0248-8.
[13] A Review of Metal Additive Manufacturing Technologies | Scientific.Net." https://www.scientific.net/SSP.278.1 (accessed January. 10, 2021).
[14] Additive Manufacturing Technologies | SpringerLink." https://link.springer.com/book/10.1007%2F978-1-4419-1120-9 (accessed January. 10, 2021).
[15] A.R. Patibandla, Effect of Process Parameters on Surface Roughness and Porosity of Direct Metal Laser Sintered Metals, University of Cincinnati, 2018.
[16] 14:00-17:00, "ISO 4287:1997," *ISO*. https://www.iso.org/cms/render/live/en/sites/isoorg/contents/data/standard/01/01/10132.html (accessed January. 10, 2021).
[17] R. Udroiu, I.C. Braga, A. Nedelcu, Evaluating the quality surface performance of additive manufacturing systems: methodology and a material jetting case study, Materials (Basel) 12 (6) (March 2019), https://doi.org/10.3390/ma12060995.
[18] A.M. Ralls, P. Kumar, P.L. Menezes, Tribological properties of additive manufactured materials for energy applications: a review, Processes 9 (1) (January 2021), https://doi.org/10.3390/pr9010031.
[19] P.L. Menezes, Kishore, S.V. Kailas, M.R. Lovell, Influence of surface texture and roughness of softer and harder counter materials on friction during sliding, Journal of Materials Engineering and Performance 24 (1) (January 2015) 393–403, https://doi.org/10.1007/s11665-014-1304-1.
[20] J. Zhao, Z.Q. Liu, Analysis and optimization of surface roughness in rotary ultrasonic burnishing of titanium alloy Ti-6Al-4V, Advanced Materials Research (2016) (accessed January. 10, 2021), https://www.scientific.net/AMR.1136.406.
[21] P. Walker, S. Malz, E. Trudel, S. Nosir, M.S.A. ElSayed, L. Kok, Effects of ultrasonic impact treatment on the stress-controlled fatigue performance of additively manufactured DMLS Ti-6Al-4V alloy, Applied Sciences 9 (22) (January 2019), https://doi.org/10.3390/app9224787. Art. no. 22.
[22] Z. Haidong, Z. Ping, M. Wenbin, Z. Zhongming, A study on ultrasonic elliptical vibration cutting of Inconel 718, Shock and Vibration (August 28, 2016). https://www.hindawi.com/journals/sv/2016/3638574/. (Accessed 10 January 2021).
[23] M. Zhang, D. Zhang, D. Geng, J. Liu, Z. Shao, X. Jiang, Surface and sub-surface analysis of rotary ultrasonic elliptical end milling of Ti-6Al-4V, Materials and Design 191 (June 2020) 108658, https://doi.org/10.1016/j.matdes.2020.108658.
[24] Z. Wang, Z. Liu, C. Gao, K. Wong, S. Ye, Z. Xiao, Modified wear behavior of selective laser melted Ti6Al4V alloy by direct current assisted ultrasonic surface rolling process, Surface and Coatings Technology 381 (January 2020) 125122, https://doi.org/10.1016/j.surfcoat.2019.125122.
[25] J. Romano, L. Ladani, M. Sadowski, Laser additive melting and solidification of inconel 718: finite element simulation and experiment, JOM 68 (3) (March 2016) 967–977, https://doi.org/10.1007/s11837-015-1765-1.
[26] S.M. Yusuf, N. Gao, Influence of energy density on metallurgy and properties in metal additive manufacturing, Materials Science and Technology 33 (11) (July 2017) 1269–1289, https://doi.org/10.1080/02670836.2017.1289444.
[27] N. Hansen, Hall–Petch relation and boundary strengthening, Scripta Materialia 51 (8) (October 2004) 801–806, https://doi.org/10.1016/j.scriptamat.2004.06.002.
[28] Z. Wang, K. Guan, M. Gao, X. Li, X. Chen, X. Zeng, The microstructure and mechanical properties of deposited-IN718 by selective laser melting, Journal of Alloys and Compounds 513 (February 2012) 518–523, https://doi.org/10.1016/j.jallcom.2011.10.107.

[29] I. Yadroitsev, I. Yadroitsava, I. Smurov, Strategy of fabrication of complex shape parts based on the stability of single laser melted track, Laser-based Micro- and Nanopackaging and Assembly V 7921 (February 2011) 79210C, https://doi.org/10.1117/12.875402.

[30] D. Wang, Y. Liu, Y. Yang, D. Xiao, Theoretical and experimental study on surface roughness of 316L stainless steel metal parts obtained through selective laser melting, Rapid Prototyping Journal 22 (4) (January 2016) 706–716, https://doi.org/10.1108/RPJ-06-2015-0078.

[31] M. Suraratchai, J. Limido, C. Mabru, R. Chieragatti, Modelling the influence of machined surface roughness on the fatigue life of aluminium alloy, International Journal of Fatigue 30 (12) (December 2008) 2119–2126, https://doi.org/10.1016/j.ijfatigue.2008.06.003.

[32] W. Yuqin, T. Jinyuan, Z. Wei, Influence of distribution parameters of rough surface asperities on the contact fatigue life of gears, Proceedings of the Institution of Mechanical Engineers, Part J: Journal of Engineering Tribology 234 (6) (June 2020) 821–832, https://doi.org/10.1177/1350650119866037.

[33] R. Klingvall Ek, L.-E. Rännar, M. Bäckstöm, P. Carlsson, The effect of EBM process parameters upon surface roughness, Rapid Prototyping Journal 22 (3) (January 2016) 495–503, https://doi.org/10.1108/RPJ-10-2013-0102.

[34] S. Yin, et al., Cold spray additive manufacturing and repair: fundamentals and applications, Additive Manufacturing 21 (May 2018) 628–650, https://doi.org/10.1016/j.addma.2018.04.017.

[35] Z. Monette, A.K. Kasar, M. Daroonparvar, P.L. Menezes, Supersonic particle deposition as an additive technology: methods, challenges, and applications, The International Journal of Advanced Manufacturing Technology 106 (5) (January 2020) 2079–2099, https://doi.org/10.1007/s00170-019-04682-2.

[36] Y. Xie, S. Yin, J. Cizek, J. Cupera, E. Guo, R. Lupoi, Formation mechanism and microstructure characterization of nickel-aluminum intertwining interface in cold spray, Surface and Coatings Technology 337 (March 2018) 447–452, https://doi.org/10.1016/j.surfcoat.2018.01.049.

[37] N. Bala, H. Singh, Fundamentals of corrosion mechanisms in cold spray coatings, in: P. Cavaliere (Ed.), Cold-Spray Coatings: Recent Trends and Future Perspectives, Springer International Publishing, Cham, 2018, pp. 351–371.

[38] M.A. Balbaa, M.A. Elbestawi, J. McIsaac, An experimental investigation of surface integrity in selective laser melting of Inconel 625, The International Journal of Advanced Manufacturing Technology 104 (9) (October 2019) 3511–3529, https://doi.org/10.1007/s00170-019-03949-y.

[39] M. Galati, P. Minetola, G. Rizza, Surface roughness characterisation and analysis of the electron beam melting (EBM) process, Materials (Basel) 12 (13) (July 2019), https://doi.org/10.3390/ma12132211.

[40] A. Sova, S. Grigoriev, A. Okunkova, I. Smurov, Potential of cold gas dynamic spray as additive manufacturing technology, The International Journal of Advanced Manufacturing Technology 69 (9) (December 2013) 2269–2278, https://doi.org/10.1007/s00170-013-5166-8.

[41] P. Richer, B. Jodoin, L. Ajdelsztajn, Substrate roughness and thickness effects on cold spray nanocrystalline Al–Mg coatings, Journal of Thermal Spray Technology 15 (2) (June 2006) 246–254, https://doi.org/10.1361/105996306X108174.

[42] T. Goyal, R.S. Walia, T.S. Sidhu, Surface roughness optimization of cold-sprayed coatings using Taguchi method, The International Journal of Advanced Manufacturing Technology 60 (5) (May 2012) 611–623, https://doi.org/10.1007/s00170-011-3642-6.

[43] R. Singh, U. Batra, Effect of cold spraying parameters and their interaction an hydroxyapatite deposition, 2004, https://doi.org/10.36884/jafm.6.04.21277.

[44] M.R. Rokni, S.R. Nutt, C.A. Widener, V.K. Champagne, R.H. Hrabe, Review of relationship between particle deformation, coating microstructure, and properties in high-pressure cold spray, Journal of Thermal Spray Technology 26 (6) (August 2017) 1308−1355, https://doi.org/10.1007/s11666-017-0575-0.

[45] V.S. Bhattiprolu, K.W. Johnson, O.C. Ozdemir, G.A. Crawford, Influence of feedstock powder and cold spray processing parameters on microstructure and mechanical properties of Ti-6Al-4V cold spray depositions, Surface and Coatings Technology 335 (February 2018) 1−12, https://doi.org/10.1016/j.surfcoat.2017.12.014.

[46] N.B. Maledi, O.P. Oladijo, I. Botef, T.P. Ntsoane, A. Madiseng, L. Moloisane, Influence of cold spray parameters on the microstructures and residual stress of Zn coatings sprayed on mild steel, Surface and Coatings Technology 318 (May 2017) 106−113, https://doi.org/10.1016/j.surfcoat.2017.03.062.

[47] R.J. Friel, R.A. Harris, Ultrasonic additive manufacturing − a hybrid production process for novel functional products, Procedia CIRP 6 (January 2013) 35−40, https://doi.org/10.1016/j.procir.2013.03.004.

[48] Influence of Ultrasonic Burnishing Technique on Surface Quality and Change in the Dimensions of Metal Shafts." https://www.hindawi.com/journals/je/2014/124247/(accessed Jan. 23, 2021).

[49] R. Teimouri, S. Amini, Analytical modeling of ultrasonic surface burnishing process: evaluation of through depth localized strain, International Journal of Mechanical Sciences 151 (February 2019) 118−132, https://doi.org/10.1016/j.ijmecsci.2018.11.008.

[50] R. Teimouri, S. Amini, A comprehensive optimization of ultrasonic burnishing process regarding energy efficiency and workpiece quality, Surface and Coatings Technology 375 (October 2019) 229−242, https://doi.org/10.1016/j.surfcoat.2019.07.038.

[51] J. Zhao, Z. Liu, Plastic flow behavior for machined surface material Ti-6Al-4V with rotary ultrasonic burnishing, Journal of Materials Research and Technology 9 (2) (March 2020) 2387−2401, https://doi.org/10.1016/j.jmrt.2019.12.071.

[52] M. Salmi, J. Huuki, I.F. Ituarte, The ultrasonic burnishing of cobalt-chrome and stainless steel surface made by additive manufacturing, Progress in Additive Manufacturing 2 (1) (June 2017) 31−41, https://doi.org/10.1007/s40964-017-0017-z.

[53] A.I. Dekhtyar, B.N. Mordyuk, D.G. Savvakin, V.I. Bondarchuk, I.V. Moiseeva, N.I. Khripta, Enhanced fatigue behavior of powder metallurgy Ti−6Al−4V alloy by applying ultrasonic impact treatment, Materials Science and Engineering: A 641 (August 2015) 348−359, https://doi.org/10.1016/j.msea.2015.06.072.

[54] B.N. Mordyuk, G.I. Prokopenko, Y.V. Milman, M.O. Iefimov, A.V. Sameljuk, Enhanced fatigue durability of Al−6Mg alloy by applying ultrasonic impact peening: effects of surface hardening and reinforcement with AlCuFe quasicrystalline particles, Materials Science and Engineering: A 563 (February 2013) 138−146, https://doi.org/10.1016/j.msea.2012.11.061.

[55] B.N. Mordyuk, G.I. Prokopenko, Fatigue life improvement of α-titanium by novel ultrasonically assisted technique, Materials Science and Engineering: A 437 (2) (November 2006) 396−405, https://doi.org/10.1016/j.msea.2006.07.119.

[56] D.A. Lesyk, S. Martinez, B.N. Mordyuk, V.V. Dzhemelinskyi, A. Lamikiz, G.I. Prokopenko, Effects of laser heat treatment combined with ultrasonic impact treatment on the surface topography and hardness of carbon steel AISI 1045, Optics and Laser Technology 111 (April 2019) 424−438, https://doi.org/10.1016/j.optlastec.2018.09.030.

[57] Y. Liu, D. Wang, C. Deng, L. Huo, L. Wang, R. Fang, Novel method to fabricate Ti−Al intermetallic compound coatings on Ti−6Al−4V alloy by combined ultrasonic impact treatment and electrospark deposition, Journal of Alloys and Compounds 628 (April 2015) 208−212, https://doi.org/10.1016/j.jallcom.2014.12.144.
[58] Z. Haidong, L. Shuguang, Z. Ping, K. Di, Process modeling study of the ultrasonic elliptical vibration cutting of Inconel 718, The International Journal of Advanced Manufacturing Technology 92 (5) (September 2017) 2055−2068, https://doi.org/10.1007/s00170-017-0266-5.
[59] G.A. Volkov, V.A. Bratov, A.A. Gruzdkov, V.I. Babitsky, Y.V. Petrov, V.V. Silberschmidt, Energy-based analysis of ultrasonically assisted turning, Shock and Vibration (2011) (accessed January 23, 2021), https://www.hindawi.com/journals/sv/2011/598106/.
[60] C. Nath, M. Rahman, K.S. Neo, A study on ultrasonic elliptical vibration cutting of tungsten carbide, Journal of Materials Processing Technology 2 09 (9) (May 2009) 4459−4464, https://doi.org/10.1016/j.jmatprotec.2008.10.047.
[61] Y. Liu, X. Zhao, D. Wang, Determination of the plastic properties of materials treated by ultrasonic surface rolling process through instrumented indentation, Materials Science and Engineering: A 600 (April 2014) 21−31, https://doi.org/10.1016/j.msea.2014.01.096.
[62] C. Liu, D. Liu, X. Zhang, S. Yu, W. Zhao, Effect of the ultrasonic surface rolling process on the fretting fatigue behavior of Ti-6Al-4V alloy, Materials (Basel) 10 (7) (July 2017), https://doi.org/10.3390/ma10070833.
[63] Z. Sun, Y. Ye, J. Xu, T. Hu, S. Ren, B. Li, Effect of electropulsing on surface mechanical behavior and microstructural evolution of Inconel 718 during ultrasonic surface rolling process, Journal of Materials Engineering and Performance 28 (11) (November 2019) 6789−6799, https://doi.org/10.1007/s11665-019-04443-y.
[64] N. Ao, D. Liu, X. Xu, X. Zhang, D. Liu, Gradient nanostructure evolution and phase transformation of α phase in Ti-6Al-4V alloy induced by ultrasonic surface rolling process, Materials Science and Engineering: A 742 (January 2019) 820−834, https://doi.org/10.1016/j.msea.2018.10.098.
[65] N. Ao, D. Liu, X. Zhang, C. Liu, J. Yang, D. Liu, Surface nanocrystallization of body-centered cubic beta phase in Ti−6Al−4V alloy subjected to ultrasonic surface rolling process, Surface and Coatings Technology 361 (March 2019) 35−41, https://doi.org/10.1016/j.surfcoat.2019.01.045.
[66] C. Liu, et al., Fretting fatigue characteristics of Ti-6Al-4V alloy with a gradient nano-structured surface layer induced by ultrasonic surface rolling process, International Journal of Fatigue 125 (August 2019) 249−260, https://doi.org/10.1016/j.ijfatigue.2019.03.042.
[67] S. Marimuthu, A. Triantaphyllou, M. Antar, D. Wimpenny, H. Morton, M. Beard, Laser polishing of selective laser melted components, International Journal of Machine Tools and Manufacture 95 (August 2015) 97−104, https://doi.org/10.1016/j.ijmachtools.2015.05.002.
[68] W. Dai, J. Li, W. Zhang, Z. Zheng, Evaluation of fluences and surface characteristics in laser polishing SKD 11 tool steel, Journal of Materials Processing Technology 273 (November 2019) 116241, https://doi.org/10.1016/j.jmatprotec.2019.05.022.
[69] W. Guo, M. Hua, P.W.-T. Tse, A.C.K. Mok, Process parameters selection for laser polishing DF2 (AISI O1) by Nd:YAG pulsed laser using orthogonal design, The International Journal of Advanced Manufacturing Technology 59 (9) (April 2012) 1009−1023, https://doi.org/10.1007/s00170-011-3558-1.

[70] Y.-D. Chen, W.-J. Tsai, S.-H. Liu, J.-B. Horng, Picosecond laser pulse polishing of ASP23 steel, Optics and Laser Technology 107 (November 2018) 180–185, https://doi.org/10.1016/j.optlastec.2018.05.025.

[71] E. Ukar, A. Lamikiz, F. Liébana, S. Martínez, I. Tabernero, An industrial approach of laser polishing with different laser sources, Materialwissenschaft und Werkstofftechnik 46 (7) (2015) 661–667, https://doi.org/10.1002/mawe.201500324.

Tribology of additively manufactured materials: fundamentals, modeling, and applications

8

Chandramohan Palanisamy [1,2] *and Raghu Raman* [1]
[1]Department of Mechanical Engineering, Sri Ramakrishna Engineering College, Coimbatore, India; [2]Department of Metallurgy, University of Johannesburg, Johannesburg, South Africa

8.1 Introduction

In the current years, additive manufacturing (AM) technique has emerged as the potential method for the manufacturing of metallic parts. AM process produces complex-shaped metal parts without wastage of material [1]. AM process involves production of metal parts in the form of layer by layer from a computer-aided design (CAD) file. Metal AM processes have mainly focused on three alloy types, especially, titanium, nickel-based superalloys, and stainless steels owing to their wide usage in aerospace, automotive, and medical sectors. Specifically, Ti6Al4V, Inconel 625, Inconel 718, 316L stainless steel, and 17-4 PH have been extensively used for numerous applications. Selective laser melting (SLM), direct metal laser sintering (DMLS), direct metal deposition (DED), electron beam melting (EBM), laser metal deposition (LMD), laser-engineered net shaping (LENS), and laser powder bed fusion (LPBF) are the processes approached successfully for fabricating such metallic alloys [2]. In numerous applications of such AM metallic alloys, tribology is important to guarantee the effectiveness and safety [3]. Wear is one of the most important root causes for failure of mechanical systems [4]. For this condition, it is highly essential to investigate tribological characteristics of metallic alloys processed by AM. The aim of this chapter is to study the tribological behavior of the Ti6Al4V, Inconel 625, Inconel 718, 316L stainless steel, and 17-4 PH metallic alloys processed by different AM processes. An attempt has been made through this chapter to review the effect of different factors such as microstructural features (dense and cellular), posttreatment (ultrasonic surface rolling, heat treatment (HT), machining, cryogenic treatment), composite development (carbides, nitrides, borides), surface modification (cladding, nitriding, mechanical attrition), lubrication condition (dry and wet) during wear test, and different counterface (soft and hard) on the wear (rotating, reciprocating, and scratch type) behavior of AM metallic alloys. In addition, the tribological behavior of the AM metallic alloys is compared with the performance of the conventionally manufactured (forged, cast, hot pressed, wrought) alloys.

Tribology of Additively Manufactured Materials. https://doi.org/10.1016/B978-0-12-821328-5.00008-1

8.2 Comparison of wear properties in cellular and dense Ti6Al4V structures

Additive manufacturing methods facilitate the manufacturing of dense and innovative metallic cellular Ti6Al4V structures with complex geometries that are unable to produce through the traditional manufacturing processes. These dense and innovative Ti6Al4V cellular structures have been produced through the AM process targeting biomedical applications. In addition, Ti6Al4V cellular structures can be made with several advantages such as pores size control, orientation, and homogeneous distribution of pores. Among several AM techniques, SLM is found to be a remarkable method for the development of Ti6Al4V dense and cellular structures. Wear is the main problem in such structures and in specific hip prostheses as these implants get in contact with the natural bone. As a result, the load is created at the interface of implant-natural bone due to micro-movements in the order of 100–200 μm [5]. Release of wear debris occurs due to these movements, resulting in pain, inflammation, and loosening of the implant. Hence, wear characteristics of Ti6Al4V cellular structures have been investigated by several researchers for their implementation in biomedical applications such as customized hip prostheses and implants [6,7].

Bartolomeu et al. [8] prepared Ti6Al4V cellular structures by SLM for biomedical applications with different cell sizes ranging from 100 to 500 μm. Reciprocating wear tests were made against hard counterface alumina with 3 N normal loads, 1 Hz frequency, 3 mm stroke length, and 60 min total test duration in phosphate-buffered saline (PBS) fluid at 37°C. The SLM machine parameters used are laser power 90 W, the scan speed of 600 mm/s, scan spacing of 80 μm, and 30 μm of layer thickness. As the open cell size increases, wall thickness reduces, leading to a reduction in the area of contact, increase in mean contact stress, and higher material loss. The increased open cell size also leads to higher range of coefficient of friction (COF), which is the measure of amount of friction existing between two surfaces. It is observed that COF increases with the drop-in wear resistance property of the implant. The mechanism of wear is plastic deformation as a result of abrasive wear.

However, to facilitate the optimum proliferation of preosteoblastic cells for scaffolds to improve the biological interaction, Bartolomeu et al. [9] produced open-cell (200 and 400 μm) structured Ti6Al4V biomedical implant specimens using SLM process (same parameters as given above) and compared it with dense (Cast/Forged) SLM-made specimens. Static COF and kinetic COF were assessed using the reciprocating sliding (0.5 mm/s) wear test by applying 50 N normal load in PBS fluid. Considerable deviation does not exist in static COF, whereas few variations exist in the kinetic COF of the specimens produced. It is lowest in the SLM-made dense implant (0.326) compared to SLM-made cellular implants (0.431, 0.459). Dense implants exhibit more wear debris than the cellular models due to their release during sliding beside the rough metallic counterparts. In total, 400 μm SLM-made cellular implant is considered to be relatively better in terms of wear resistance.

8.2.1 Wear studies of Ti6Al4V hybrid cellular structures

To enhance the tribocorrosion performance of the Ti6Al4V cellular structures, incorporation of biocompatible polymers in the cellular structures would be a preferable choice. Among various biocompatible polymers, poly-ether-ether-ketone (PEEK) is a suitable material since it has a high chemical stability, corrosion, and wear resistance [10]. PEEK is being reinforced and used as the bearing materials in finger joints, tibial inserts, and acetabular cups [11]. Specifically for hip implants made of Ti6Al4V cellular structures, combining it with PEEK aids in the betterment of tribological properties [12]. Hence, the same research group produced Ti6Al4V−PEEK hybrid cellular structures (open-cells—400 μm and walls thickness—300 μm) using the same SLM/hot pressing (HP) processes. By applying 6 N load with all other conditions similar, the specific wear rate was tested using the reciprocating sliding wear test and compared with the previously produced dense (forged), dense (SLM), open-cell structured Ti6Al4V. Ti6Al4V−PEEK multimaterial cellular structures exhibited the lowest kinetic COF (0.362) among other structures (0.389, 0.395, and 0.417). Its wear resistance has improved in the order of increased specific wear rate by 450% compared to conventional implants due to the presence of PEEK material. Its corrosion resistance is also better than the other dense or cellular specimens [13].

In continuation by the same research group, Ti6Al4V−PEEK multimaterial structures were fabricated for hip implants under all similar conditions of SLM and sliding wear tests, except for a load of 50 N. These wear resistance behavior of structures were compared with cast/wrought Ti6Al4V. It is found that the Ti6Al4V−PEEK structures had 40% lower mass loss (0.75×10^{-3} g) and slightly lower COF (0.359) than cast/forged structures (1.2×10^{-2} g and 0.393), respectively, due to their cellular structures. The better wear resistance of cellular structures is due to the self-lubricating ability of PEEK, which is also evident in fine-scale polishing wear mechanism in worn-out surfaces of Ti6Al4V−PEEK multimaterial structures [14]. Buciumeanu et al. [15] prepared Ti6Al4V−PEEK hybrid cellular structures using SLM for biomedical applications with different open-cell sizes (Fig. 8.1—350, 400, 450, 500 μm) of 6 mm diameter and 4 mm height. PEEK was impregnated as filler material in the cellular structure of the implant by HP and subjected to reciprocating sliding wear test against hard counterface alumina. PEEK and titanium do not bond chemically, rather as micromechanical interlocking due to their rough surface. Maximum cellular size of 500 μm responds better wear resistance due to the presence of the higher amount of PEEK. The wear mechanism is abrasive, and more PEEK impregnation averts the release of the titanium metal ion.

Another set of competitive biomaterials to PEEK that can be reinforced with Ti6Al4V is hydroxyapatite or β-tricalcium phosphate. These bioactive materials possess greater biocompatibility, osteoconductivity, and cell adhesion, which results in enhanced bone tissue formation and greater bonding between the bone and the implant [16].

Costa et al. [17] constructed a multimaterial Ti6Al4V cellular structure produced by the SLM process impregnated with bioactive materials using press and sintering techniques. Wear sliding tests were conducted with reciprocating alumina contact ball,

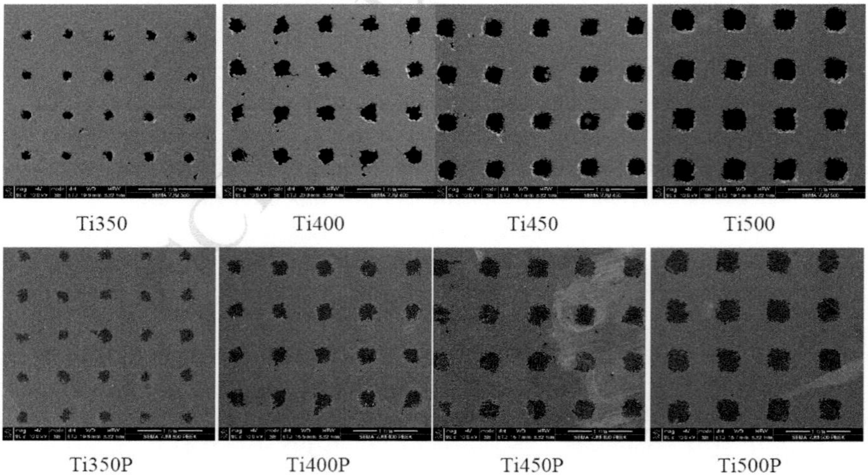

Figure 8.1 SEM images of the Ti6Al4V and Ti6Al4V–PEEK hybrid cellular structures [15].

50 N load, and 1 HZ frequency in PBS lubrication fluid. It is concluded that the hydroxyapatite and β-tricalcium phosphate structures had a decreased weight loss of 20.5% and 27.4% compared to the unreinforced structure, which translates to higher wear resistance resulting due to introduction of bioactive materials. Hence, it is apparent from various research findings that in terms of wear resistance, small cellular structures (100 μm) are favored. However, if larger-sized cellular structures are to be selected (400 μm and above) for osseointegration requirement, Ti6Al4V hybrid cellular structures made using SLM can be a preferred choice.

8.3 Effect of posttreatments on wear of additive-manufactured Ti6Al4V parts

Wear resistance properties of AM build Ti6Al4V parts should be enhanced for their extensive usage in the wear-related field as the Ti6Al4V parts exhibit low shear strength and low work hardening coefficient [18]. Hence, various researchers paid attention in imparting wear resistance to titanium alloys through different posttreatment techniques.

8.3.1 Effect of ultrasonic surface rolling on Ti6Al4V parts

Ultrasonic surface rolling process (USRP) is one of the posttreatment methods that induce intensive strain and refine the grain significantly up to several microns in the surface. This process produces ultrasonic frequency as well as deep rolling, which lead to severe plastic deformation and impart wear resistance, corrosion resistance, and fatigue strength [19].

USRP was carried out in SLM-made Ti–6Al–4V alloy. Reciprocating wear tests were made in polished and rolled samples against silicon nitride ball counterface with 5 N normal load, 8 Hz frequency, and for different sliding times (3, 6, 9, 15 min). The outcome indicates that COF dropped from 0.74 (polished surface) to 0.64 (USRP-treated surface) and wear volume from 0.205 to 0.195 mm^3. The improved wear resistance is due to increased hardness since severe work hardening has taken place in the compressed layer, resulting in refined grains, deformation twins, and dislocation walls, as shown in Fig. 8.2 [20].

Even though the USP process enhances the wear resistance of SLM-made Ti6Al4V parts through work-hardening, further improvement in wear resistance is achieved through the modified USRP method with electropulsing [21].

Three different posttreatments, namely, HT, USRP, and direct current–assisted USRP (DC-USRP), were studied on SLM-made Ti6Al4V alloy samples for its wear performance with 25 N load and hydraulic oil lubrication [22]. In HT, specimens were heated to 1000°C, holding it for 2 h, and cooled. The following DC-USRP processing parameters were employed: spindle speed—45 rpm, feeding rate—0.08 mm/rev, static force—1000 N, vibration amplitude—10 μm, vibration frequency—30 kHz, current—200 A, and target temperature—300°C. The lowest wear rate of 1×10^{-6} mm^3/Nm is reported for

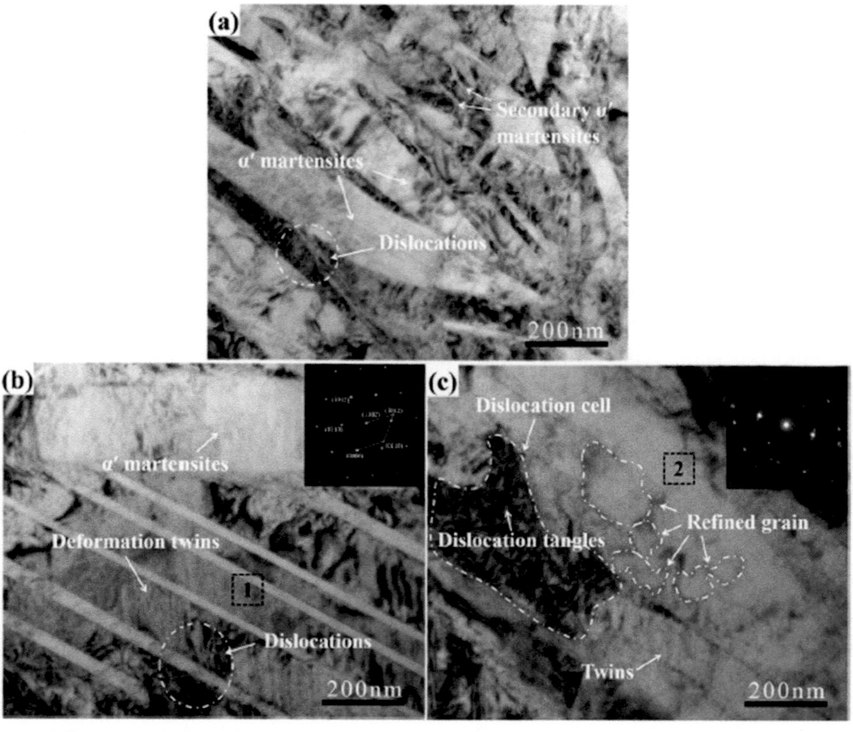

Figure 8.2 Transmission electron microscope (TEM) images of (A) top surface before USRP (B) and (C) after 1000 N 6 passes of USRP [20].

the DC-USRP-treated sample followed by USRP-treated samples (5×10^{-6} mm^3/Nm), heat-treated samples (2×10^{-5} mm^3/Nm), and as-built samples (4×10^{-5} mm^3/Nm). The COF also reduces in the same order. Severe abrasive wear and delamination is observed in as-built and heat-treated samples while mild abrasive wear is observed in the USRP- and DC-USRP-treated samples. This enhanced wear resistance in DC-USRP-, USRP-treated samples, and mild abrasive wear in as-sintered, heat-treated samples are due to more compressive residual stress in the former and more tensile residual stress in later. An altered lubrication mechanism has also contributed to higher wear resistance in DC-USRP- and USRP-treated samples [22].

8.3.2 Effect of heat treatment on Ti6Al4V parts

Apart from different posttreatments, the wear performance of the AM-made Ti6Al4V parts are greatly affected by the parameters employed during building and post HTs. This is due to rapid heating and cooling rates during the building process, which significantly influences microstructural characteristics, mechanical properties, and wear properties. Hence, several researchers paid greater attention to controlling these parameters [23].

Kaya et al. [24] investigated the effects of parameters during production and postproduction HTs on Ti6Al4V alloy produced by the SLM process. In the HT process, samples were heated to 840°C for 4 h followed by 2 h holding and then furnace cooling. In general, the COF values of the Ti6Al4V castings are lesser than the parts built in the SLM method. The highest densification rate of 99.5% is obtained at 60 μm hatch spacing and 45 degrees build orientation. The wear rate of SLM-made parts is 26% lower than parts made by the casting method. This is attributed to the $\alpha + \beta$, α' (fine needle—shaped acicular martensite) $+\beta$ and α/α'(coarse martensite grain) microstructures formed in the cast, SLM-made, and SLM heat-treated parts, respectively [24].

In a specific study [25], build directions of SLM parts and further modification of structures through different HT procedures considering different cooling rates were attempted to enhance the wear resistance of SLM-made Ti6Al4V parts. Chandramohan et al. [25] fabricated Ti6Al4V specimens through the SLM process in vertical (VB) and horizontal (HB) directions. Three types of HTs were made. HT1: heating to 1100°C and holding for 1 h followed by furnace cool. HT2: heating to 900°C and holding for 1 h followed by furnace cooling. The same sample is heated to 650°C and held for 3 h followed by air cooling. HT3: heating to 1100°C and holding for 2 h followed by water quenching. The same sample is heated to 900°C and held for 3 h followed by air cooling. Rotary dry wear under varying loads of 5, 15, and 25 N at 25 m/s has been carried out. Wear volume loss of both HB and VB is lower in HT two specimens than in HT 1, which can be attributed to better grain refinement. The low wear volume is due to the presence of compact oxide particles on the surface of the HT 2-VB specimen. The wear mechanism is of abrasive and adhesive nature [25]. In addition to HT of SLM-made Ti6Al4V parts, EBM process parts are also taken into consideration since it is used for making of large-scale titanium-based components. EBM process offers a vacuum work environment, high material deposition rate, and better mechanical properties [26]. The EBM-made Ti6Al4V alloy was subjected to three types of HTs and tested for its fretting wear resistance. HT1—stress-relieving at 600°C, HT2—subtransus HT at 950°C, and HT3—supertransus HT at 1050°C. The subtransus HT (HT2) yielded improved mechanical and antifretting properties due to homogeneous and coarser $\alpha + \beta$ lamellar structure along with fine secondary α-phases in β lath and nanoscale dispersoids precipitated in α lath. This microstructure also leads to a minimum wear rate of 1.52×10^{-15} mm^3/Nm with a lower COF of 0.63 compared to the other treatments [26]. It is identified that performing machining before the HT process on the EBM parts have a significant effect on the wear resistance properties [27]. Hence, Bruschi et al. [28]studied the influence of coupling machining (Lathe machine) with HT on the wear performance of EBM Ti6Al4V alloy. In HT, the material is heated to 980°C, followed by holding and cooling to RT at the rate of 20°C. Reciprocating sliding wear tests were conducted in the presence of saline solution (0.9% NaCl in distilled water) against the cobalt alloy CoCrMo counterpart. The stroke length of 500 μm, frequency of 10 Hz, and the normal load of 7 N were maintained in the wear test. The heat-treated cylinders exhibited less wear volume ($0.5 \pm 0.8 \times 10^{-3}$ mm^3) and lowered COF compared to the non-HT samples ($7.6 \pm 1 \times 10^{-3}$ mm^3). This is due to the higher microhardness imparted during the HT through the formation of a martensitic structure near the surface. The synergistic effect of coupling machining and HT was the reason for the overall enhancement of EBM Ti6Al4V wear resistance. Wear mechanisms such as adhesive and abrasive wear were observed predominantly in the worn-out surfaces of the samples [28].

8.4 Wear studies of additive-manufactured Ti6Al4V composite parts

Ti6Al4V metal matrix composites (MMCs) possess greater specific strength, wear resistance, and elevated temperature strength compared to the unreinforced alloy. AM techniques have the ability to develop MMCs with homogeneous dispersion of nanoparticles through the rapid melting and solidification manner [29]. Several research studies were focused on developing the Ti6Al4V composites and investigated on its wear performance.

Cai et al. [30] synthesized TiB (0−3 wt%)/Ti6Al4V (97−100 wt%) ceramic−matrix nanocomposites in the SLM process using TiB and Ti6Al4V powders to study the wear-resistant properties. It is reported that the TiB-free sample exhibited twofold wear loss as that of the 3 wt% TiB_2 sample, with the average COF being 0.44, 0.39, 0.36, and 0.35 for 0, 1, 2, and 3 wt% TiB, respectively. This is attributed to the grain refinement, complete conversion of TiB_2 particles into needle-like nanoscale TiB phase, and enhanced hardness. The microstructure of composites is poised with mutual parallel strip architecture containing enriched TiB in a few regions but very few in the remaining regions. The wear mechanism is primarily linked with oxidation and adhesion wear for TiB−Ti system composites.

Ti6Al4V/TiC composites were produced through the LMD process by delivering Ti6Al4V and TiC powders into the laser-melted Ti6Al4V substrate [31]. The effect of different scanning velocities (0.015 and 0.105 m/s) with an interval of 0.01 m/s on its wear resistance under dry conditions was studied. The composite was made at a laser power of 3.2 kW and a gas flow rate of 2 L/min. Wear tests were conducted on the developed specimens using at the load of 25 N. The wear volume of the samples decreased up to 0.05 mm^3 with increasing scanning velocity of up to 0.065 m/s. With a further increase in scanning velocity, the wear volume displays an increasing trend. Less amount of unmelted carbide particles were formed at low scanning velocity, which contributes to low wear resistance. As the scanning velocity increased above 0.065 m/s, large size unmelted carbide was formed in less quantity. These particles tend to produce much damage to the composite surface and consequences in low wear performance [31].

The Ti6Al4V/TiC composite was prepared by LMD with a laser power of 2.5 KW, scanning speed of 0.01 m/s, and gas flow rate of 2 L/min [32]. The lowest wear volume of 0.021 mm^3 was reported in built composite and the highest wear volume of 0.120 mm^3 in the Ti6Al4V substrate material. The formation of the hard intermetallic compound Ti_3Al has greatly enhanced the wear resistance of the composite. Adhesion, abrasion, and plastic deformation were the predominant type of wear in the substrate without reinforcement [32].

Few Ti6Al4V/TiC composites were produced through the LMD process with the powder flow rates of 2 and 4 g/min and laser powers of 1.5 and 3 kW [33]. Wear tests performed convey that wear volume gets improved with the increase in the powder flow rate. The lowest wear volume at a lower powder flow rate was due to proper melting of powders using sufficient laser power at this powder flow rate. In addition, the formation of in-situ Ti_3Al during processing was found hard and brittle, which aids in the increase of wear resistance. The increased powder flow rate led to incomplete melting of powders. On the contrary, the wear resistance got increased with an increase in laser power [33]. At low laser power, the size of unmelted carbides is found larger that induces wear by a three-body wear mechanism. As the laser power increases, the size of unmelted carbides shrinks with better Ti_3Al formation that aids in increased wear resistance. As a result, the laser power of 3 kW and powder flow rate of 2 g/min were found as optimum parameters to obtain lowest wear volume (0.000024 mm^3) [33].

Hence, it can be inferred that the optimum process parameters for producing TiC/Ti6Al4V composites through LMD is with a laser power of 3 KW, powder flow rate of 2 g/min, scanning speed of 0.05 m/s, and gas flow rate of 2 L/min.

8.5 Wear studies on surface-modified and coated additive-manufactured Ti6Al4V parts

Various surface modification methods were handled by several researchers for enhancing the wear properties of the AM-made Ti6Al4V alloy. Such surface-modified AM-made Ti alloys have been extensively utilized in various applications [34].

8.5.1 Wear studies on laser-cladded Ti6Al4V composite coatings

In recent years, laser cladding is being largely employed for enhancing the wear resistance of the metals as its economic one [35]. The laser cladding process offers dense coatings, higher cooling rate, defect-free layer, limited dilution ratio, narrow heat-affected zone, and better bonding of coating with the substrate [36]. However, the counterparts in contact with this laser cladding materials experience wear at elevated temperatures during operation. To overcome this wear of laser-cladded materials at elevated temperatures, the parts should possess self-lubrication at elevated temperatures. Single lubricants and composite lubricants have been employed for applications depending on the temperature range exposed [37].

The laser-cladded Ti6Al4V matrix with metal sulfides was subjected to the high-temperature wear test at various temperatures (20, 300, 600, 800°C) against a 5 mm-diameter ceramic material Si_3N_4 ball (high hardness—1700 HV) [38]. N1 (Ni60−16.8TiC-23.2WS_2) and N2 (Ni60−19.6TiC-20.4WS_2) coating effects were studied, and it was found that the bond of N1 with the matrix is superior to N_2 with the matrix. In N1 at 800°C, the oxide layer formation is continuous, leading to lower COF and wear rates. The predominant wear mechanisms of N1 coating are reported as oxidation wear and adhesive wear. In the N2 coating, partial decomposition of some lubrication films and larger adhesion force than binding force lead to higher COF and wear rates. The COF and the wear rate of the substrate decline with an increase in temperature due to the formation of a metal oxide layer that acts as self-lubricating and antiwear. The synergetic effect of metal oxide, metal sulfides, and hard phases improved these properties at 800°C. The predominant wear mechanisms of Ti6Al4V alloy are plastic deformation, adhesive wear, and abrasive wear, whereas at an elevated temperature of 800°C, it is oxidation wear, delamination of oxide film, and plastic deformation [38].

It is observed that TiC is one of the best reinforcements for improving hardness and low COF in Ti alloys. Reinforcing it ex-situ results in brittle features and defect at the matrix/ceramic interface. Hence, attempts have been made in synthesizing TiC reinforcement in-situ by adding ceramic powders to the Ti alloy [39]. Ti6Al4V alloy surface was laser cladded with Ti and SiC particles (90%Ti+10%SiC and 80%-Ti+20%SiC), and a wear test was performed with 2 N load and 50 Hz frequency against 10 mm stainless steel counterface. The residual Ti-rich phase volume has been identified to drop the wear resistance. For 10% SiC reinforcement, the residual Ti-rich phase volume was measured as 87.12% and 32.49% for 20% SiC reinforcement. This leads to the 0.1 COF and 15 μm wear depth in 10% SiC reinforcement, 0.4 COF and 7 μm wear depth in 20% SiC reinforcement, and 0.5 COF and 0.4 μm wear depth in the substrate. The difference in the track depth is due to the increased hardness from 339.1 to 932.2 HV, which is offered by the hard, brittle intermetallic compounds such as TiC and Ti_5Si_3 (Fig. 8.3) formed by incorporation of Ti and SiC particles [40].

Other than carbide-based reinforcements, several other reinforcements based on nitrides (TiN) and borides (TiB) have also been added ex-situ and also formed in-situ to assess the tribological behavior.

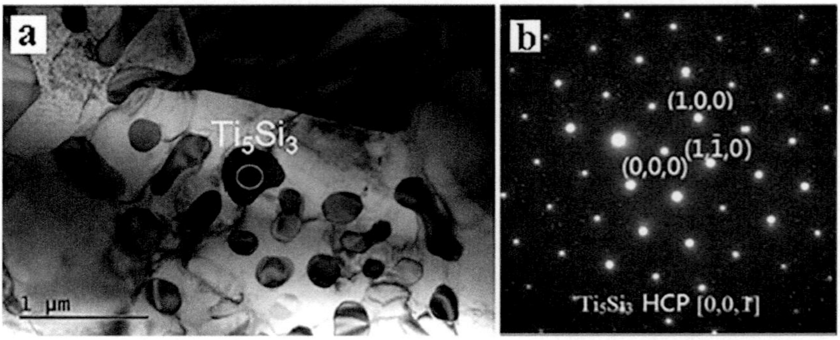

Figure 8.3 TEM bright field image showing (A) Ti_5Si_3 particles and (B) corresponding selected electron diffraction analysis [40].

The Ti6Al4V alloy surface was deposited with TiN for different concentration (10, 15, 20 and 40 wt%) through the LENS process and its tribological performance is investigated [41]. Reciprocating tribological tests were performed in the simulated body fluid (SBF) for the distance of 1000 m for load of 10 N against 3 mm-diameter Al_2O_3 ball. The alloy reinforced with TiN coatings has greater wear performance than that of the unreinforced alloy. Unreinforced alloy exhibits a wear rate of 1.03×10^{-3} mm^3/Nm, and it reduces with an increase in the fraction of TiN. The Ti6Al4V alloy reinforced with 40 wt% TiN exhibited lowest wear rate of 3.74×10^{-6} mm^3/Nm displaying shallow worn tracks on its surface. On the other hand, similarly processed CoCrMo alloy exhibited a wear rate of 1.04×10^{-5} mm^3/Nm, which has been utilized for the acetabular ball joint implants. In addition, SEM analysis of the TiN-reinforced alloy revealed that there is no full removal of reinforcement from the alloy surface after the tribological test [41]. Ti6Al4V powder and calcium phosphate were premixed and coated on Ti6Al4V samples for biomedical application through LENS process. It produces a metal ceramic composite coating of titanium nitrides and calcium titanate. In comparison to as-sintered sample, the composite wear rate reduced by 91% with a lower COF due to its increased surface hardness by 148% (868 ± 9 HV). In addition, Ti ion release has reduced from 12.45 to 3.17 ppm accounting to 70% drop between coated and uncoated [42].

In-situ TiB−TiN was formed in Ti6Al4V load-bearing implants by feeding Ti6Al4V powder with 5 and 15 wt% BN through LENS process [43]. Set parameters of laser power 300−400 W, scan speed 10−20 mm/s were used. A constant linear oscillatory motion Wear test was done with load of 5 N for a distance of 1000 m against chrome steel ball. Wear resistance of laser processed composite coatings over Ti substrate was found to enhance with increase in BN. The average wear rate of 5 wt% BN lie from 1.51×10^{-4} to 4.26×10^{-5} mm^3/Nm and for 15 wt% BN, lie from 6.20×10^{-6} to 1.90×10^{-6} mm^3/Nm, respectively. This is due to higher fraction of TiB and TiN phase formation upon increase in the BN concentration. The TiB whiskers reinforce composites significantly through a load-transfer mechanism, and the in situ−generated TiB nanorods prevent brittle TiN phase dendritic development [43].

Effects of trace boron (0.07 wt%) on dry sliding wear behavior of electron beam freeform (EBM)−fabricated Ti6Al4V alloy were investigated [44]. Reciprocating tribological tests were conducted in as formed Ti6Al4V and in Ti6Al4V−0.07B using 2 N load and frequency of 5 Hz against 4.5 mm-diameter GGr15 sphere. Addition of trace boron has minimum effect on COF but declines the wear rate. As formed Ti6Al4V shows lesser wear rate of 3.5×10^{-13} mm^3/Nm in longitudinal (Z) direction as against 5.2×10^{-13} mm^3/Nm in transverse (Y) direction. In contrary, Ti6Al4V−0.07B exhibits lowest wear rates in both longitudinal (2.8×10^{-13} mm^3/Nm) and transverse directions (2.0×10^{-13} mm^3/Nm). This is because boride considerably refines prior-βgrain, α-lath, and eliminates grain boundary in the Ti6Al4V alloy's microstructure. TiB whiskers are practically organized in parallel along the build direction and at the grain boundaries of the prior-β grains in Ti6Al4V−0.07B alloy. Wear mechanisms such as adhesion, oxidation, and abrasive wear were found in the worn out surface of the EBM samples [44].

Other than carbide-, nitride-, and boride-based surface modification of titanium alloys, another preferential choice of material for surface modification of titanium is the Co-containing alloys especially for sliding wear application [45]. This is owing to formation of the hard Ti_2Co precipitates upon adding cobalt to the titanium alloy. In addition, Co-containing alloys have inherent hardness and strength at higher temperatures. Adesina and Popoola [46] fabricated To-Co binary coatings on Ti6Al4V substrate by the laser surface cladding technique. Reciprocating wear test was performed against tungsten carbide counterface ball. The set parameters included laser outputs of 750 and 900 W, a scan speed of 1.2 m/min, a beam size of 3 mm, and an argon shield gas flow rate of 1.2 L/min. The low COF of 0.05 was reported in 900 W (high energy input) coated specimens with high hardness due to the efficient melting of cobalt particles, formation of hard intermetallic phases like $CoTi_2$, CoTi, and its even distribution. The improved wear resistance was attributed to the combined factors of premixed Ti, Co powders, and high energy input cladding [46].

8.6 Studies on lubrication and counterface materials during wear test

Ti6Al4V alloy has been largely used as implants for artificial knee joints, artificial hip joints, fracture fixation screws, pacemakers, bone plates, cardiac valve prostheses, and artificial hearts [47]. In addition, Ti6Al4V alloy is also utilized for dentistry applications such as crowns, over dentures, and bridges [48]. While the Ti implant is put in operation in these applications, they tend to get small-scale sliding contact with surfaces having different hardness in various physiological environments [49]. Exposure to such environment leads to progressive damage on the interface, which may shorten the lifetime of the implant. Progressive damage releases the debris that migrates and causes inflammation at the surrounding soft tissues. Hence, before putting in operation, performance of Ti6Al4V alloy in several environmental conditions has to be investigated. Therefore, several researchers studied its performance against different counterfaces and in different physiological environments.

8.6.1 Impact of lubrication and counterface materials in SLM-made Ti6Al4V parts

A conventionally processed (CP) Ti6Al4V sample, an SLM-made Ti6Al4V sample, and a heat-treated SLM Ti6Al4V sample were produced [50]. In the HT process, samples were heated between 720 and 740°C followed by 2 h holding and then furnace cooling. The samples were subjected to friction and wear tests with L-HM 46 hydraulic oil lubrication having kinematic viscosity of 48.2 mm^2/s at 40°C and viscosity index of 96.3. These samples were made to be in contact with a relatively soft H68 brass material (110 HV) and hard 38CrMoAl material (1001 HV).With the soft brass, it is found that CP sample develops a 2–3 μm thin tribo-oxide layer during the wear test without any plastic deformation leading to more wear, whereas it is not so with SLM samples. Altogether, mild oxidative wear is reported with brass. In case of wear with hard 38CrMoAl, a severe oxidative, abrasive, and delamination wear has been recorded for CP and SLM samples. Heat-treated SLM Ti6Al4V sample revealed a lowest wear rate without any plastic deformation due to the protective tribo oxide layer formation. The lubricant has effectively reduced the wear and friction [50].

8.6.2 Impact of lubrication and counterface materials in EBM-made Ti6Al4V parts

Tribological properties of Ti6Al4V alloy manufactured using the EBM process were studied and compared with conventionally manufactured Grade 5 Ti6Al4V samples. The wear rate of the samples was measured at a sliding speed (3 cm/s) and load (1 N) against 100Cr6 steel ball counterface with and without Hank's solution lubrication. In general, the friction and wear of both the samples decrease with lubrication. Average specific wear rates of the conventionally manufactured and EBM-produced Ti6Al4V samples are 93.9×10^{-14} and 78.4×10^{-14} mm^3/Nm, respectively, which conveys the relatively higher wear resistance of EBM-produced sample. This higher wear resistance is attributed to the heavily twinned and acicular martensitic microstructure resulting in higher surface hardness compared to commercially available Ti6Al4V alloys [51].

Ti6Al4V specimens were manufactured for implant application by the EBM process under different building directions, EBM-x (build direction perpendicular to longitudinal axis) and EBM-z (build direction parallel to longitudinal axis) [52]. Rotary wear tests were performed in three simulated synovial fluids that replace the natural articular cartilage in total joint replacement. The three fluids are PBS (NaCl 9 g/L + KH_2PO_4 0.144 g/L + $Na_2HPO_4{-}7H_2O$ 0.795 g/L), PBS + 19 g/L bovine serum albumin +3 g/L hyaluronic acid, and PBS + 48 g/L bovine serum albumin +3 g/L hyaluronic acid. To study the effect of body fluid (lower pH level up to pH 2.0 and lower protein concentration up to 60 g/L) lubrication on anisotropic wear behavior, the protein concentration was changed in the lubrication fluids taken for study [52]. The synovial fluids with higher protein decline surface energy and improve the viscosity and lubrication effect at the joint interface. PBS had highest surface energy, second fluid with intermediate energy, and third fluid with the least energy. The wear rate of the samples was measured on z-planes of both the samples at a rotary speed of 60 rpm, normal load of 30 mN, and 30-min test duration against the spherical tungsten carbide pin counterface. It was found that the wear rate of EBM_X is greater than EBM_Z in all three simulated synovial fluids. EBM_Z specimens exhibited greater wear resistance. This is due to its columnar grain structure as against equiaxed grains. Both the samples exhibited a Widmanstatten or basket-wave morphology microstructure, consisting alternate layers of acicular α phase divided by thin layers of retained β [52].

The same research group performed reciprocating wear tests on EBM-made samples and the mill-annealed Ti6Al4V samples. The tests were conducted in three environments, that is, dry environment, PBS with pH 7.4 (passivating environment), and PBS + 19 g/L bovine serum albumin +3 g/L hyaluronic acid (synovial fluid) against 1 mm diameter CP grade-2 titanium sphere. The measured dynamic COF values (lateral/normal force ratio) under dry, PBS, and synovial fluid conditions in mill-annealed sample are 0.75, 0.6, and 0.23 respectively. In EBM-z along X-direction, the COF values are 0.65, 0.53, and 0.26, respectively. In EBM-z along diagonal, the COF values are 0.69, 0.56, and 0.25, respectively. In EBM-x along X-direction, the COF values are 0.6, 0.55, and 0.24, respectively. In EBM-x along Y-direction, the COF values are 0.7, 0.57, and 0.255, respectively. The calculated wear volume values under dry, PBS, and synovial fluid conditions in mill-annealed sample are 843, 287, and 23 μm^3, respectively. In EBM-z along X-direction, it is 463 μm^3 for dry, 593 μm^3 for PBS, and 29 μm^3 for synovial fluid. In EBM-z along diagonal, it is 519 μm^3 for dry, 505 μm^3 for PBS, and 33 μm^3 for synovial fluid. In EBM-x along X-direction, it is 269 μm^3 for dry, 519 μm^3 for PBS, and 28 μm^3 for synovial fluid. In EBM-x along Y-direction, it is 685 μm^3 for dry, 611 μm^3 for PBS, and 26 μm^3 for synovial fluid. In the dry environment, the wear performance of EBM-made samples performed better than the mill-annealed samples. This is attributed to the accumulation of dislocation in mill-annealed samples. The uninterrupted breakage and restoration of oxide layer and consistent growth and breakdown of adhesive junction are typical wear mechanisms. In EBM-built specimens, PBS environment considerably decreased the effect of anisotropic wear. The synovial fluid containing high protein improves the wear resistance and electrochemical dissolution of all the samples due to its absolute liquid film lubrication and decreasing surface fatigue response. Furthermore, the formation of fine lamellar α-phase in large β grains of EBM specimens facilitated reduced wear volume [53].

8.7 Comparison of additive-manufactured and diverse processing routes of Ti6Al4V parts

Ti6Al4V parts can be produced through various manufacturing processes, that is, either by additive or traditional manufacturing processes [54]. Among AM processes, SLM process uses laser beam for producing parts, whereas the EBM process uses the high-energy electron beam as the source [55]. SLM has the processing ability to handle wide range of alloys, whereas the EBM can process limited range of alloys. Spot size and the particle size employed in these processes vary, which results in the variation of the melt pool shape and size [54]. As these two AM processes only possess minimal differences, however, it will result in significant variation in microstructural variations and properties of the parts [56]. In general, AM-produced Ti6Al4V parts show unique microstructure and properties compared to casting, forging, machining, and HP [57]. Microstructural and property variation, especially wear resistance, greatly relies on the processing steps involved [58]. Wear property of the Ti6Al4V is one of the significant properties for its extensive usage in numerous applications [59]. Hence, various research studies were focused on comparing the wear of AM Ti6Al4V parts with the traditional manufacturing process [59].

Ti6Al4V were produced through forging, SLM, EBM, and subjected to rotary wear test against high-carbon chromium-bearing steel GCr15 ring (hardness ~630 HV) counterface with a load of 50 N and relative humidity ranging from 50% to 60% [60]. The COF of forged, SLM, and EBM have been reported as 0.6, 0.4, and 0.5, which are also evidenced with more tribooxide layer formation in SLM—(10.43%) and EBM—(12.37%) made specimens when compared to the forged (5.73%) specimen. The hardness values measured are 399 HV in SLM, 383 HV in EBM, and 368 HV in forged specimen. Higher hardness in SLM-made specimens is due to the finer acicular martensite (α') in the prior columnar β microstructure [60]. In spite of moderate hardness in EBM-made specimen, its wear rate is the lowest ($16 \pm 4.2E-5$) in comparison with forged ($23.9 \pm 4.6E-5$) and SLM—($19 \pm 3.7E-5$) made specimens. This is due to the lesser delamination in EBM-made specimens supported by the presence of very few horizontal cracks and vertical cracks as against several horizontal cracks in SLM specimens, shown in Fig. 8.4 [60].

Different processing routes such as casting, HP, and SLM were employed to manufacture and study their effect on the tribological behavior of Ti6Al4V alloy implants. It is reported that the casting samples had 75% β phase, whereas HP samples had 50% β phase and SLM-made samples had 25% β phase, respectively. A reciprocating wear test was performed with PBS lubrication. The wear rate was found to be lowest in the SLM process–built sample compared to HP—and casting-made samples. It is commonly known that hardness of the sample is linearly proportional to the wear resistance, which is ensured by the maximum hardness value of 388 VHN obtained in the

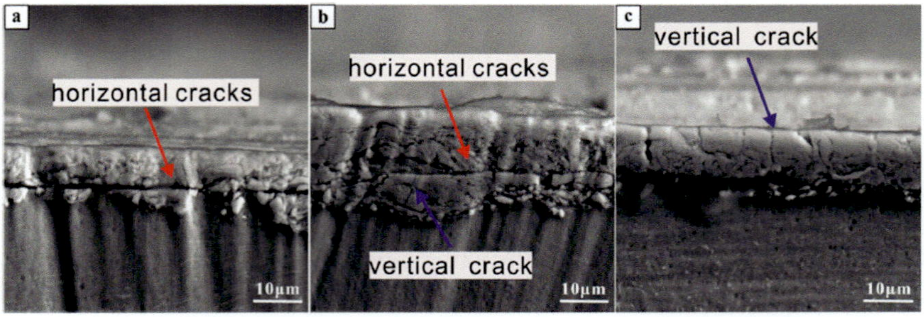

Figure 8.4 Cross-section comparison of Ti6Al4V alloy (A) forging (B) SLM and (C) EBM [60].

SLM-made sample as against 360 and 342 VHN in HP and casting processes, respectively. The lowest wear rate in SLM the sample is attributed to the formation of α and α' acicular martensite (supported by faster solidification), which imparts hardness to the Ti6Al4V sample. Plastic deformation as a result of adhesive wear and abrasive wear has been identified as the wear mechanism in the worn out surfaces [61].

Buciumeanu et al. [62] studied the tribocorrosion behavior of Ti6Al4V alloy manufactured by casting, HP, and LENS method for biomedical implants. The reciprocating wear test was carried out by applying a normal load of 1 N, frequency of 1 HZ, and stroke length of 3 mm against a 10 mm alumina ball counterface in PBS lubrication fluid. The wear rate of cast and hot-pressed alloys was more than 8E−04, whereas the LENS-made specimen yielded a value of about 6E−04, respectively. This is due to the lamellar microstructure of $\alpha + \beta$ in HP Ti−6Al−4V samples and columnar prior β grain microstructure with a Widmanstatten structure (α laths surrounded by retained β grain boundaries) between layers. However, no significant change was recorded for the change in COF for all samples [62].

Machining (turning) operation was performed on wrought and EBM-made Ti6Al4V biomedical specimens in dry and cryogenic cooling conditions [63]. The wear test was carried out by applying a normal load of 7 N, frequency of 10 HZ, and maximum contact pressure of 90 MPa against CoCrMo counterface plates parallel to their axis in saline environment. It is found that cryogenic cooling induces lower COF (0.36 and 0.37) in EBM Ti6Al4V than COF (0.41 and 0.4) of wrought product at a lower feed rate (0.1 mm/rev) and lower cutting speed (50 m/min), respectively. Similar trend is reported in case of higher feed rate (02 m/rev) with lower cutting speed (50 m/min), whereas it is otherwise for lower feed rate (0.2 m/rev) and higher cutting speed (80 m/min). It is due to the two main factors that determine the friction, that is, microhardness and roughness of rubbing surfaces. Both the alloy surfaces get harder due to cryogenic cooling and rougher due to increased feed rate. Under cryogenic cooling, adhesive wear is recognized as the dominant wear mechanism in both EBM and wrought Ti6Al4V, but abrasive wear with greater wear debris is identified in dry machining [63].

Machining (turning) operation was performed on wrought and EBM-made Ti6Al4V biomedical specimens in dry, wet, and cryogenic cooling conditions [64]. The wear test was carried out by applying a normal load of 45 N, frequency of 1 HZ, and maximum contact pressure of 90 MPa against ZrO_2 flat-plate in artificial saliva environment. The generated wear volume is inversely proportional to the hardness of the softest contacting surface, according to the Archard's wear rate law. In wrought and EBM-made samples under dry condition, the wear volume almost remains the same (75×10^{-3} mm^3). But, the average wear volume in wrought samples is lesser in case of wet (32.3×10^{-3} mm^3) and cryogenic-cooled (34.6×10^{-3} mm^3) samples. In general, the wear volume of machined sample in dry condition is nearly twice as that of wet and cryogenic-machined samples. Cryogenically cooled samples had 60% lesser stress leading to better surface finish of the samples. Irrespective of route and cooling strategies, abrasive and adhesive wear mechanisms were reported [64].

Investigation was carried out on wear resistance of Ti6Al4V alloys fabricated using the wrought process, wrought + heat treated process, and SLM processes. In HT, specimens were heated to 960°C, held for 1 h, cooled down to 525°C, held for 6 h, and air cooled. Reciprocating wear tests were carried out by applying a normal load of 2 N, reciprocating stroke of 10 mm, and frequency of 1 Hz in artificial saliva solution using 4 mm-diameter Al_2O_3 ceramic ball (1300 HV). The COF of wrought sample is 0.425, and wrought + heat-treated is 0.410. The range of COF for SLM-made samples is 0.346–0.393. The lowest wear rate of 0.35×10^{-3} mm^{-3}/N/min has been reported in SLM (horizontal built) sample. In wrought- and wrought + heat-treated samples, higher wear rates of 0.047 and 0.044×10^{-3} mm^{-3}/N/min have been reported. This is due to the observed microstructure in the samples made through different processes. Microstructure of wrought- and wrought + heat–treated samples contain both α' martensite and β phases. SLM-made samples contain only α' martensite phase (higher hardness of ~410 HV). Wear mechanisms such as plastic deformation, abrasive, and adhesive wear were observed in the worn out surfaces [65].

8.8 Wear rate comparison of additive-manufactured Ti6Al4V with other additive-manufactured alloys

Several metallic alloy parts were produced through AM processes in the recent decade [66]. For some specific application, few alternative material parts were made through AM method with distinct microstructural features and wear properties [67]. Hence, it is highly important to understand wear behavior of the AM-made parts so that the part with superior property can be selected for the specific application. Therefore, several researchers produced different materials through AM routes and compared their wear behavior.

Various processes like SLM, SLS, and conventional tool making were compared to develop a wear-resistant material. The SLM process was employed to build parts using Concept Stainless steel (17HRC), Concept Tool steel (31HRC), EOS SS 17-4 (19HRC), Ti–6Al–4V, and Co–Cr–Mo. SLS process was employed to produce parts using LaserForm (60% steel and 40% bronze with 83.4HRB) and DirectSteel (Fe 60%, Ni 31% and Cu_3P 9%, C 0.08% with 89.6HRB). Conventional process was employed to make tool steel UHB 11 (C 0.5, Si 0.3, Mn 0.6, S 0.04, Fe rest) by heating it to 850°C and hardened by water quenching. Fretting wear tests were performed using 2 and 6 N loads in SLM-made samples and 2, 4 N loads in SLS-made samples. Furthermore, a constant sliding distance of 200 μm, 10 Hz frequency, 10,000 cycles and 52% humidity were maintained in the test against 30 mm chrome steel ball as counter face. Among SLM made steels, wear volume of Concept tool steel is maximum (198×10^3 and 462×10^3 μm^3 for 2 and 6 N, respectively) and wear of EOS SS is more (39×10^3 and 66×10^3 μm^3 for 2 and 6 N, respectively) than Concept SS (15×10^3 and 51×10^3 μm^3, respectively). Unlike regular concept, here wear resistance of steel-based SLM materials decreases with an increase in hardness. In SLM made, least wear volume was recorded for Ti–6Al–4V (7362×10^3 and 9337×10^3 μm^3 for 2 and 6 N, respectively) and Co–Cr–Mo (1474 and 2252×10^3 μm^3 for 2 and 6 N, respectively) parts. The wear volume of SLS-made alloy-LaserForm is lesser (10.7, 29.6 × 10^3 μm^3) than all the SLM-made steels, which are supported by the low-friction element (copper) in the alloy. The wear volume of conventionally made hardened tool steel was much higher (3899 × 10^3 μm^3) than both the SLM- and SLS-made parts. Hence, material chemistry determines the wear resistance property than the type of processing [68].

Ti6Al4V and CoCr samples were produced by EBM for biomedical applications [69]. Rotary wear test was carried out for a load of 1 N, circular path radius of 1.5 mm, linear sliding velocity of 2 cm/s for 30,000 laps against 6 mm 100Cr6 steel ball counterface. The COF of Ti6Al4V and CoCr specimens have been reported as 0.68 and 0.81, respectively. Higher COF in EBM-made specimen is due to the presence of more carbides through which the counter ball has made efforts to shear rather than in the matrix. The specific wear rate in Ti6Al4V is higher (137.8 × 10^5 mm^3/Nm) than CoCr (1.8 × 10^5 mm^3/Nm) in spite of its lower COF. This is due to the contact of the counter ball with irregular shaped carbides in CoCr specimens, whereas it is not so in Ti6Al4V specimens. Moreover, in the same Ti6Al4V specimens of 1 and 20 mm sizes, 1 mm specimen exhibits lower specific wear rate (122.4 × 10 mm^3/Nm) than the 20 mm specimen (137.8 × 10 mm^3/Nm) due to its higher hardness and shear strength. The higher hardness in 1 mm Ti6Al4V specimen is due to its acicular martensite (α') in the prior columnar β microstructure as against $\alpha + \beta$ structure on 20 mm Ti6Al4V specimen [69].

8.9 Wear studies on additive-manufactured Inconel alloy parts

In the past 4 decades, nickel-based superalloys were extensively developed and received enormous attraction toward various applications [70]. Specifically, Inconel 718 is used in applications of combustion chambers and nuclear reactors due its greater creep resistance and oxidation resistance [71]. Inconel 625 has been widely used in sea water application [72]. It is being utilized as propeller blades in auxiliary propulsion motors and submarine quick-disconnect fittings. In addition, seals, springs, fasteners, and oceanographic instrument components have been made using Inconel 625 [73]. Inconel alloy developed through cast, wrought, and powder metallurgy process possesses reasonable microstructural features and properties [74]. As the Inconel 718 alloy acts as the core part of hot structural components, it should withstand its properties for the wide temperature range. Conventionally produced Inconel 718 superalloys retain its properties through solid solution and precipitation strengthening [75]. However, structures with high precision and elevated temperature performance are higher requirement [76]. Inconel 625 superalloy has been highly used in moving assemblies, which get subjected to critical environment and suffer through friction and wear.

Hence, the AM processing technologies evolved for processing nickel superalloys to produce net-shaped complex parts with higher performance for the critical conditions. Systematic research on the wear behavior of the AM-made Inconel parts was the need due to its severe working conditions. Hence, several researchers focused on investigating the wear behavior of AM-made Inconel 625 and 718 parts by conveniently modifying the process parameters such as scan speed, laser power, layer thickness, scan strategy, scan spacing, and so on.

8.9.1 Wear studies of SLM-made Inconel parts

Jia and Gu [77] manufactured Inconel 718 alloy parts through SLM process under different laser power and scan speed. Wear tests were carried out against steel GCr15 ball counterface. SLM machine parameters used are laser power of 180–330 J/m and scan speeds of 400 and 600 mm/s. With an increase in laser power from 180 to 330 J/m, COF drops from 0.62 to 0.58 and wear rate dropped from 9.12×10^{-4} to 4.64×10^{-4} mm^3/Nm. This is due to dramatic increase in micro hardness from 331.9 $HV_{0.2}$ to 395.8 $HV_{0.2}$ supported by the formation of grain refinement, densification behavior (98.4% at 330 J/m), intermetallic precipitation phase γ'—Ni_3 (Al, Ti), and incorporation of nickel element in columnar dendrite γ (Ni—Cr—Fe) matrix (Fig. 8.5).

During the layer-by-layer addition and fusion of Inconel 718 alloy powder in SLM process, heat will be accumulated due to frequent laser radiation on powder bed. The preceding solidified layer will experience an aging treatment upon subsequent increment of layers. This leads to the precipitation of γ' [77].

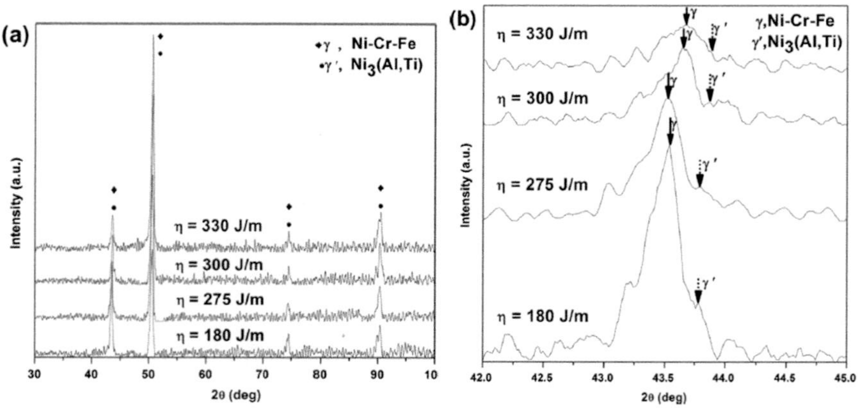

Figure 8.5 XRD spectra of SLM-processed Inconel 718 parts at different processing conditions obtained in a wide range of $2\theta = 30-100°$ (A) and obtained in a small range of $2\theta = 42-45°$ [77].

8.9.2 Wear studies of EBM-made Inconel parts

Electron beam surface processing of Inconel 718 alloy was made for varied combination of heat input (48, 32, 24 J/mm), scan speed (500, 750, 1000 mm/min), voltage of 40 kV, and beam current of 10 mA. A fretting wear resistance test was conducted for a load of 10 N and oscillating frequency of 10 Hz against tungsten carbide ball. Wear rates are 8.2, 3.9, 9, and 5.2×10^{-7} mm^3/Nm. The minimum wear rate of 3.9×10^{-7} mm^3/Nm is achieved in the combination of 48 J/mm and 500 mm/min, which is attributed to its initial faster kinetics of wear for a little longer duration and later becoming negligible. Moreover, due to its higher heat input, grain refinement has taken place leading to a maximum microhardness of 700 HV as against 300, 350, and 400 HV in the subsurface of the specimens made through remaining combinations [78].

Gao and Zhou [79] investigated the fretting behavior of Inconel 625 alloy manufactured by the electron beam selective melting (EBSM) process. Fretting wear test was performed under normal loads of 40, 62, 84, and 106 N and frequency of 2 Hz for 20,000 cycles against the 42 $CrMo_4$ stainless steel flat. The COFs measured were found to 0.39, 0.42, 0.46, and 0.5 for the loads selected from 40 to 106 N. The wear depth of conventional nickel-based materials varies from 40 to 140 μm per 200 cycles, and EBSM-Inconel 625 alloy fretting surface remains less than 2.4 μm after 2×10^4 cycles at the load of 106 N. The microstructure of the worn-out surface reveals continuous and stable tribo-layers consisting of Fe_2O_3, Fe_3O_4, Cr_2O_3, and Mn_2O_3, which is confirmed through high-resolution XPS (Fig. 8.6). This layer improves the fretting wear characteristics of the EBSM-Inconel 625 alloy [79].

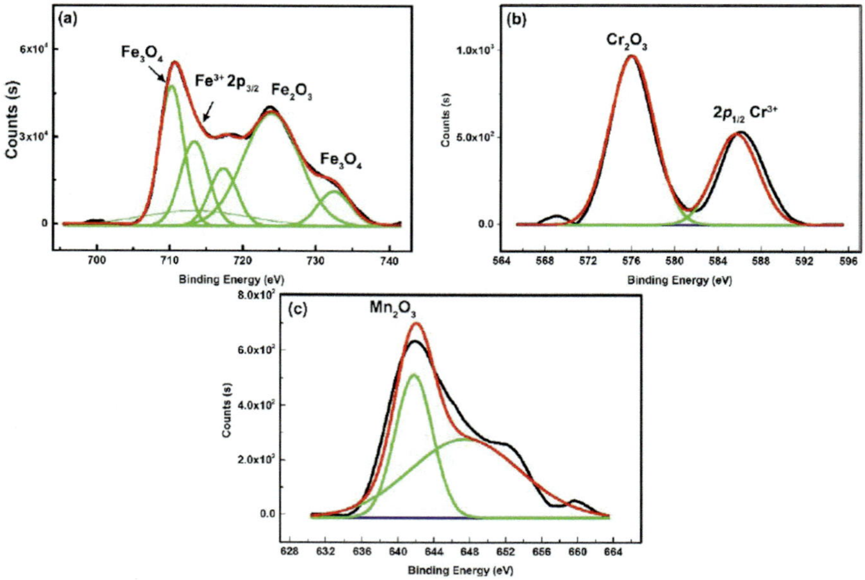

Figure 8.6 XPS spectra on the wear tracks of AM-Inconel 625 after fretting wear [79].

8.10 Effect of posttreatments on wear of additive-manufactured Inconel parts

Inconel parts with complex shapes and high precision are being made through different AM routes that are difficult to produce through conventional manufacturing method [80]. Production of such parts is cost-effective, which is carried out in a fast manner to meet out the fluctuation demand. However, these AM-made parts have to overcome fewer limitations such as high surface roughness, low ductility, and residual stress formation. These are addressed to some extent by conveniently modifying the parameters such as scan speed, laser power, layer thickness, scan strategy, scan spacing, and so on. Therefore, optimizing the parameters alone does not fulfill the requirements; hence, there is a need for postprocessing treatment to be performed on the AM build Inconel parts. Postprocessing treatments such as HT, machining, electrochemical polishing, laser polishing, shot-peening etc., can be carried out to improve properties of the AM parts. Few researchers have employed HT as the tool for enhancing the wear behavior of the AM-made Inconel parts.

8.10.1 Effect of heat treatment on SLM-made Inconel parts

Karabulut et al. [81] manufactured Inconel 718 alloy by SLM process, heat treated it, and examined its wear performance in comparison with a wrought Inconel 718 alloy product. Dry sliding wear tests were performed for 40 min using tungsten carbide ball under a contact load of 5 N. Two types of HTs were carried out. HT1: Heating the as-built specimens to 600°C with 2 h holding and air cooling and HT2: heating the as-built specimens to 1100°C with 2 h holding and air cooling. As-built parts yield poor wear resistance (27 μm wear depth) when compared to the wrought product (23 μm wear depth). The wear rate of wrought specimen, as-sintered specimen, HT1 specimen, and HT2 specimen was 4.4, 4.6, 4.1, and 4.7 mm^3/Nm, respectively. HT1specimen yields better wear resistance due to its improved microhardness of 340 HV. This is attributed to its cellular dendritic structures and precipitation hardening of δ-phase that inhibits the dislocation motion. An increase in wear rate and reduction in hardness (268 HV) of HT2 specimen are due to its microstructural features like formation of grains and grain boundaries, complete dissolution of strengthening phases (γ and γ') in high temperature, and removal of residual stress in as-sintered and in HT1 specimens [81].

Zhanyong Zhao et al. [82] fabricated IN718 alloy specimens through the SLM process. Three types of HTs were made. HT1: Solutionizing—940°C × 1.5 h, followed by water quenching (WQ) plus double aging—720°C × 8 h (FC, 50 °C/h) + 620°C × 8 h (AC), HT2: Solutionizing—980°C × 1.5 h (WQ) plus double aging—720°C × 8 h (FC, 50°C/h) + 620°C 8 h (AC), HT3: Solutionizing—1020 °C × 1.5 h (WQ) plus double aging—720°C × 8 h (FC, 50°C/h) + 620°C × 8 h (AC). Dry wear tests have been carried out using a normal load of 30 N, friction speed of 0.02 m/s, and stroke length of 5 mm against a friction pair of 52,100 alloy balls. The wear rate of as-sintered specimen was evaluated as 265 × 10^3 mm^3/Nm due to its fine cellular dendritic sub-structure with laves ((Ni, Fe, Co)$_2$(Nb, Ti, Mo)) and MC phases in the interdendritic regions. After HT1, the wear rate dropped to 152 $\times$ 10^3 mm^3/Nm due to the large-sized hard precipitates such as a γ' (Ni$_3$(Al, Ti)) phase (FCC) and γ'' (Ni$_3$Nb) phase (BCT) distribution in the matrix. After HT2, the wear rate further dropped to 80 $\times$ 10^3 mm^3/Nm, which is attributed to γ'' and δ phases that acted as abrasives in the wear test. After HT3, the wear resistance further dropped to 70 $\times$ 10^3 mm^3/Nm since the size of nano-scale γ' and γ'' phases in the γ phase matrix increased. Altogether, during solid solutionizing, a fraction of the Laves phase dissolves into the matrix and during double aging, the γ'' phase and the equilibrium phase of the δ phase (Ni$_3$Nb) precipitates. After solutionizing and double aging, nano-scale γ' and γ'' phases homogeneously gets distributed in the matrix, that improves the wear resistance of the SLM-made IN718 alloy. The same trend has been recorded with reducing COFs in as-sintered, HT1, HT2, and HT3 as 0.457, 0.342, 0334, and 0.256, respectively [82].

8.10.2 Effect of machining on SLM-made Inconel parts

Wear resistance of the AM Inconel parts has been improved upon carrying out different HT procedures [83]. On the other side, the AM-made part's surface quality is being an issue due to the surface roughness caused by partially melted powders [84]. Hence, the influence of finish milling operation on the wear resistance of AM part has been investigated.

Machining (Finish-Milling), a posttreatment operation, was carried out in SLM-made Inconel 625 specimens. The effect of different feed rates (0.05, 0.1, and 0.15 mm) on wear resistance was investigated for a constant cutting speed of 60 m/min and axial and radial depth of cut of 0.3 and 6 mm, respectively. Dry reciprocating wear tests were done for 40 min against 6 mm diameter tungsten carbide ball counterpart. The wear rate of as-sintered specimens in build and scan directions are 6.25 and 6.5 mm^3/Nm, respectively. Wear of machined specimens with the feed rates of 0.05, 0.1, and 0.15 mm are 5, 3.5, and 3 mm^3/Nm, respectively, which records up to 50% drop of wear rate as that of as-sintered specimen. This trend is in line with the observed maximum hardness of 414 HV in 0.15 mm feed specimen as against lower hardness of 319 HV in as-built and 369 HV in 0.05 mm feed specimens. Higher feed rate has caused work hardening owing to compressive residual stress and increased dislocation density in surface and subsurface. Surface roughness increases with increase in the feed rate, but its influence on wear rate is not significant [85].

8.11 Wear studies of additive-manufactured Inconel composite parts

In the recent years, several studies were focused on high-performance tribological applications with the aid of composite development [86]. Conventional manufacturing of nickel-based MMCs possess various limitations such as irregular microstructure, agglomeration, lower densification, and wettability problem. Evolving of AM techniques aids in production of nickel-based composite parts without problems that met in conventional manufacturing of composite parts [87]. Inconel 718 parts were reinforced with hard ceramic particles through AM methods that enhance the wear performance due to its load bearing ability. In specific, SLM process involves nonequilibrium, rapid melting, and solidification characteristics, which have the potential to address the aforementioned problems, particularly, the interfacial bonding between the matrix and reinforcement [88].

8.11.1 Influence of WC on Inconel 718 composite

WC/Inconel 718 composites were produced through SLM process with a laser power of 110W and varied laser scan speeds of 350, 450, 550, and 650 mm/s [89]. Dry sliding wear tests were carried out using 3 mm diameter bearing steel GCr15 ball (HRC60) at load of 6 N and by rotating friction unit at 560 rpm for 15 min. The lowest wear rate of 2.5×10^{-4} mm^3/Nm was recorded for scan speed of 450 mm/s, and highest wear rate of 4.1×10^{-4} mm^3/Nm was recorded for the scan speed of 650 mm/s. As the laser scan speed reduces, the densification behavior of composite part increases from 85.7% to 97.8% and microhardness increases from 317.5 to 393.2 HV. At a scan speed of 450 mm/s, the morphology of gradient interface become regular and ordered. Mean thickness of graded interface increases to 0.27 μm and a diffusion layer exists. The existence of the gradient interface plays a vital role in improvement of wear properties (Fig. 8.7). As the

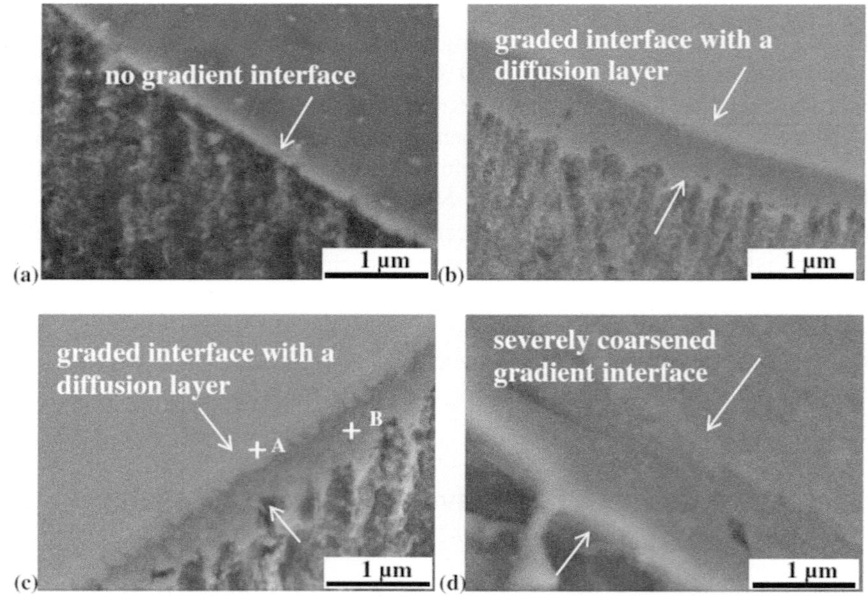

Figure 8.7 Field emission scanning electron microscope (FE-SEM) images showing graded interfacial layer between reinforcing particles and matrix at different laser scanning speeds: (A) v = 650 mm/s; (B) v = 550 mm/s; (C) v = 450 mm/s; and (D) v = 350 mm/s [89].

scan speed is further reduced to 350 mm/s, extreme heat gets accumulated leading to coarsened microstructure, and hence, its friction and wear performance drop down. The diffusion layer disappears in this scan speed and the mean thickness reduces to 1.07 μm. As the laser scan speed reduces from 650 to 350 mm/s, the wear mechanism changes constantly from severe abrasive wear to adhesive wear [89].

Conventional Inconel 718 parts and WC/Inconel 718 composites were produced through the SLM process [90]. WC/Inconel 718 composites were made by mixing Inconel 718 powder with varied WC particle mean sizes of 21 μm, 10.5 μm, and 5.25 μ mat; the weight ratio of WC and Inconel 718 was 0.25:0.75. Dry sliding wear tests were carried out on the developed specimens for a test load of 430 g, friction unit rotation of 560 rpm for 15 min with a rotation radius of 2 mm against 3 mm steel GCr15 ball (HRC60) as counterface material. COF of conventional part is 0.43, and it decreases from 0.40 to 0.21 as the WC particle size reduces from 21 to 5.25 μm. The wear resistance of WC/Inconel composite outperforms with smaller-sized WC (5.25 μm) due to its better interfacial bonding. In its microstructure, fine dendrites were almost perpendicular to the edge of the particle formed under the impact of temperature gradient (4700°C/mm) around the WC [90]. Hence, the contact area between the matrix and particle increases, leading to improvement in interfacial bonding [90].

8.11.2 Influence of boron nitride on Inconel 718 composite

Apart from WC reinforcement addition to Inconel 718 alloy, hexagonal boron nitride has been a potential reinforcement owing to its greater thermo-mechanical properties. BN delivers lubricating behavior to the Inconel alloy 718 and offers good wear and antifrictional properties [91].

Kim et al. [92]fabricated BN-reinforced IN718 composites with different volume ratios (0, 6, and 12 vol%) of hexagonal boron nitride using the laser power bed fusion method. A dry sliding-wear test was conducted by rotating the friction unit at 200 rotation/min for 2400 s and by applying a load of 5.0 N against 5.4 mm diameter tungsten carbide ball counterface. The average wear depths were 6.5, 5.1, and 3 μm during the entire 2400 s process, and COFs were 0.43, 0.35, and 0.31 for the composite with 0, 6, and 12 vol% BN. This is because the BN has an hexagonal structure and surface exfoliation, which greatly decrease abrasion by facilitating high-sliding, lubricating characteristics of the IN718 matrix. More volume of BN addition supports this through reduction in composite density from 8.12 to 7.03 g/cm^3 and increased specific hardness from 38.7 to 41.7 $Hv_{0.5}$ cm^3/g [92].

8.12 Wear studies on surface-modified and coated additive-manufactured Inconel parts

A competitive method to develop Inconel composites is the laser-manufactured coating over the substrates [93]. Even the failed component can be surface engineered through laser AM, which saves the cost and eventually improves the wear resistance and life of the component.

A10 mm-thick IN718 alloy wrought plate was laser cladded with 12 layers of IN718 superalloy to attain a clad thickness about 2.4 mm and heat treated [94]. Both the as-received wrought and cladded plates were subjected to HT. The plates were heated to 980°C, held for 1 h, and air cooled to room temperature. Then, the samples were further heated to 720°C, held for 8 h, and furnace cooled at the rate of 100°C/h to 620°C. The samples were held at this temperature for 8 h and air cooled to room temperature. The wear test was performed on laser-manufactured IN718 superalloy coating with different applied loads (75 N, 100, and 125 N), rotational speeds (90, 120, and 150 RPM), and lubricating mediums (dry and wet-distilled water) against the Al_2O_3 ceramic ring counterpart. The loss of material due to wear in heat-treated coated plate is found to be higher (4.408×10^{-4} g/m) than the sheer-coated plate (1.1703×10^{-4} g/m) and heat-treated wrought plate (1.949×10^{-4} g/m) [94]. This is attributed to the higher microhardness of the heat-treated coated plate with 480–500 HV and heat-treated wrought plate with 460–480 HV, which is about twice higher than that of the shear-coated plate [94]. Homogeneous distribution of brittle Laves phase contributed for hardness improvement and retarded the wear rate of the coating. The lubricating medium decreases the COF and improves the wear resistance considerably, while the influence of load on wear rate is more different than that of the rotational speed. Fatigue wear along with slight adhesive wear type of failure is observed predominantly in the sample [94].

8.13 Wear studies on additive-manufactured stainless steel parts

Stainless steel has been extensively utilized in the several industries, as structural materials and especially as medical implants [95]. Since the stainless steel comprises of the metals like Cr, Ni, and Mo, there was a great shift toward the AM processing of stainless steel parts in the recent years to avoid the loss of expensive materials and also to build near net-shaped structures. As the AM processes offer greater flexibility on the product shape and production time [96], several researchers produced stainless steel parts through different AM technologies and investigated its tribological properties since tribology plays a significant role in several applications [97].

8.13.1 Wear studies on SLM-made stainless steel

316L stainless steel was produced in Renishaw AM250 SLM system with the 70 μm beam diameter and 1071 nm wavelength [98]. Stainless steel powder of 15–45 μm with a spherical shape was used to build the specimen under the nitrogen atmosphere. Laser power of 200 W, scanning speed of 750 mm/s, and hatch space of 110 μm were set for building the samples in different scanning strategies (zigzag, parallel remelting, and perpendicular remelting). Scratch tests were performed in automatic scratch tester (WS-2005) using the diamond indenter with the tip radius of 200 μm [98]. Sliding speed of 10 mm/min, sliding distance of 5 mm, and different loads of 1, 3, and 5 N were applied for performing the wear test. The scratch depth was found deeper in case of the zigzag built specimen (3.5 μm), whereas the shallow scratches were found in remelting—(2.5 μm) and perpendicular remelting-built specimen (2 μm), which indicates the higher wear resistance. In general, SLM samples undergo rapid heating and cooling, which forms cellular and columnar subgrains. Implementing of different remelting strategies aids in grain refinement and formation of high density of subgrain boundaries [98].This leads to increase in dislocation density that resists the plastic deformation and contributes in wear resistance [98].

316L stainless steel samples was produced in the SLM machine (Renishaw AM400) using the stainless steel powder of size 15–45 μm [99]. The laser wavelength of 1070 nm, beam diameter of 70 mm, hatch distance of 110 μm, and layer thickness of 50 μm were maintained for different laser powers of 100, 150, 200, and 300 W. Sliding wear tests were conducted in linear reciprocating ball-on-plate tribometer for the sliding distance of 1000 m, sliding stroke of 10 mm, frequency of 4 Hz, and load of 10 and 20 N. Hardened steel (E-52,100, 58 HRC) of 10 mm diameter with surface roughness 0.04 μm was used as the counterface. An increase in the load increased the wear rate of the stainless steel samples. The wear rate of the samples produced at lower (1.21×10^{-6} mm^3/Nm) and higher power (1.23×10^{-6} mm^3/Nm) were found merely similar, which indicates that there is no significant influence of laser power on the wear resistance. This was due to debris released during the wear test and pressed into the porous structure of the stainless steel samples. Pores produced in the samples under low laser power generally attract the debris produced during the sliding wear test [99]. Porosity produced under different laser powers is shown in Fig. 8.8. The debris attachment in the pores increases the hardness of the sample and eventually increases the wear resistance [99].

8.13.2 Wear studies on LMD-made stainless parts

17-4 PH stainless steel was produced through LMD process using the laser power (1000–2600 W) and wavelength of 1.06 μm [100]. Stainless steel powder (45–90 μm) was utilized for producing specimens of 100 × 100 × 10 mm. Wear tests were carried out on the samples using the ball-on-disc configuration in ANTON PAAR-tribometer as per the ASTM G 99-05 [100]. Alumina ball with diameter of 6 mm is used as the counterface, and the load of 10 N, acquisition rate of 100 Hz, time of 16 min 40 s, and sliding distance of 31.406 m were set for conducting the wear test. The wear rate for the samples produced at different laser power of 1400, 1800, and 2600 W was 3.6828×10^{-9}, 3.6491×10^{-9}, and 3.3306×10^{-9} mm^3/Nm, respectively. The wear rate was found to decrease as the laser power during deposition was increased, which was due to the higher microhardness and formation of higher fraction of the martensitic structure along with coarser delta ferrite [100].

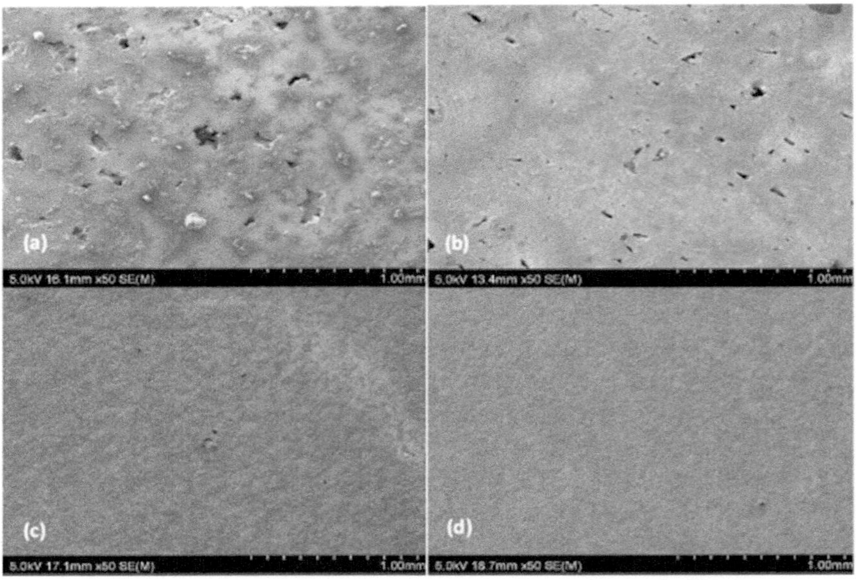

Figure 8.8 Porosity at different laser powers: (A) 100 W, (B) 150 W, (C) 200 W, and (D) 300 W [99].

8.13.3 Wear studies on DED-made stainless steel parts

316L stainless steel was produced through laser-assisted direct energy deposition (DED) using the powders size of 45–105 μm [101]. Laser power of 2-KW was used for producing the samples [101]. Laser power (1000, 1200, and 1400 W), scanning speed (0.4, 0.5, and 0.6 m/min), and powder feed rate (3, 6, and 8 g/min) were varied for each experiments. Tribology tests were conducted on the samples (5 × 27 mm) using the pin-on-disc configuration according to ASTM-G99 standard [101]. Stainless steel EN 31 disc is the counterface for performing wear test for sliding distance of 1000 m under varied loads (10, 20, and 30 N) and sliding velocities (0.5, 0.75, and 1 m/s). The wear rate gets increased from 0.18557×10^{-4} to 0.295×10^{-4} mm^3/Nm with an increase in sliding velocity from 0.5 to 1 m/s due to material softening by generation of higher frictional heat. Similarly, the wear rate increased with an increase in the normal load (10–30 N) from 0.0247×10^{-4} to 0.07022×10^{-4} mm^3/Nm due to severe plastic deformation as a result of adhesive wear. Wear rate gets decreased for the samples made under higher laser scanning speed even at higher loads. This is due to the action of higher loads, which compress the pores and decrease the level of crack initiation. Scratches and grooves were identified as the major form of wear in case of lower sliding velocity, whereas the deeper grooves and plastic deformation were noticed at higher velocities. This is due to the higher temperature developed at the interface that leads to material softening and more material removal [101].

8.14 Influence of posttreatment on the wear of additive-manufactured stainless steel parts

Although the advantages of laser-based AM of stainless steel are quite enormous, there are also drawbacks in laser processing of stainless steel such as formation of higher stress concentration, higher surface roughness, and residual stresses. In addition, defects such as internal voids, partially melted powders, and thermal stress occur, which leads to unfavorable properties of the part [102]. More specifically, the wear property of stainless steel parts is affected due to the abovementioned defects that arise in the parts during manufacturing. Optimization of processing parameters for stainless steel parts is one option for attaining the functional properties to meet the standards in critical applications. Another option is posttreatment such as HT (high and low temperatures) that has a greater effect on the tribological behavior of the AM-made stainless steel parts [84]. Hence researchers approached both elevated-temperature and sub-zero-temperature HT for the enhancement of the wear behavior of the AM-made stainless steel parts.

8.14.1 Influence of high-temperature heat treatment

316L stainless steel samples were produced by the SLM process in argon atmosphere using the powder size of 14–45 μm in diameter [103]. Building of samples was carried out at laser power of 200 W, layer thickness of 50 μm, hatch distance of 110 μm, and point distance of 60 μm. Different HTs, HT-1 (600°C/2 h/air cooling), HT-2 (850°C/2 h/air cooling), and HT-3 (1100°C/2 h/air cooling) were performed on the as-built SLM samples to investigate the effect of different HT temperatures on wear behavior. Wear tests were carried out using reciprocating tribometer for 40 min at the load of 15 N. Tungsten carbide ball of diameter 6 mm was used as the counterface material. A wear depth of the as-built SLM sample was found as 52 μm, whereas the wear depth of HT-1, HT-2, and HT-3 was 62, 52, and 45 μm respectively. The wear rate of the as-built specimen was 6.4 mm^3/Nm, and it gets increased in the HT-1 condition and further decreased by 5% and 14% in the post-HT-2 and HT-3 conditions, respectively. Wear resistance of the SLM samples were observed to be increasing with a decreasing porosity level of the samples (HT-1 (0.38%), HT-2 (0.29%), and HT-3 (0.08%)) as shown in Fig. 8.9.

This discrepancy is attributed to the porosity levels that present on the SLM samples after the different HT temperature, which indicates that porosities have a prominent role over the wear resistance than that of hardness. HT temperature of 850°C was essential to modify the microstructure into homogeneous, and an increase in HT temperature was found to decrease the porosity in the SLM sample [103].

8.14.2 Influence of sub-zero temperature treatment

The influence of cryogenic treatment on the wear behavior of DMLS 316L stainless steel was investigated [104]. Samples were manufactured under the laser power of

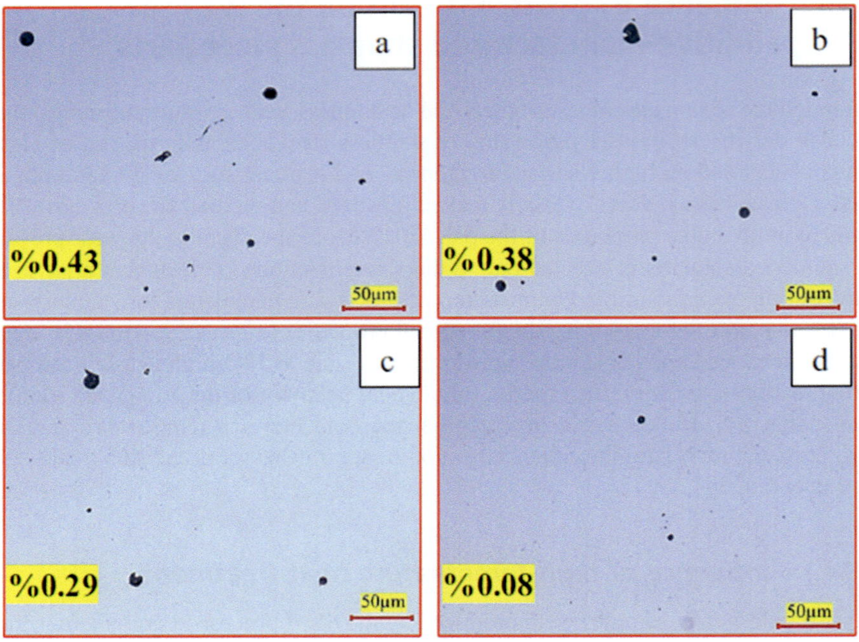

Figure 8.9 Porosity in (A) as-built specimen, (B) HT-1 specimen (600°C), (C) HT-2 specimen (850°C), and (D) HT-3 specimen (1100°C) [103].

200–400 W in the EOSINTM280 machine. Cryogenic treatment was performed at temperature of −196°C for the holding time of 24 h under liquid nitrogen atmosphere. The cooling was performed with the temperature drop at the rate of 0.62°C/min. Porosity of the specimens was found decreased drastically after the cryogenic treatment. Reciprocating wear tests were performed on the specimens using the DUCOM TR-285-M1 machine according to the standard ASTM G133-05. Cryogenic-treated and non-cryogenic-treated specimens of size 30 × 20 × 10 mm were tested against the zirconia ball for the time of 30 min. A frequency of 2 Hz, load of 5 N, and sliding velocity of 40 mm/s were maintained during the wear test. The COF plot (Fig. 8.10) showed that the specimen exhibited three distinct phases such as running-in period, transition period, and steady-state period. COF was found as 0.41 for the untreated sample, whereas the cryogenic-treated sample possessed the COF as 0.32, which is attributed to the formation of martensitic structure upon cryogenic treatment. Worn-out surfaces of untreated samples revealed the presence of wear debris as it contains micropores in the surface, whereas the debris was absent in the cryogenic-treated sample due to sealing of pores during cryogenic treatment. The presence of loose debris in the untreated sample aids in more contact pressure and consequences in severe wear compared to the treated sample [104].

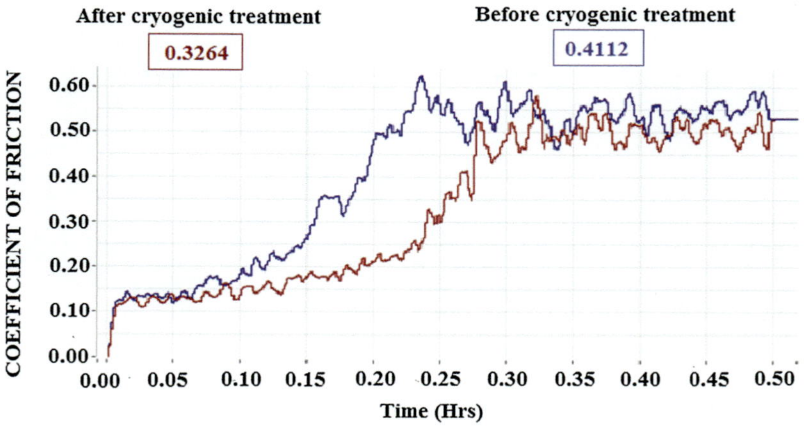

Figure 8.10 COF plot for 316L SS before and after cryogenic treatment [104].

8.15 Wear studies of additive-manufactured stainless steel composite parts

AM-made 316L stainless steel parts have received greater attention toward its application in aviation and petrochemical industries owing to its better toughness, weldability, plasticity, and oxidation resistance [95]. However, wear performance of the 316L stainless has to be improved equal to martensitic steels for extensive applications [105]. As a result, 316L composite parts were developed with ceramic particle reinforcement, and its tribological behavior has been assessed [106]. Wear of the 316L stainless steel and 316L stainless steel/TiC composite manufactured through SLM process was investigated. Stainless steel powders with the size of 45 μm and TiC powders of the size 2–5 μm were utilized for building the specimens in the SLM machine (Renishaw AM 250). Reciprocating wear tests were performed for a sliding speed of 60, 80, and 100 mm/min at the load of 15, 25, and 35 N; GCr15-bearing steel was used as a counterface to slide against the stainless steel for the time and distance of 30 min and 5 mm, respectively. The wear rate of stainless steel and composite decreased with an increase in the sliding speed. An increase in the sliding speed causes more frictional heat and oxidation of the materials, which tend to form an oxide layer that acts as a protective layer and inhibits further wear. Worn-out surfaces revealed the wear mechanism at a low sliding speed as abrasive and oxidation, whereas it turned into severe oxidation at higher sliding speed. The wear rate gets increased with an increase in the normal load for both the unreinforced and composite. However, the wear rate of the composite was found to be lower than the unreinforced stainless steel at a high-load condition [106]. This is due to hardness and work hardening behavior of composites compared to the unreinforced stainless steel [106]. TiC particles hinder the movement of dislocation during the plastic deformation of the material, which

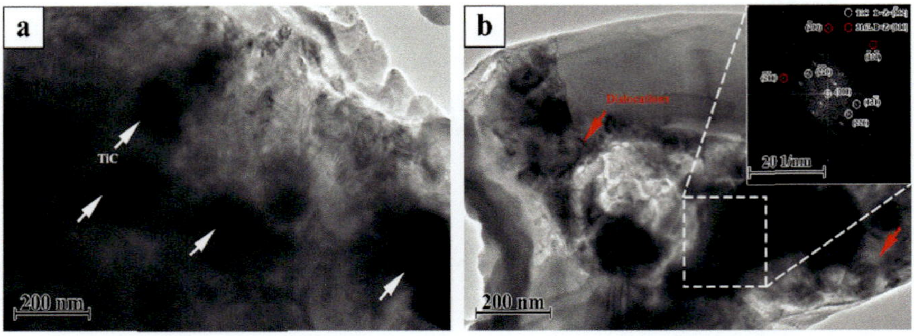

Figure 8.11 TEM micrographs of (A) and (B) TiC/316L stainless steel composite [106].

eventually results in work hardening that increases the hardness (335 ± 8 HV) and wear resistance of composite. The presence of TiC particles and the dislocations were shown in Fig. 8.11 [106].

8.16 Wear studies on surface-modified additive-manufactured stainless steel parts

Austenitic stainless steels are widely accepted for its greater corrosion resistance due to the presence of chromium and nickel [95]. These metals tend to form stable and passive oxide layers (Cr_2O_3) on surface and protects the surface from corrosion [107]. Specifically, 316L stainless steel is utilized in food and medical equipment. SLM-made alloys usually possess higher surface roughness in the range of 10 and 20 μm depending on build parameters, build orientation, and the material [99]. Higher surface roughness is not desirable for critical applications since it causes stress concentration, causes crack initiation, and consequently, results in wear [108]. Almost many components require the surface roughness below 0.8 μm for a better tribological performance [108]. Thus, certain works were focused on improving the surface integrity and wear performance of AM-processed 316L stainless steel parts through nitriding and mechanical attrition treatment.

8.16.1 Effect of nitriding on stainless steel parts

316L stainless steel was additive-manufactured through LPBF process in the EOSINT M280 (400 W) machine using the powders with size of 15–45 μm [109]. Laser power of 285 W, laser speed of 960 mm/s, hatching distance of 0.11 mm, and laser diameter of 0.055 mm were maintained during fabricating of the samples under the nitrogen atmosphere (99.5%). Additive-manufactured samples were subjected to solution annealing at 1060°C for 30 min using the horizontal furnace followed by air cooling. Additive-manufactured samples (both as-built and solution treated) and commercially available 316L stainless steel were subjected to low-temperature plasma nitriding surface treatment in the Metaplas Ionon HZIW 600/1000 reactor under the 75 vol.% H_2: 25 vol % N_2 gas mixture [109]. Wear tests were carried out on the samples under dry condition using the ball-on-flat configuration. Counterface of 100 Cr6 ball with diameter of 32 mm is utilized for sliding against the samples at the load of 20 N, sliding speed of 0.12 m/s, amplitude of 4 mm, frequency of 15 Hz, time of 833 s, and sliding distance of 100 m. Conventional 316L stainless steel possessed a wear volume of 5.4×10^{-2} mm^3, additive-manufactured as-built samples has the wear volume of 7.2×10^{-2} mm^3, whereas the solution-treated sample has the wear volume of 6.6×10^{-2} mm^3. Solution-treated sample exhibited better wear performance than as-built sample; however, it is not better than that of conventional 316L stainless steel. The improvement in resistance of wear through solution treatment is owing to the decrease in the dislocation density and nanosegregation. Nitrided additive-manufactured samples possessed lowest wear volume of 1.0×10^{-2} mm^3 compared to other samples due to the inducing compressive residual stresses through nitriding process [109].

8.16.2 Effect of mechanical attrition treatment on stainless steel parts

316L stainless steel specimens were produced through SLM process using the Renishaw machine [110]. Stainless steel powders of size ranging from 15 to 45 μm were utilized for building the samples under the controlled argon atmosphere. Laser power of 200 W, scan speed of 480 mm/s, and layer thickness of 50 were set for the building of the samples. The built sample surfaces were subjected to surface mechanical attrition treatment (SMAT) process to enhance the smoothness of the surface. Stainless steel balls of diameter 6 mm having surface roughness of 0.04 μm and hardness of 840 HV were made to impact on the surface of the samples. The attrition treatment was carried out for different times (10–80 min) at the frequency of 40 Hz and amplitude of 7 mm. The tribological performance of the samples was tested using the reciprocating tester under dry condition [110]. A stainless steel ball was used as the counterbody in the test, which has the diameter of 6 mm, surface finish of 0.04 μm, and hardness of 840 HV. Stainless steel samples were made to slide against this stationary counterface steel ball at the amplitude of 8 mm, frequency of 1 Hz, and time of 3600 s. Sliding speed of 0.016 m/s, sliding distance of 57.6 m, and loads of 2 N, 5, and 10 N were applied. Mechanically attrition samples possess lower COF (0.4) compared to as-built (0.6), which is due to the smoothing of the surface. The longer the attrition time, the smoother the sample surface, hence, the lower the friction [110]. Strain hardening induced by mechanical attrition tends to reduce the friction as it offers higher resistance toward the plastic deformation during the wear test. Wear volume of the as–built and attrition sample at the maximum load of 10 N were found as 0.5 and 0.2 mm^3, respectively. Wear track on the sample subjected to attrition appeared shallow and narrow than the as-built sample. Samples subjected to attrition for 10 min showed 50% lesser wear rate than the as-built sample. This is attributed to the strain hardening and surface smoothening (Fig. 8.12) produced through the mechanical attrition process. Wear mechanisms such as abrasive wear and delamination were found in both the as-built and attrition samples [110].

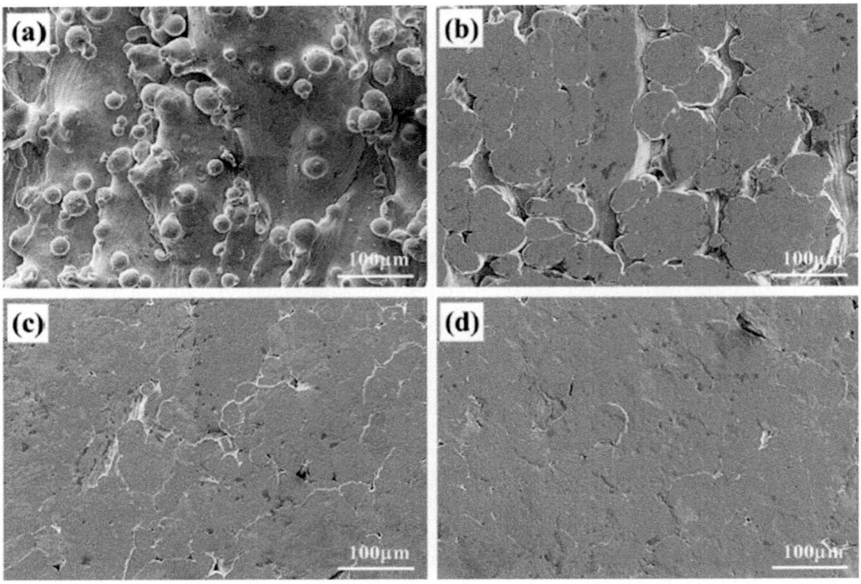

Figure 8.12 SEM images showing the as-SLM surface (A) and the SLM surfaces after SMAT for (B) 10 min, (C) 30 min, and (D) 80 min [110].

8.17 Comparison of additive-manufactured and diverse processing routes of stainless steel parts

AM-made stainless steel parts are being utilized extensively for orthopedic implants and prosthesis in which tribological properties are highly essential [95]. AM-made parts are fully dense, and it possess distinct microstructural features, which lead to unique properties compared to the parts produced through conventional manufacturing routes. Distinct microstructural characteristics and properties were due to the higher cooling rate (10^3-10^8 K/s) involved in the AM process [111]. In addition, AM process has a reduced environmental impact and low wastage of material [111]. 316L stainless steel prosthesis were also been produced through casting as well as forging process [112]. However, application of AM would produce the prosthesis in an easy and faster manner for a specific individual patient that cannot be attained through traditional manufacturing process. Numerous research studies have focused on comparing the tribological properties of the AM-made parts with CP parts (HP, casting, and wrought). The processing method differs in each process, which has a significant influence on the wear behavior of the stainless steel parts.

Wear behavior of 316L stainless steel manufactured through three different routes such as SLM, HP, and conventional casting was investigated to study the effect of the processing route [113]. In the SLM process, a laser spot size of 87 μm, laser power of 70 W, scan rate of 417 mm/s, scan spacing of 0.07 mm, and layer thickness of 0.03 mm were maintained to produce the samples. Wear tests were conducted in reciprocating ball-on-plate configuration using the tribometer (Bruker-UMT-2). A load of

3 N, frequency of 1 Hz, stroke length of 2 mm, and sliding time of 60 min were set for performing the test against the Al_2O_3 counterface (10 mm diameter) [113]. Samples were immersed in the PBS solution during the test. Specimens produced through SLM process have the lowest wear rate of 3.5×10^{-5} mm^3/Nm, whereas the HP and cast specimens has the high wear rate of 4.3×10^{-5}, and 4.8×10^{-5} mm^3/Nm, respectively. Worn-out surface of the samples revealed sliding grooves and abrasive type of a wear mechanism [113]. Cast specimen has deeper grooves compared to the HP and SLM samples. This trend of wear rate and wear mechanism is ascribed to the hardness of the samples attained through different processing routes. Hardness of the SLM sample (229 HV) is found to be higher than the HP (176 HV) and cast (165 HV) sample. The higher hardness in the SLM sample is due to formation of finer grain size (13 ± 4 μm). Higher cooling rate of SLM process (10^3–10^8 K/s) leads to significant grain refinement compared to the HP (3 K/s) and cast process (0.5 K/s) [113].

High-temperature wear behavior of the 316 stainless steel produced through the SLM process was investigated [114]. Specimens were produced in EOS INT M270 system (EOS GmbH, Germany) using the laser power of 195 W, scanning speed of 800 mm/s, line spacing of 0.1 mm, and layer thickness of 20 μm [114]. Wear tests were performed using the Rtec Universal Tribometer (Rtec, San Jose, USA) through ball-on-disc configuration. Alumina balls having diameter of 9.3 mm with hardness of 1650 HV was used as the counterface. A load of 5 N, sliding speed of 0.031 m/s, and sliding duration of 1 h were set for performing the wear test at the different elevated temperatures of 300, 400, and 600°C. The COF for SLM 316L sample was found as 0.5 at room temperatures, and it gets increased up to 0.6 for the elevated temperature of 600°C. Attainment of higher COF values in SLM 316L with an increase in temperature can be attributed to the microstructure enhancement with respect to high-temperature mechanical characteristics. Conventionally manufactured samples possessed COF of 0.73 at room temperature, whereas it gets decreased to 0.52 at the temperature of 600°C. The lower COF of the conventional stainless steel was due to the softening and lower hardness upon increase in temperature. Higher COF in SLM was due to its hierarchical and fine microstructure. The wear rate of the conventionally manufactured sample and SLM sample was found as $6.6 \times 10^{-4}(\pm 0.6)$ and $6.4 \times 10^{-4}(\pm 0.7)$ mm^3/Nm, respectively. The lower wear rate of SLM sample was due to the formation of oxide layer that protects the surface wear and also the hierarchical microstructure resulted through the SLM process. The presence of thermally cellular sub-grains was found to increase the hardness of SLM sample (325 HV), which eventually hinders the deformation of sub-surface and increased the wear resistance compared to the conventionally produced stainless steel (175 HV). Wear tracks of the SLM sample at 600°C revealed abrasive ploughing type of wear with the average groove depth of 11.0 ± 0.8 μm due to its higher hardness. In case of conventionally produced sample, oxide glaze layer removal occurred and the removed oxide acted as the debris which induced abrasive wear in the sample with the wider groove of depth 27.2 ± 4.0 μm. Overall, the enhanced wear performance of SLM stainless steel at elevated temperatures was due to its hierarchical microstructure, high-temperature hardness, less softening, and stable oxide layer formation [114].

316L stainless steel samples were produced through the SLM technique in Renishaw 125 machine using powders of size 30 μm [115]. Samples were built at the laser power of 150 W for different scanning speeds ranging from 125 to 200 mm/s and for the layer thickness of 50 μm. Wear experiments were conducted on stainless steel samples using the pin-on-disc configuration against the steel ball (6 mm dia, 840 HV) [115]. Tests were performed at the rotational speed of 120 rpm, time of 3600 s, track diameter of 7 mm, load of 10 N, and sliding speed of 4.4 cm/s. SLM samples revealed wider and deeper wear tracks (60 μm) compared to the wrought 316L steel (10 μm). On contrary, the hardness of the SLM samples is found higher (230−240 HV) than wrought sample (185 HV), which indicates that hardness alone has not significant influence over the wear resistance, whereas the porosities (1.7%−6.7%) in SLM samples has predominant effect. It is indicated that proper HT after SLM will aid in relieving the residual stress as well optimizing the microstructural features. The predominant wear mechanism in the SLM sample was found fracture type as the pores in the SLM sample tends to act as the crack initiation location and results in material removal [115].

17-4 PH stainless steel was produced through LB-PBF process in EOS M290 machine under argon atmosphere [116]. Laser power of 220 W, scanning speed of 755.5 mm/s, hatching distance of 100 μm, and layer thickness of 40 μm were set for building the specimen using the particles of size 15−45 μm. Specimens were subjected to solution HT (1050°C for 30 min followed by air cooling) and held at 482°C for 1 h and then air cooled to room temperature. A ball-on-disc tribometer (Bruker UMT-3) was used to determine the wear rate of the samples against the high carbon chrome steel balls (10 mm diameter). In dry sliding condition, sliding velocity of 0.6 m/s and time of 4000 s were set for the different load condition of 10 and 30 N. LB-PBF samples possesses greater wear resistance (0.005 mm^3/Nm) compared to the wrought 17-4 PH samples (0.007 mm^3/Nm) even at higher load condition (30 N). This is attributed to the higher hardness (417 ± 21 HV) in LB-PBF samples than the wrought samples (392 ± 24 HV). Higher hardness was due to the grain refinement as the consequence of faster cooling rate in the LB-PBF process [116]. Grain refinement increases the density of dislocation near the grain boundaries, which contribute in enhancing the hardness as well as wear resistance. Adhesive type of wear mechanism was found as the dominant wear in the worn-out surfaces due to dry sliding condition [116].

8.18 Conclusion

Ti6Al4V cellular structures with small cell size are preferable for higher wear resistance application, whereas cellular structures with large cell size are desirable for osseo-integration. Among different posttreatment for Ti6Al4V, HT of parts is appropriate to increase the wear resistance since it forms martensitic structure irrespective of the AM process. However, coupling ultrasonic rolling or machining with HT of parts further enhances the wear resistance through increase in hardness. Apart from posttreatment, reinforcement incorporation in Ti6Al4V enhances its usage for several applications. Ti6Al4V metallic alloys are mostly reinforced with TiC through LMD process. Controlling of parameters such as scanning velocity and laser power aids in formation of less amount unmelted carbide particles with hard intermetallic Ti_3Al and contributes in better wear performance. Rather than incorporation of reinforcements throughout the part, cladding of hard reinforcements on Ti6Al4V is found desirable for sliding wear applications as it is economical one. However, in both composite and cladding development on Ti6Al4V, formation of hard intermetallic phases in the part significantly enhanced the wear resistance. Even though the part produced has better resistance to wear, the working environment should possess lubrication for effective reduction in wear and friction irrespective of the AM process. Specifically, exposure of additively manufactured Ti6Al4V part to synovial fluid environment experiences low wear loss due to absolute liquid film lubrication and decreasing surface fatigue response. Overall, the additively manufactured Ti6Al4V part possesses greater wear resistance due to the finer acicular martensite (α') phase compared to the α' martensite and β phases in the part produced through any conventional manufacturing processes. It is noteworthy to mention that material chemistry determines the wear resistance property than the type of processing.

In case of nickel-based superalloy Inconel 718 processed through SLM, EBM, and laser bed fusion, formation of finer grains, and intermetallic phase ((Ni_3(Al, Ti))) γ' has resulted in higher hardness and consequently higher wear resistance. Various posttreatments on the additively manufactured nickel-based superalloy have been carried out to assess the improvement of wear resistance. HT of Inconel 718 alloy further increased the wear resistance by formation and homogeneous dispersion of nanoscale γ' ((Ni_3(Al, Ti))) phase (FCC) and γ'' (Ni_3Nb) phase (BCT) in the matrix. Performing cladding on the heat-treated Inconel 718 parts decreased the wear rate further through hardness improvement and homogeneous distribution of laves phase. To improve the surface integrity as well as wear resistance, machining is performed as post treatment in Inconel 625 produced in SLM process. However, Inconel 718/tungsten carbide composites developed through SLM process has extensive potential for numerous applications since it possess better wear resistance than as built Inconel 718 alloy. As the size of tungsten carbide gets decreases, wear resistance of the composite increased due to better interfacial bonding. Upon incorporation of hexagonal boron nitride in the Inconel 718, antifrictional properties get improved due to its lubricating behavior and hexagonal structure.

Stainless steels are produced through different processes such as SLM, LMD, and DED by controlling different process parameter and found that each parameter influences differently on the wear behavior of stainless steels. In 316L stainless steel produced by SLM process, remelting strategy has been found as an influencing parameter on the wear rate, whereas laser power is found to be least influential. In contrary, an increase in the laser power during laser deposition of 17-4 PH stainless steel decreases the wear rate due to higher formation of martensitic structure along with coarser delta ferrite. An increase in the laser scanning speed in DED of 316L stainless steel decreases the wear rate significantly. Several posttreatments were found to be effective in improving the wear performance of the additive-manufactured as-built stainless steel in different manner. High-temperature HTs improve its wear performance by decreasing the porosity levels. On the other hand, sub-zero treatment of the stainless steel parts improves wear performance by the formation of martensitic structure. As there is significant development in AM of stainless steel, composites are manufactured through incorporation of hard ceramic reinforcements that improve wear resistance to a greater extent. Reinforcing of titanium carbide in the stainless steel through SLM imparts hardness and work hardening, which result in lower wear loss. To improve the wear resistance along with the surface integrity, surface modification process such as nitriding and mechanical attrition treatment is found to be useful. Both the processes impart compressive residual stresses and lower the friction by surface smoothening. As a whole, the additive-manufactured stainless steel outperformed in wear resistance compared to the conventionally manufactured stainless steel. Several factors such as finer grains formation, hierarchical microstructure, high temperature hardness, less softening, and stable oxide layer formation through AM process contributed to improve wear resistance.

References

[1] P. Chandramohan, R. Raman, B. Ravisankar, Anodizing and its effects on mechanical properties and corrosion resistance of laser additive manufactured Ti-6Al-4V alloy, International Journal of Material Research 111 (8) (2020) 654–660.

[2] P. Chandramohan, S. Bhero, B. Abiodun Obadele, P. Apata Olubambi, B. Ravisankar, Effect of built orientation on direct metal laser sintering of Ti-6Al-4V, Indian Journal of Engineering and Materials Sciences 25 (February 2018) 69–77.

[3] M. Lorusso, Tribological and Wear Behavior of Metal Alloys Produced by Laser Powder Bed Fusion (LPBF), Friction, Lubrication and Wear, IntechOpen, 2019.

[4] C.G. Cordovilla, N. Narciso, E. Louis, Abrasive wear resistance of aluminum alloy/ceramic particulate composites, Wear 192 (1996) 170–177, https://doi.org/10.1016/0043-1648(95)06801-5.

[5] M. Viceconti, R. Muccini, M. Bernakiewicz, M. Baleani, L. Cristofolini, Large-sliding contact elements accurately predict levels of bone–implant micromotion relevant to osseointegration, Journal of Biomechanics 33 (2000) 1611–1618.

[6] J. Wieding, A. Wolf, R. Bader, Numerical optimization of open-porous bone scaffold structures to match the elastic properties of human cortical bone, Journal of the Mechanical Behavior of Biomedical Materials 37 (2014) 56–68.

[7] N. Taniguchi, S. Fujibayashi, M. Takemoto, K. Sasaki, B. Otsuki, T. Nakamura, T. Matsushita, T. Kokubo, S. Matsuda, Effect of pore size on bone ingrowth into porous titanium implants fabricated by additive manufacturing: an in vivo experiment, Materials Science and Engineering: C 59 (2016) 690–701.

[8] F. Bartolomeu, M. Sampaio, O. Carvalho, E. Pinto, N. Alves, J.R. Gomes, F.S. Silva, G. Miranda, Tribological behavior of Ti6Al4V cellular structures produced by selective laser melting, Journal of the Mechanical Behavior of Biomedical Materials 69 (2017) 128–134, https://doi.org/10.1016/j.jmbbm.2017.01.004.

[9] F. Bartolomeu, M.M. Costa, J.R. Gomes, N. Alves, C.S. Abreu, F.S. Silva, G. Miranda, Implant surface design for improved implant stability—a study on Ti6Al4V dense and cellular structures produced by selective laser melting, Tribology International 129 (2019) 272−282, https://doi.org/10.1016/j.triboint.2018.08.012.

[10] M. Sampaio, M. Buciumeanu, B. Henriques, F.S. Silva, J.C.M. Souza, J.R. Gomes, Tribocorrosion behaviour of veneering biomedical PEEK to Ti6Al4V structures, Journal of the Mechanical Behavior of Biomedical Materials 54 (2016) 123−130, https://doi.org/10.1016/j.jmbbm.2015.09.010.

[11] S.M. Kurtz, J.N. Devine, PEEK biomaterials in trauma, orthopedic, and spinal implants, Biomaterials 28 (32) (November 2007) 4845−4869, https://doi.org/10.1016/j.biomaterials.2007.07.013.

[12] R.M. Cowie, A. Briscoe, J. Fisher, L.M. Jennings, Wear and friction of UHMWPE-on-PEEK OPTIMA, Journal of the Mechanical Behavior of Biomedical Materials 89 (2019) 65−71, https://doi.org/10.1016/j.jmbbm.2018.09.021.

[13] F. Bartolomeu, M. Buciumeanu, M.M. Costa, N. Alves, M. Gasik, F.S. Silva, G. Miranda, Multi-material Ti6Al4V and PEEK cellular structures produced by selective laser melting and hot pressing: a tribocorrosion study targeting orthopedic applications, Journal of the Mechanical Behavior of Biomedical Materials 89 (2019) 54−64, https://doi.org/10.1016/j.jmbbm.2018.09.009.

[14] F. Bartolomeu, C.S. Abreu, C.G. Moura, M.M. Costa, N. Alves, F.S. Silva, G. Miranda, Ti6Al4V-PEEK multi-material structures—design, fabrication and tribological characterization focused on orthopedic implants, Tribology International 131 (March 2019) 672−678, https://doi.org/10.1016/j.triboint.2018.11.017.

[15] M. Buciumeanu, S. Almeida, F. Bartolomeu, M.M. Costa, N. Alves, F.S. Silva, G. Miranda, Ti6Al4V cellular structures impregnated with biomedical PEEK—new material design for improved tribological behavior, Tribology International 119 (March 2018) 157−164, https://doi.org/10.1016/j.triboint.2017.10.038.

[16] T.A. Dantas, M.M. Costa, G. Miranda, F.S. Silva, C.S. Abreu, J.R. Gomes, Effect of HAp and β-TCP incorporation on the tribological response of Ti6Al4V biocomposites for implant parts, Journal of Biomedical Materials Research Part B: Applied Biomaterials (2017b) 1−7, https://doi.org/10.1002/jbm.b.33908.

[17] M.M. Costa, F. Bartolomeu, N. Alves, F.S. Silva, G. Miranda, Tribological behavior of bioactive multi-material structures targeting orthopedic applications, Journal of the Mechanical Behavior of Biomedical Materials 94 (2019) 193−200, https://doi.org/10.1016/j.jmbbm.2019.02.028.

[18] S. Yerramareddy, S. Bahadur, The effect of laser surface treatments on the tribological behavior of Ti-6Al-4V, Wear 157 (1992) 245−262.

[19] C. Ye, A. Telang, A.S. Gill, Gradient nanostructure and residual stresses induced by ultrasonic nano-crystal surface modification in 304 austenitic stainless steel for high strength and high ductility, Materials Science and Engineering: A 613 (2014) 274−288.

[20] Z. Wang, Z. Xiao, C. Huang, L. Wen, W. Zhang, Influence of ultrasonic surface rolling on microstructure and wear behavior of selective laser melted Ti-6Al-4V alloy, Materials 10 (2017) 1203−1217, https://doi.org/10.3390/ma1010120.

[21] H. Wang, G. Song, G. Tang, Evolution of surface mechanical properties and microstructure of Ti-6Al-4V alloy induced by electropulsing-assisted ultrasonic surface rolling process, Journal of Alloys and Compounds 681 (2016) 146−156, https://doi.org/10.1016/j.jallcom.2016.04.067.

[22] Z. Wang, Z. Liu, ChaofengGao, K. Wong, S. Ye, Z. Xiao, Modified wear behavior of selective laser melted Ti6Al4V alloy by direct current assisted ultrasonic surface rolling

process, Surface and Coatings Technology 381 (2020) 125122, https://doi.org/10.1016/j.surfcoat.2019.125122.

[23] H. Attar, M. Calin, L.C. Zhang, et al., Manufacture by selective laser melting and mechanical behavior of commercially pure titanium, Materials Science and Engineering: A 593 (2014) 170–177.

[24] G. Kaya, F. Yildiz, I. Hacisalihoglu, Characterization of the structural and tribological properties of medical Ti6Al4V alloy produced in different production parameters using selective laser melting, 3D Printing and Additive Manufacturing 6 (2019) 5, https://doi.org/10.1089/3dp.2019.0017.

[25] P. Chandramohan, S. Bhero, B. Abiodun Obadele, P. Apata Olubambi, Laser additive manufactured Ti-6Al-4V alloy: tribology and corrosion studies, The International Journal of Advanced Manufacturing Technology 92 (2017) 3051–3061, https://doi.org/10.1007/s00170-017-0410-2.

[26] X. Tao, Z. Yao, S. Zhang, L. Zhong, Y. Xu, Correlation between heat-treated microstructure and mechanical and fretting wear behavior of electron beam freeform-fabricated Ti6Al4V alloy, JOM 71 (7) (2019), https://doi.org/10.1007/s11837-019-03469-w.

[27] L.E. Murr, E.V. Esquivel, S.A. Quinones, S.M. Gaytan, M.I. Lopez, E.Y. Martinez, et al., Microstructures and mechanical properties of electron beam-rapid manufactured Ti-6Al-4V biomedical prototypes compared to wrought Ti-6Al-4V, Materials Characterization 60 (2009) 96–105.

[28] S. Bruschi, R. Bertolini, A. Ghiotti, Coupling machining and heat treatment to enhance the wear behaviour of an Additive Manufactured Ti6Al4V titanium alloy, Tribology International 116 (2017) 58–68, https://doi.org/10.1016/j.triboint.2017.07.004.

[29] S.C. Tjong, Y.W. Mai, Processing-structure-property aspects of particulate- and whisker-reinforced titanium matrix composites, Composites Science and Technology 68 (3) (2008) 583–601.

[30] C. Cai, C. Radoslaw, J. Zhang, Y. Qian, S. Wen, B. Song, Y. Shi, In-situ preparation and formation of TiB/Ti-6Al-4V nanocomposite via laser additive manufacturing: microstructure evolution and tribological behaviour, Powder Technology 342 (January 15 , 2019) 73–84, https://doi.org/10.1016/j.powtec.2018.09.088.

[31] R.M. Mahamood, E.T. Akinlabi, M. Shukla, S. Pityana, Scanning velocity influence on microstructure, microhardness and wear resistance performance of laser deposited Ti6Al4V/TiC composite, Materials and Design 50 (2013) 656–666, https://doi.org/10.1016/j.matdes.2013.03.049.

[32] R.M. Mahamood, E.T. Akinlabi, Laser metal deposition of functionally graded Ti6Al4V/TiC, Materials and Design 84 (November 2015) 402–410, https://doi.org/10.1016/j.matdes.2015.06.135.

[33] R.M. Mahamood, E.T. Akinlabi, Effect of laser power and powder flow rate on the wear resistance behaviour of laser metal deposited TiC/Ti6Al4V composites, Materials Today: Proceedings 2 (2015) 2679–2686, https://doi.org/10.1016/j.matpr.2015.07.233.

[34] S. Kumar, A. Mandal, K.D. Alok, R.D. Amit, Parametric study and characterization of AlN-Ni-Ti6Al4V composite cladding on titanium alloy, Surface and Coatings Technology 349 (2018) 37–49 (H).

[35] L. Zhu, P. Xue, Q. Lan, G. Meng, R. Yuan, Z. Yang, P. Xu, Z. Liu, Recent research and development status of laser cladding: a review, Optics and Laser Technology 139 (2021) 1–26.

[36] K. Jayakumar, T.S. kumar, B. Shanmugarajan, Review study of laser cladding processes on Ferrous substrates, International Journal of Advanced Multidisciplinary Research 2 (6) (2015) 72–78.

[37] M.R. Torresa, Ripolla, B. Prakash, Tribological behaviour of self-lubricating materials at high temperatures, International Materials Reviews 63 (2018) 309–340.
[38] Z.-Y. Zhou, X.-B. Liu, S.-G. Zhuang, X.-H. Yang, M. Wang, C.-F. Sun, Preparation and high temperature tribological properties of laser in-situ synthesized self-lubricating composite coatings containing metal sulfides on Ti6Al4V alloy, Applied Surface Science 481 (2019) 209–218, https://doi.org/10.1016/j.apsusc.2019.03.092.
[39] L.J. Huang, L. Geng, H.Y. Xu, et al., In situ TiC particles reinforced Ti6Al4V matrix composite with a network reinforcement architecture, Materials Science and Engineering: A 528 (2011) 2859–2862.
[40] N. Li, Y. Xiong, H. Xiong, G. Shi, J. Blackburn, W. Liu, R. Qin, Microstructure, formation mechanism and property characterization of Ti+SiC laser cladded coatings on Ti6Al4V alloy, Materials Characterization 148 (2019) 43–51, https://doi.org/10.1016/j.matchar.2018.11.032.
[41] V. Krishna Balla, A. Bhat, S. Bose, B. Amit, Laser processed TiN reinforced Ti6Al4V composite coatings, Journal of the Mechanical Behaviour of Biomedical Materials 6 (2012) 9–2 0, https://doi.org/10.1016/j.jmbbm.2011.09.007.
[42] H. Sahasrabudhe, A. Bandyopadhyay, In situ reactive multi-material Ti6Al4V-calcium phosphate-nitride coatings for bio-tribological applications, Journal of the Mechanical Behavior of Biomedical Materials 85 (September 2018) 1–11, https://doi.org/10.1016/j.jmbbm.2018.05.020.
[43] M. Das, K. Bhattacharya, S.A. Dittrick, C. Mandal, V. Krishna Balla, T.S. Sampath Kumar, B. Amit, I. Manna, In situ synthesized TiB–TiN reinforced Ti6Al4V alloy composite coatings: microstructure, tribological and in-vitro biocompatibility, Journal of the Mechanical Behaviour of Biomedical Materials 29 (2014) 259–271, https://doi.org/10.1016/j.jmbbm.2013.09.006.
[44] H. Wang, Z. Yao, X. Tao, S. Zhang, D. Xu, M. Oleksander, Role of trace boron in the microstructure modification and the anisotropy of mechanical and wear properties of the Ti6Al4V alloy produced by electron beam freeform fabrication, Vacuum 172 (February 2020) 109053, https://doi.org/10.1016/j.vacuum.2019.109053.
[45] Y. Xue, H.M. Wang, Microstructure and dry sliding wear resistance of CoTi intermetallic alloy, Intermetallics 17 (2009) 89–97.
[46] O.S. Adesina, A.P.I. Popoola, A study on the influence of laser power on microstructural evolution and tribological functionality of metallic coatings deposited on Ti-6Al-4V alloy, Tribology—Materials, Surfaces and Interfaces 11 (2017) 7, https://doi.org/10.1080/17515831.2017.1367150.
[47] N. Eliaz, Corrosion of metallic biomaterials: a review, Materials (Basel) 12 (3) (February 2019) 407, https://doi.org/10.3390/ma12030407.
[48] C.N. Elias, J.H.C. Lima, R. Valiev, M.A. Meyers, Biomedical applications of titanium and its alloys, JOM 60 (3) (2008) 46–49.
[49] J. Alipal, T.C. Lee, P. Koshy, H.Z. Abdullah, M.I. Idris, Evolution of anodised titanium for implant applications, Heliyon 7 (7) (July 2021) e07408, https://doi.org/10.1016/j.heliyon.2021.e07408.
[50] Y. Zhu, X. Chen, J. Zou, H. Yang, Sliding wear of selective laser melting processed Ti6Al4V under boundary lubrication conditions, Wear 368–369 (December 15 , 2016) 485–495, https://doi.org/10.1016/j.wear.2016.09.020.
[51] W. Khun, Q. Toh, X.P. Tan, E. Liu, S.B. Tor, Tribological properties of three- dimensionally printed Ti–6Al–4V material via electron beam melting process tested against 100Cr6 steel without and with Hank's solution, Journal of Tribology 140 (6) (2018) 061606, https://doi.org/10.1115/1.4040158 (8 pages.

[52] M. Qasim Riaz, M. Caputo, M.M. Ferraro, J.J. Ryu, Influence of process-induced anisotropy and synovial environment on wear of EBM built Ti6Al4V joint implants, Journal of Materials Engineering and Performance 27 (2018) 3460−3471, https://doi.org/10.1007/s11665-018-3458-8.
[53] J.J. Ryu, S. Shrestha, M. Guha, J.K. Jung, Sliding contact wear damage of EBM built Ti6Al4V: influence of process induced anisotropic microstructure, Metals 8 (131) (2018) 1−16, https://doi.org/10.3390/met8020131.
[54] S. Liu, C.S. Yung, Additive manufacturing of Ti6Al4V alloy: a review, Materials and Design 164 (February 15 , 2019) 107552.
[55] S.L. Sing, J. An, W.Y. Yeong, F.E. Wiria, Laser and electron-beam powder-bed additive manufacturing of metallic implants: a review on processes, materials and designs, Journal of Orthopaedic Research 34 (2016) 369−385.
[56] D. Herzog, V. Seyda, E. Wycisk, C. Emmelmann, Additive manufacturing of metals, Acta Materialia 117 (2016) 371−392.
[57] J.A. Tamayo, M. Riascos, C.A. Vargas, M. Libia, Baena, Additive manufacturing of Ti6Al4V alloy via electron beam melting for the development of implants for the biomedical industry, Heliyon 7 (5) (May 2021) e06892, https://doi.org/10.1016/j.heliyon.2021.e06892.
[58] B. Song, X. Zhao, S. Li, C. Han, Q. Wei, S. Wen, J. Liu, Y. Shi, Differences in microstructure and properties between selective laser melting and traditional manufacturing for fabrication of metal parts: a review, Frontiers of Mechanical Engineering 10 (2015) 111−125.
[59] M. Fellah, L. Mohamed, A. Omar, L. Dekhil, A. Taleb, H. Rezag, A. Iost, Tribological behavior of Ti-6Al-4V and Ti-6Al-7Nb alloys for total hip prosthesis, Advances in Tribology (2014) 1−13, https://doi.org/10.1155/2014/451387.
[60] W. Zhang, P. Qin, Z. Wang, C. Yang, L. Kollo, D. Grzesiak, K. Gokuldoss Prashanth, Superior wear resistance in EBM-processed TC4 alloy compared with SLM and forged samples, Materials 12 (782) (2019) 1−11, https://doi.org/10.3390/ma12050782.
[61] F. Bartolomeu, M. Buciumeanu, E. Pinto, N. Alves, F.S. Silva, O. Carvalho, G. Miranda, Wear behaviour of Ti6Al4V biomedical alloys processed by selective laser melting, hot pressing and conventional casting, Transactions of Nonferrous Metals Society of China 27 (2017) 829−838, https://doi.org/10.1016/S1003-6326(17)60060-8.
[62] M. Buciumeanu, A. Bagheri, N. Shamsaei, S.M. Thompson, F.S. Silva, B. Henriques, Tribocorrosion behavior of additive manufactured Ti-6Al-4V biomedical alloy, Tribology International 119 (March 2018) 381−388, https://doi.org/10.1016/j.triboint.2017.11.032.
[63] S. Bruschi, R. Bertolini, A. Bordin, F. Medea, A. Ghiotti, Influence of the machining parameters and cooling strategies on the wear behavior of wrought and additive manufactured Ti6Al4V for biomedical applications, Tribology International 102 (2016) 133−142, https://doi.org/10.1016/j.triboint.2016.05.036.
[64] R. Bertolini, S. Bruschi, A. Ghiotti, L. Pezzato, M. Dabal, Influence of the machining cooling strategies on the dental tribocorrosion behaviour of wrought and additive manufactured Ti6Al4V, Biotribology, 11(2017), 60-68, DOI: 10.1016/j.biotri.2017.03.002.
[65] J. Ju, Y. Zhou, K. Wang, Y. Liu, J. Li, M. Kang, J. Wang, Tribological investigation of additive manufacturing medical Ti6Al4V alloys against Al_2O_3 ceramic balls in artificial saliva, Journal of the Mechanical Behavior of Biomedical Materials 104 (April 2020) 103602, https://doi.org/10.1016/j.jmbbm.2019.103602.
[66] W.E. Frazier, Metal additive manufacturing: a review, Journal of Materials Engineering and Performance 23 (2014) 1917−1928, https://doi.org/10.1007/s11665-014-0958-z.

[67] T.D. Ngo, A. Kashania, G. Imbalzano, T. Kate, Q. Nguyen, D. Hui, Additive manufacturing (3D printing): a review of materials, methods, applications and challenges, Composites B: Engineering 143 (June 15 , 2018) 172−196.
[68] S. Kumar, Microstructure and Wear of SLM Materials, International Solid Freeform Fabrication Symposium, 2008.
[69] W. Toh, Z. Sun, X. Tan, E. Liu, S.B. Tor, K. Chee, Comparative study on tribological behaviour of Ti-6Al-4V and CoCrMo samples additively manufactured with electron beam melting, in: Proceedings of the 2nd International Conference on Progress in Additive Manufacturing, Research Publishing, Singapore, 2016.
[70] T.M. Pollock, S. Tin, Nickel-based superalloys for advanced turbine engines: chemistry, microstructure and properties, Journal of Propulsion and Power 22 (2) (2006) 361−374, https://doi.org/10.2514/1.18239.
[71] S.H. Chang, In situ TEM observation of γ', γ'' and δ precipitations on Inconel 718 superalloy through HIP treatment, Journal of Alloys and Compounds 486 (2009) 716−721.
[72] A. Ismail, Corrosion performance of Inconel 625 in high sulphate content, IOP Conference Series: Materials Science and Engineering 131 (2016) 012010, https://doi.org/10.1088/1757-899X/131/1/012010.
[73] C.P. Paul, P. Ganesh, S.K. Mishra, P. Bhargava, J. Negi, A.K. Nath, Investigating laser rapid manufacturing for Inconel-625 components, Optics and Laser Technology 39 (2007) 800−805.
[74] C.M. Kuo, Y.T. Yang, H.Y. Bor, C.N. Wei, C.C. Tai, Aging effects on the microstructure and creep behavior of Inconel 718 superalloy, Materials Science and Engineering A 510 (2009) 289−294.
[75] B. Izquierdo, S. Plaza, J.A. Sanchez, I. Pombo, N. Ortega, Numerical prediction of heat affected layer in the EDM of aeronautical alloys, Applied Surface Science 259 (2012) 780−790.
[76] L. Zheng, M.C. Zhang, J.X. Dong, Hot corrosion behavior of powder metallurgy Rene95 nickel-based superalloy in molten $NaCl-Na_2SO_4$ salts, Materials and Design 32 (2011) 1981−1986.
[77] Q. Jia, D. Gu, Selective laser melting additive manufacturing of Inconel 718 superalloy parts: densification, microstructure and properties, Journal of Alloys and Compounds 585 (2014) 713−721, https://doi.org/10.1016/j.jallcom.2013.09.171.
[78] S.K. Sharma, K. Biswas, J. Dutta Majumdar, Wear behaviour of Electron beam surface melted Inconel 718, Procedia Manufacturing 35 (2019) 866−873, https://doi.org/10.1016/j.promfg.2019.06.033.
[79] Y. Gao, M. Zhou, Superior mechanical behaviour and fretting wear resistance of 3D-printed Inconel 625 superalloy, Applied Science 8 (2018) 1−15, https://doi.org/10.3390/app8122439, 2439.
[80] B. Dubiel, S. Jan, Precipitates in additively manufactured Inconel 625 superalloy, Materials 12 (2019) 1144, https://doi.org/10.3390/ma12071144.
[81] Y. Karabulut, E. Tascioglu, Y. Kaynak, Heat Treatment Temperature-Induced Microstructure, Microhardness and Wear Resistance of Inconel 718 Produced by Selective Laser Melting Additive Manufacturing, 2019, p. 163907, https://doi.org/10.1016/j.ijleo.2019.163907. Optik.
[82] Z. Zhao, H. Qu, P. Bai, L. Jing, L. Wu, P. Huo, Friction and wear behaviour of Inconel 718 alloy fabricated by selective laser melting after heat treatments, Philosophical Magazine Letters 98 (12) (2019) 547−555, https://doi.org/10.1080/09500839.2019.1597991.

[83] S. Marian Zaharia, L. Antoneta Chicos, C. Lancea, M.A. Pop, Effects of homogenization heat treatment on mechanical properties of Inconel 718 sandwich structures manufactured by selective laser melting, Metals 10 (2020) 645, https://doi.org/10.3390/met10050645.

[84] D. Bourell, J.P. Kruth, M. Leu, G. Levy, D. Rosen, A.M. Beese, A. Clare, Materials for additive manufacturing, CIRP Annals 66 (2017) 659−681.

[85] E. Tascioglu, Y. Kaynak, O. Poyraz, A. Orhangul, S. Oren, The effect of finish-milling operation on surface quality and wear resistance of Inconel 625 produced by selective laser melting additive manufacturing, in: S. Itoh, S. Shukla (Eds.), INCASE 2019, LNME, 2020, pp. 263−272.

[86] J.M. Wilson, Y.C. Shin, Microstructure and wear properties of laser-deposited functionally graded Inconel 690 reinforced with TiC, Surface and Coatings Technology 207 (2012) 517−522.

[87] Z. Chen, W. Pei, S. Zhang, B. Lu, L. Zhang, X. Yang, K. Huang, Y. Huang, X. Li, Q. Zhao, Graphene reinforced nickel-based superalloy composites fabricated by additive manufacturing, Materials Science and Engineering: A 769 (January 2 , 2020) 138484.

[88] D.D. Gu, Laser Additive Manufacturing of High-Performance Materials, first ed., Springer-Verlag, Berlin Heidelberg, Germany, 2015.

[89] T. Rong, D. Gu, Q. Shi, S. Cao, M. Xia, Effect of tailored gradient interface on wear properties of WC/Inconel 718 composite using selective laser melting, Surface and Coatings Technology 307 (2016) 418−427, https://doi.org/10.1016/j.surfcoat.2016.09.011.

[90] Q. Shi, D. Gu, K. Lin, W. Chen, M. Xia, D. Dai, The role of reinforcing particle size in tailoring interfacial microstructure and wear performance of selective laser melting WC/Inconel 718 composites, Journal of Manufacturing Science and Engineering 140 (11) (2018) 111019, https://doi.org/10.1115/1.4040544 (12 pages).

[91] S.P. Rawal, Metal-matrix composites for space applications, JOM 53 (2001) 14−17.

[92] S. Hoon Kim, G.-H. Shin, B.-K. Kim, K. Tae Kim, D.-Y. Yang, C. Aranas Jr., J.-P. Choi, J.-H. Yu, Thermo-mechanical improvement of Inconel 718 using ex situ boron nitride-reinforced composites processed by laser powder bed fusion, Scientific Reports 7 (2017) 14359, https://doi.org/10.1038/s41598-017-14713-1.

[93] M.Z. Ghodsi, S. Khademzadeh, E. Marzbanrad, M.H. Razmpoosh, N. De Marchi, E. Toyserkani, Development of Yttria-stabilized zirconia reinforced Inconel 625 metal matrix composite by laser powder bed fusion, Materials Science and Engineering: A 827 (October 19 , 2021) 142037.

[94] Y. Zhnag, Q. Pan, L. Yang, R. Li, J. Dai, Tribological behaviour of IN718 superalloy coating fabricated by laser additive manufacturing, Lasers in Manufacturing and Materials Processing 4 (2017) 153−167, https://doi.org/10.1007/s40516-017-0043-1.

[95] M. Odnobokova, A. Belyakov, N. Enikeev, R. Kaibyshev, R.Z. Valiev, Microstructural changes and strengthening of austenitic stainless steels during rolling at 473 K, Metals 10 (2020) 1614, https://doi.org/10.3390/met10121614.

[96] A.J. Sedriks, Corrosion resistance of austenitic Fe-Cr-Ni- Mo alloys in marine environments, International Metals Reviews 27 (1982) 321−353.

[97] H. Chen, D. Gu, Effect of metallurgical defect and phase transition on geometric accuracy and wear resistance of iron-based parts fabricated by selective laser melting, Journal of Materials Research 31 (10) (2016) 1477−1490.

[98] Y. Yang, Y. Zhu, H. Yang, Enhancing wear resistance using selective laser melting (SLM): influence of scanning strategy, Jurnal Tribologi 23 (2019) 113−124.

[99] H. Li, M. Ramezani, M. Li, C. Mac, J. Wang, Effect of process parameters on tribological performance of 316L stainless steel parts fabricated by selective laser melting, Manufacturing Letters 16 (2018) 36–39, https://doi.org/10.1016/j.mfglet.2018.04.003.
[100] A. Adeyemi, E.T. Akinlabi, R.M. Mahamood, Influence of laser power on the microhardness and Wear resistance properties of laser metal deposited 17-4 ph stainless steel, Annals of "Dunarea De Jos" University of Galati 29 (2018), https://doi.org/10.35219/awet.2018.08.
[101] K. Benarji, Y. Ravi Kumar, C.P. Paul, A.N. Jinoop, K.S. Bindra, Parametric investigation and characterization on SS316 built by laser-assisted directed energy deposition, Proceedings of IMechE L: Journal of Materials: Design and Applications 234 (3) (2019) 452–466, https://doi.org/10.1177/1464420719894718.
[102] M. Yakout, M. Elbestawi, S.C. Veldhuis, On the characterization of stainless steel 316L parts produced by selective laser melting, The International Journal of Advanced Manufacturing Technology 95 (2018) 1953–1974.
[103] E. Tascioglu, Y. Karabulut, Y. Kaynak, Influence of heat treatment temperature on the microstructural, mechanical, and wear behavior of 316L stainless steel fabricated by laser powder bed additive manufacturing, The International Journal of Advanced Manufacturing Technology 107 (2020) 1947–1956, https://doi.org/10.1007/s00170-020-04972-0.
[104] M. Sugavaneswaran, A. Kulkarni, Effect of cryogenic treatment on the wear behavior of additive, manufactured 316L stainless steel, Tribology in Industry 41 (1) (2019) 33–42, https://doi.org/10.24874/ti.2019.41.01.04.
[105] Z. Zhao, J. Li, P. Bai, H. Qu, M. Liang, H. Liao, L. Wu, P. Huo, H. Liu, J. Zhang, Microstructure and mechanical properties of TiC-reinforced 316L stainless steel composites fabricated using selective laser melting, Metals 9 (2019) 267.
[106] L. Jing, Z. Zhao, P. Bai, H. Qu, M. Liang, H. Liao, L. Wu, P. Huo, Tribological behavior of TiC particles reinforced 316Lss composite fabricated using selective laser melting, Materials 12 (2019) 950, https://doi.org/10.3390/ma12060950.
[107] Q. Chao, V. Cruz, S. Thomas, N. Birbilis, P. Collins, A. Taylor, P.D. Hodgson, D. Fabijanic, On the enhanced corrosion resistance of a selective laser melted austenitic stainless steel, Scripta Materialia 141 (2017) 94–98.
[108] E. Brinksmeier, G. Levy, D. Meyer, A.B. Spierings, Surface integrity of selective laser melted components, CIRP Annals Manufacturing Technology 59 (2010) 601–606.
[109] M. Godec, C. Donik, A. Kocijan, B. Podgornik, D.A. Skobir Balantic, Effect of post-treated low-temperature plasma nitriding on the wear and corrosion resistance of 316L stainless steel manufactured by laser powder-bed fusion, Additive Manufacturing 32 (March 2020) 101000, https://doi.org/10.1016/j.addma.2019.101000.
[110] Y. Sun, R. Bailey, A. Moroz, Surface finish and properties enhancement of selective laser melted 316L stainless steel by surface mechanical attrition treatment, Surface and Coatings Technology 378 (November 25 , 2019) 124993, https://doi.org/10.1016/j.surfcoat.2019.124993.
[111] K. Antony, N. Arivazhagan, K. Senthilkumaran, Numerical and experimental investigations on laser melting of stainless steel 316L metal powders, Journal of Manufacturing Processes 16 (2014) 345–355.
[112] Y. Okazaki, E. Gotoh, Comparison of metal release from various metallic biomaterials in vitro, Biomaterials 26 (2005) 11–21.
[113] F. Bartolomeu, M. Buciumeanu, E. Pinto, N. Alves, O. Carvalho, F.S. Silva, G. Miranda, 316L stainless steel mechanical and tribological behavior—a comparison between

selective laser melting, hot pressing and conventional casting, Additive Manufacturing 16 (August 2017) (2010) 81–89, https://doi.org/10.1016/j.addma.2017.05.007.

[114] S. Alvi, K. Saeidi, F. Akhtar, High temperature tribology and wear of selective laser melted (SLM) 316L stainless steel, Wear 448–449 (15) (2020) 203228, https://doi.org/10.1016/j.wear.2020.203228.

[115] Y. Sun, A. Moroz, K. Alrbaey, Sliding wear characteristics and corrosion behaviour of selective laser melted 316L stainless steel, Journal of Materials Engineering and Performance 23 (2) (2014) 518–526, https://doi.org/10.1007/s11665-013-0784-8, 23.

[116] S. KC, P.D. Nezhadfar, C. Phillips, M.S. Kennedy, N. Shamsaei, R.L. Jackson, Tribological behavior of 17–4PH stainless steel fabricated by traditional manufacturing and laser-based additive manufacturing methods, Wear 440–441 (15 December 2019) (2019) 203100, https://doi.org/10.1016/j.wear.2019.203100.

Tribology of additively manufactured titanium alloy for medical implant

9

Rasheedat M. Mahamood[1,2], *Tien-Chien Jen*[2], *Stephen A. Akinlabi*[3], *Sunil Hassan*[3] *and Esther T. Akinlabi*[2,4]
[1]Department of Material and Metallurgical Engineering, University of Ilorin, Ilorin, Nigeria; [2]Department of Mechanical Engineering Science, University of Johannesburg, Auckland Park Kingsway Campus, Johannesburg, South Africa; [3]Department of Mechanical Engineering, Butterworth Campus, Walter Sisulu University, Butterworth, Eastern Cape, South Africa; [4]The Directorate, Pan Africa University for Life and Earth Sciences Institute, Ibadan, Nigeria

9.1 Introduction

Titanium and titanium alloys are among the most important advanced materials with good exciting of properties and are used in a number of application areas. Some of these properties include low density, high strength-to-weight ratio, and excellent corrosion resistance [1,2]. The strength-to-weight ratio of titanium is more than any other metallic materials and it also has a better corrosion resistance property than the best grade of stainless steels [3,4]. The commercially available pure titanium is about 99.2% pure and its strength is comparable to the strength of some steels, but it weighs less than steel, of about 45% lighter than steel [5]. Titanium has a relatively high melting point of about 1675 °C and is about 60% denser than aluminum, but it is two times stronger than aluminum [6]. These impressive properties have made titanium a high sort after material in numerous applications that include the aerospace, biomedicals, marine, and automobile industries [7]. Titanium alloys are grouped into four, namely alpha alloy, near alpha, alpha-beta, and beta alloys and are designated with grades [4,8]. Titanium grades 1−4 are known as commercially pure titanium (CP), while the other grades are referred to as titanium alloys. The titanium grade 5 is a titanium alloy that consists of an alpha-beta phase (also called Ti6Al4V, Ti−6Al−4V, Ti 6−4, or Ti64) [4]. Ti6Al4V is called the workhorse on the industry and it is the most widely used of all titanium alloys. The reason for its popularity is as a result of its better properties [4]. It is heat treatable to increase its strength and it can be welded and employed at high service temperatures. It accounts for about 50% of the titanium alloys used in the industries worldwide [4]. Some industries that use Ti6Al4V include the following: the aerospace, marine, chemical processing industries, and medical industries. Titanium grade 23 is the highest purity version of Ti6Al4V and it is referred to as Ti6Al4V ELI or surgical titanium. It is preferred

Tribology of Additively Manufactured Materials. https://doi.org/10.1016/B978-0-12-821328-5.00009-3

when high strength, light weight, high toughness, and good corrosion resistance properties are required, and it has a superior toughness than other titanium alloys [7. 9]. It has good fatigue strength, low modulus of elasticity, and its biocompatibility made it number one choice in biomedical applications and it is used in surgical procedures such as orthopedic cables, screws and orthopedic pins, in joint replacements, orthodontic appliances, ligature clips, cryogenic vessels, springs, surgical staples, and bone fixation devices [9].

The biocompatibility (nontoxic to the human body) of titanium and its inherent ability of osseointegration is responsible for its use as surgical implements as well as dental and orthopedic implants including dental implants that can last for more than 20 years and for hip balls and sockets implant [9]. Titanium has low modulus of elasticity, making it to be relatively close to the human bone that makes it possible to be used as implant in conjunction with human bone. This allows the skeletal loads to be evenly shared between human bone and the implant which helps to reduce the bone degradation that is caused by induced stress leading to bone fractures. Titanium is also nonferromagnetic which makes the titanium implants to be safely examined with magnetic resonance imaging or X-ray. The stiffness of titanium is more than twice that of the human bone which could cause the adjacent bone to bear a reduced load which may cause it to deteriorate [10]. This problem is often prevented by making the implant to be porous [11]. All these properties are the bulk properties of titanium that meets the requirement as biomaterials. However, the interaction between the surface of the implant and the biological environment, the human body, creates some biological reactions which make it important to have the surface of the implant modified in order for the implant to behave as required during its service life. Some of the manufacturing processes employed to produce the implant usually result in some surface properties that are not desirable in biological environments. Such surface properties include oxidized surface, contaminated surface that are stress, plastically deformed, and nonuniform [13]. These types of surface properties are not suitable for biomedical applications and the surface of the implant needs to be modified. Depending on the area where the implant will be used in the body, some specific surface properties may be desired such as good wear resistance properties where the implant may come in sliding contact with the bone or other organs in the body. Good corrosion resistance may be desired where the implant will be in contact with body fluid that is corrosive. Other application need could be the necessity to have to have proper organ integration. The proper surface modification is needed not only to satisfy the operational requirement of the implant but also to enable the bulk material properties to be retained [13]. There are several surface modification methods that are used for titanium implants which include sand blasting, plasma spray deposition, sputtering deposition, and cathodic arc deposition. The use of laser for surface modification of implant is new and promising with a lot of advantages [12,14,15].

In this chapter, the surface modification of titanium alloy implants using additive manufacturing technology—laser metal deposition process—is presented and some of the research works in this field are reviewed. The organization of this chapter is as follows: the brief background of titanium and its alloys are presented is section 2; properties of titanium and its biomedical applications are discussed in section 3; types of surface modifications performed on titanium implants and techniques used to achieve them are discussed in section 4; section 5 presents the use of laser metal deposition for titanium alloy implants with some of research work in this area. The summary is presented in section 6 and some future research directions are highlighted.

9.2 Brief background of titanium and its titanium alloys

Titanium was first discovered in England by William Gregor in 1791 and it is one of the most important metals in use in the industries today [16]. Titanium alloys may be classified into four groups as α, near α, $\alpha + \beta$, and β [16]. The alpha phase is strong and less ductile, while the beta phase is more ductile. Alloying elements are added to the pure titanium to either stabilize the alpha grain's structure or to stabilize the beta phase. The alpha stabilizing elements include aluminum, nitrogen, and carbon and the beta stabilizing elements include vanadium, iron, and manganese. All these elements are added to improve the properties of titanium. Alpha-titanium alloys exhibits better corrosion resistance; however, it exhibits low temperature strength. Alpha + beta alloys on the other hand exhibit better strength due to the presence of both. The proportion of the alpha and beta phases in titanium alloy determines the properties that the titanium alloys exhibit. The superior corrosion resistance is exhibited by beta titanium alloys and low elastic modulus [17]. The excellent properties of titanium and its alloys made them to be material of choice for many industries, but their applications are limited because of the high cost of the materials [17]. The aerospace and petrochemical industries are interested in the high strength to weight ratio of titanium alloys due to the weight saving that the alloys offer which is key in these industries. The corrosion resistance properties are also another attractive property that made them to be favored in several industries [18]. The low modulus and biocompatibility are the key properties that make them attractive to the medical industry. The aerospace industry is also interested in titanium due to its cryogenic temperature capabilities. Some of these properties are explained as follows:

The high strength to weight ratio offers a great advantage in the transportation industry such as aerospace and automobile [16,17]. The high strength with low density offers a weight saving advantage for those industries which will in turn help to reduce the fuel consumption and hence help to reduce the carbon footprint of these industries. Example of where titanium has helped to save weight in the aerospace industry is the replacement of the low alloy steel with titanium alloy in the landing gear on the Boeing 777 aircraft [18]. The replacement of this steel with titanium alloy produced a weight saving of about 580 kg. Titanium and its alloys provide many opportunities for weight savings. Apart from the weight saving advantage, the high corrosion resistance of the titanium alloy helps to keep the maintenance cost low considerably [7]. Also, the use of titanium for the reciprocating parts like connecting rods in automotive engines will reduce the engine weight significantly, especially in the heavy-duty engines. Titanium also offers a great advantage where space limitation is a problem [7].

The ability of titanium and its alloy to retain its properties at elevated temperatures is another property that makes titanium and its alloys to be attractive in engine structure and exhaust areas. Titanium can maintain its properties even at a temperature above 500°C. This made them to be preferred to nickel-based alloys which are heavier than titanium in applications below 600°C. Titanium is also preferred for impellors for rocket engines for cryogenic temperatures application [7].

Titanium forms strong passive oxide coating when exposed to the atmosphere. The titanium oxide formed creates a strong protection layer on the surface of the bulk material that prevents further environmental attack on the material. The passive oxide layer is only a few nanometers thick even after several years [9]. This gives the titanium the characteristic high corrosion resistance property which is close to that of platinum. The protective layer of this oxide of titanium can withstand some corrosive media such as dilute sulfuric and hydrochloric acid and chloride solution [19]. Titanium is corrosion resistance to human body fluids which is one of the reasons why it is favoured as biomedical implants.

Titanium has low modulus of elasticity which is very close to that of the human bone [3,8]. This property is responsible for the reason why titanium is used for orthopedic implants. Also, the low modulus of elasticity of titanium is one of the attractive properties that make them to be used as springs. Replacing the steel spring with titanium will amount to a weight saving of about 70% because the modulus of titanium is about half of that of steel and the weight is lighter; only about half of the number of coils are required from the titanium spring [8]. The main limitation of titanium is its high cost and that is attributed to the limited use of the material and is responsible for its use where the cost justified the intended use. The high cost of titanium is from the high energy required for its production as well as high cost of machining, because titanium is classified as a difficult to machine material [20]. The biocompatibility property of titanium and its biomedical application is discussed in the next section.

9.3 Titanium biocompatibility

The corrosion resistance property is one of the most important properties that is desirable from a biomedical material because the body fluids such as blood and saliva can cause corrosion when they are in contact with metals. Also, some bodily fluids contain protein and amino acid which tend to aggravate corrosion process [21]. If metallic material that is used for medical implant corrodes, the corrosion product of the material (metal ion) is released into the body fluid and causes toxicity and allergic reaction in patients. Therefore, the material to be used as a medical implant needs to have high corrosion resistance property. When titanium comes in contact with body fluids, the corrosion rates are extremely low, and the oxide formed on the outer layer of the material prevents further corrosion activity [22] on the material which is why titanium and its alloys are widely being used for biomedical applications.

The high specific strength exhibited by titanium and its alloys and low modulus of elasticity is responsible for the attractiveness of titanium and its alloys as orthopedic implants. Another very important property of titanium that makes them to be preferred in the biomedical industry is its biocompatibility. Biocompatibility is the ability of a material not to cause potential toxicity because of its coming in contact with a human body. That is, the ability of a material to not cause any reactive response when it comes in contact with the body. Titanium and its alloys are relatively not reactive and do not corrode significantly in biological environment such as body fluid. Titanium has been proven to support cell growth and differentiation and readily absorbs proteins from the body fluid [23–25]. All these properties are responsible for the reason why titanium and its alloys are the preferred biomedical material. The areas of application of titanium in the medical industry are highlighted in the next subsection.

9.3.1 Biomedical applications of titanium and its alloys

Titanium and its alloys are widely employed in biomedical applications which include medical implants as well as medical surgery equipment. Titanium and its alloys characterized by good corrosion resistance, high specific strength, low modulus, and biocompatibility are the key driving factors for the increased use of titanium in biomedical applications. Commercially pure titanium, Ti6Al4V, and some new developed titanium-based alloys are the commonly used titanium used for medical applications. The biomedical applications of titanium and its alloys are classified based on their areas of application in the biomedical industries as follows:

The low modulus of elasticity of titanium is the key property because they are being used as bone and dental implants. Titanium is used in the production of hip implant (see Fig. 9.1), in knee joint replacements, and as dental implant.

Titanium and its alloys are also used in cardiovascular implants because of excellent properties. Prosthetic heart valves, protective cases in pacemakers, and circulatory devices are made using titanium. The high specific strength and high corrosion resistance are the key properties of titanium that make them to be attractive in cardiovascular applications. Also, the titanium is used in the mechanical components of the pump in the heart. Titanium and its alloys are often used in osteosynthesis, such as bone fracture-fixation. Osteosynthesis is a method of treating the bone fracture by surgical means which include bone screws and bone plates. Despite all the exciting properties of titanium, the wear resistance behavior is poor [26]. The poor wear resistance property among other things of titanium and its alloys has necessitated the need for surface modification of titanium used for medical implants.

Titanium has low wear and abrasion resistance properties, and in applications where they will be rubbing other surfaces, their surfaces need to be altered in order to improve the wear resistance behavior. For example, in hip and knee joint implant, the surface of the titanium implant is required to be optimized for better wear resistance property. Also, in order to achieve proper integration of the implant and the body tissue, the surface of the titanium implant needs to be modified in order to assist in biological bonding of the implant with the bone or other tissues. The surface of the titanium implant plays an important role in the way the human body will react with the implant

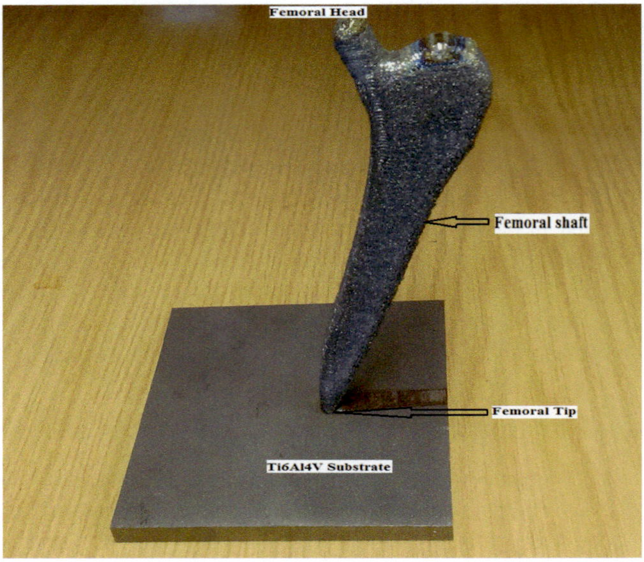

Figure 9.1 Pictorial diagram of a model hip implant made with Ti6Al4V printed at CSIR, Pretoria, South Africa.

and when the reaction may start [13]. The manufacturing methods for the implant also play an important role on the type of the surface property that the implant will have. For example, titanium implant made through a high temperature method may produce surfaces that are oxidized. During the forming process, the surface of the implant may also be plastically deformed, nonuniform, and with residual stresses. This type of surfaces needs to be modified before being used. If the implant is to be used in blood-contacting environment such as artificial heart valves, good corrosion resistance in such medium is necessary and hence the surface needs to be optimized for such application. The surface of titanium implant is required to be modified based on the clinical needs or the service requirement from the implant. Different types of surface modifications performed on titanium implants for the desired clinical requirements are discussed in the next section.

9.4 Types of surface modifications and methods used in surface modification of titanium and its alloys for biomedical applications

Before discussing the types of surface modifications that are performed on titanium for biomedical applications, there is a need to clearly understand the meaning of biomaterials. According to Williams [27], biomaterials are materials that are used to treat, augment, or replace tissue or organ in human body without causing any undesirable interaction with the human body. There are different types of materials used as biomaterials which include metals and ceramics; of all these materials, titanium and its alloys are the most significant materials used for biomedical applications. The excellent properties exhibited by titanium and its alloys, especially the biocompatibility, are responsible for their extensive use in this area. Commercially pure titanium is more favored as dental implants, while the Ti6Al4V are mostly used as orthopedic implants [28,29]. Biocompatibility is the ability of a material to perform as desired in the biological environment (e.g., human body) without causing any local or systematic undesirable reaction to the body but produce beneficial clinical response [28]. Titanium and its alloys are found to remain unchanged when implanted in human body because of their excellent corrosion resistance and they are termed as biostable or biologically inert materials [28]. Although the properties desired from the bulk material are different from those desired from the surface of the implant, the bulk material needs to have the desired properties for the biomechanical properties which is important for the biological performance of the implant such as the strength, weight, and modulus. The properties required from the surface of the implant are different because it is this surface that will be in contact with working environment as well as interact with other organs or devices that they may need to work together. The surface property of the implant should be such that a stable bond can be formed between the implant and the surrounding tissues and also to be able to prevent wear if the implant is required to have a sliding motion against another surface. The type of surface modification will depend on the intended implant location, the containing environment, and the service requirement of the implant. For instance, if the implant will be located where it will be surrounded by blood all the time, then the surface of the implant should have the properties that will prevent undesirable interaction of the implant with the blood such as causing blood coagulation. For bone implants or dental implants, the surface property that will promote osseointegration, are always desired. Also, for implants that will be in sliding motion against its host or other implants, the surface property should be such that the wear resistance property will be improved. In order to enhance the performance of the implant and to prevent implant rejection by the body as well as to reduce implant failure rate (which could be due to bacterial adhesion), there is need for proper surface modification of the surface of the implant for the intended use. There are different types of surface modifications performed on titanium-based implants and they are discussed in the next subsection.

9.4.1 Types of surface modification performed on titanium implants

Corrosion resistance is an important property that is desirable of an implant because the body fluids consist of electrochemical systems which are aggressive corrosion media to the implant. Surface finish of an implant is a key parameter for the behavior of material in corrosive media for implants such those used in the heart or blood vessel region. Surface roughness increases the surface area of a material that is exposed to the corrosive media which will in turn increase the total amount of corrosion. In order to improve the corrosion resistance of the titanium implants that will be used in the location where it will be surrounded by corrosive media, the surface finishing of the implant should be improved [30]. On the other hand, bone implants need to have rough surface and even porosity to promote bone integration with the surrounding tissues [31]. If the implant is unable to properly integrate with the surrounding tissues it can lead to implant failure. The porous coating of the surface of the implant with pore size range between 100 and 200 μm is found to improve osseointegration of the implant [29]. The property of the surface of the implant at the nano level (nano roughness) has also been proved to be advantageous in osseointegration. The higher the surface roughness of the implant, the higher the surface electrostatic charge density produced and hence the higher the adhesion energy of the surface [32]. The increase in surface energy also promotes cell adhesion and also helps to improve wear resistance [32]. Introduction of porosity on the surface of the implant will not only promote proper bone integration but it will also help to further lower the modulus of elasticity which will further help to prolong the life of the implant and prevent the loss of bone mass which could be a potential problem for the neighboring bone [33].

For titanium implants that will be used at the joints, the wear resistance property of the implant needs to be improved. This is key to the service life of the implant as well as reducing the adverse reaction that can result from having the wear debris taken to other areas of the body or organs. Surface modification could also be in form of changing the surface morphology or in form of adding coating on the surface of the titanium implant with other materials such as hydroxyapatite and calcium phosphate in order to improve blood compatibility, corrosion, and wear resistance property of the implant. The combination of surface morphology and coating can also be applied to achieve the desired properties. Different methods are used to achieve surface modification of titanium implants and they are discussed briefly in the next subsection.

9.4.2 Surface modification techniques for titanium and its alloys for biomedical applications

There are different methods used for the surface modification of titanium-based implants; they are basically coating technique or coating deposition method and noncoating technique or surface treatment method. Coating techniques use another material to cover the surface of the bulk material to improve the surface property of the implant. The coating technique involves the introduction of thin film of material onto the surface of the implant thereby substituting the new layer with the original surface of the bulk material. The new surface has different physical and chemical properties from that of the bulk material and there is an interface between the layer and the bulk material. The main advantage of this method is that a wide range of surface properties can be achieved; however, because of the interface that exists between the bulk material and the deposited layer, delamination can occur leading to the failure of the coating. In the noncoating technique, on the other hand, surface treatment technique achieves surface modification without deposition of another material, but the treatment allows the surface properties of the material to be altered by altering the surface roughness or morphology. The range of properties that is achievable with the noncoating technique is limited when compared to the coating method. Although the main advantage of noncoating technique is that there is no problem of surface failure (delamination) as it is the case with the coating technique. Also coating may be undesirable in some applications where the coating could result in creating an isolation layer between the implant and the surrounding organism where interaction is necessary especially in protein absorption. In such an instance, one may consider noncoating method. There are different types of coating techniques that are employed on titanium-based implants such as physical, chemical, electrochemical, and biochemical method. The different types of noncoating techniques include mechanical, chemical, and electrochemical methods. Table 9.1 below summarized the different surface modification methods used for titanium-based implants. The use of additive manufacturing for surface modification is fairly a new method of surface modification for titanium and its alloy for biomedical applications. Additive manufacturing technique has the advantage of producing the implant no matter the complexity and at the same time achieves the desired surface property in the same additive manufacturing process [34]. The surface modification methods presented in Table 9.1 are applied after the implant has been produced using series of manufacturing processes depending on the complexity of the implant. The additive manufacturing method can build the implant using information produced by the digital image of the implant to be made and with the desired surface properties [34]. In the next section, application of laser metal deposition process for the surface modification of titanium and its alloys for biomedical application is presented.

9.5 Surface modification of titanium using AM technology

The different types of surface modification methods mentioned earlier have various limitations [34]. The thermal spray coatings, for example, possess low coating density and low bond strength exists between the coating and the surface of the implant which is the major cause of implant failure [13,34]. Also, some of these methods produce lower bonding strength and interfacial defect that results in the premature failure of the implant. Additive manufacturing technology has revolutionized the way products are made. The main advantage of using laser in surface modification of implant is that it forms a strong metallurgical bonding between the surface of the implant and the deposited material.

Laser is an acronym that is used to describe the operation of Laser [35]. LASER means Light Amplification by Simulated Emission of Radiation [35]. Laser is produced from light source and the light is amplified in a way that is similar to how a microphone amplifies sound. The light amplification is achieved by a process called simulated emission which is also known as optical amplification [33,36]. The light that is emitted from the light source is used to create an excitation in the atoms in the lasing medium; this is called gain amplification [37]. The optical amplification is achieved through the arrangement of mirrors in the gain chamber (containing solid,

Table 9.1 Surface modification method of titanium-based implants [13].

Surface modification methods	Modified layer	Objective
Mechanical methods Machining, grinding, polishing, blasting Chemical methods, chemical treatment, acidic treatment, Alkaline treatment Hydrogen peroxide treatment Sol-gel Anodic oxidation Chemical vapor deposition (CVD)	Rough or smooth surface formed by subtraction process <10 nm of surface oxide layer 1 μm of sodium titanate gel 5 nm of dense inner oxide and porous outer layer 10 mm of thin film, such as calcium phosphate, TiO2, and silica 10 nm−40 μm of TiO2 layer, adsorption, and incorporation of electrolyte anions 1 μm of TiN, TiC, TiCN, diamond, and diamond-like carbon thin film	Produce specific surface topographies, clean, and roughen surface; improve adhesion in bonding Remove oxide scales and contamination Improve biocompatibility, bioactivity, or bone conductivity Improving biocompatibility, bioactivity, or bone conductivity Improve biocompatibility, bioactivity, or bone conductivity Produce specific surface topographies; improved corrosion resistance; improve biocompatibility, bioactivity, or bone conductivity Improve wear resistance, corrosion resistance, and blood compatibility
Biochemical methods	Modification through silanized titania, photochemistry, self-assembled monolayers, protein resistance, etc.	Induce specific cell and tissue response by means of surface-immobilized peptides, proteins, or growth factors
Physical vapor deposition (PVD) Evaporation, ion plating, Sputtering	1 μm of TiN, TiC, TiCN, diamond, and diamond-like carbon thin film	Improve wear resistance, corrosion resistance, and blood compatibility
Ion implantation and deposition Beam-line ion, implantation, PIII	10 nm of surface modified layer and/or 1 μm of thin film	Modify surface composition; improve wear, corrosion resistance, and biocompatibility
Glow discharge plasma treatment	1 nm−100 mm of surface modified layer	Clean, sterilize, oxide, nitride surface; remove native oxide layer

liquid, or gas); the excited atoms bounce back and forth between these mirrors thereby causing a powerful, amplified, and coherent beam of light called "Laser" [35,38]. The excited atoms emit a coherent type of light. It is the optical amplification that has generated countless images of the light through the mirror arrangement. Laser light is characterized by a single wavelength, same phase position, and they spread out in parallel lines which are known as monochromaticity, coherency, and low divergence respectively [33,35]. The advantage of these characteristics makes the intensity of the laser beam to be high and the high intensity is concentrated only at the point of interest. The advantages of laser have been utilized for surface modification process especially for titanium implant in the recent past [39]. This is because the energy can be directed and easily controlled as required without causing damage to the bulk material due to low heat affected zone it created. There are different types of lasers with different wavelengths, but not all of them are suitable for material processing. The most common types of lasers used in materials processing are the Nd-YAG laser and Co_2 laser, while Nd-YAG is the most commonly used for surface modification of implants [39–42].

Laser cladding has been widely used for the surface modification of titanium-based implants [41–45]. Laser cladding is achieved by preplacing powder on the surface of the implant and uses the energy from the laser to scan the powder. The laser melts the powder, and the melt-pool is rapidly solidified and forms a strong coating on the surface of the implant. Laser cladding has been used to coat titanium with improved tribological properties and high-temperature oxidation resistance [46]. Laser cladding has also been used to perform surface alloying by using the laser to melt the preplaced materials as well as the surface of the material with the intention to cause the preplaced material to mix with the surface of the implant to form the desired alloy on the surface of the material [41,43]. The main disadvantage of using this method for surface modification for medical implant is that the powder needs to be mixed with chemical binder especially for complex shaped implants [46]. The chemical binder is evaporated during the cladding process which could create gas porosity in the implant [47]. Also, the dilution depth is not easy to be controlled with the preplaced laser cladding. Laser metal deposition process does not require chemical binder because the powder is deposited directly from coaxial nozzle and not preplaced. Also, the dilution zone is controllable by controlling the processing parameters. Complex and intricate surfaces can also be easily modified using laser metal deposition process.

The process of performing surface modification on metal implants using lasers has a number of advantages that ranged from increased speed of production to the ability to reach intricate areas that cannot be reached by any other manufacturing process [43]. That is, the ease of coating complex geometry and with the required tailoring of the surface composition. This process is more cost effective in that it has high speed of production and low processing time which reduces the lead time to market of a new product [48]. Laser processing is also used to create a good surface property that can help in osseointegration through its ability to produce implant with required porosity and sufficient surface roughness [49,50]. Laser metal deposition (LMD) process is an additive manufacturing process that is capable of producing three-dimensional (3D) implant directly from the 3D computer aided design (CAD) model of the implant by adding material layer after layer [33,51]. Additive manufacturing has revolutionized the manufacturing industry and the impact continues to be felt in every sector [52–59]. The LMD process can also be used to produce the implant with the desired surface properties in a single manufacturing process because the process can handle more than one material simultaneously [34]. This is one of the key advantages of this process unlike the conventional manufacturing process where the implant needs to be first made through one or more processes and then the surface modification is performed using another process. Also, LMD can be used to perform surface modification on an existing implant and it can also be used to effect repair on damaged component [60]. Laser surface modification has been applied in a number of titanium medical implants and some of the research works in this area are presented in the next subsection.

9.5.1 Additive manufacturing technology for surface modification in titanium-based implant

Additive manufacturing, an advanced manufacturing process, has the capability to produce components such as implants directly from the 3D CAD image of the implant by building up the solid component through the addition of materials layer by layer which is against metal removal as was the practice in subtractive manufacturing process [33,61]. The use of the conventional manufacturing process of implants usually involves the construction of the implant from different manufacturing process and it is made using different materials in order to have an implant with the desired properties. The different materials include structural materials with better mechanical properties such as titanium and its alloys, a porous material, and a corrosion or wear resistance material such as tantalum and cobalt or ceramic materials, respectively. These different materials are mechanically bonded to the base material. One of the problems with fabricating implant material from the conventional manufacturing process is that the mismatch of the mechanical properties of this material and the sharp interface that exists between them will result in premature failure of the implant [34,62]. Galvanic corrosion is another problem with this type of arrangement. Additive manufacturing can offset most of these problems in the sense that some additive manufacturing technology such as laser metal deposition process can handle multiple materials simultaneously making it possible to produce an implant of this nature in a single manufacturing process and with sound metallurgical bonding of different materials [48,51]. The sharp interface between different materials can also be eliminated by making the composite a functionally graded one.

LMD process, an additive manufacturing technology, is capable to perform surface modification on medical implants especially titanium and its alloys because of their ability to process multiple materials at the same time making it possible to deposit composite materials on the implant that are impossible to be made through other processes [33,63]. LMD uses a highly focused laser beam to create a melt pool on the surface of the substrate and the melt pool created is used to melt the powder or wired materials deposited in this melt pool. The heat-affected zone generated in this process is very low because the substrate material acts as a heat sink which results in the rapid cooling of the deposited material. This manufacturing process enables novel materials to be made which due to thermodynamic limitation cannot be produced using other methods [33,51]. A number of research works have been undertaken to enable this important additive manufacturing process to be used for the production of medical implants as well as for surface modification of already produced implants [33,64].

Singh et al. [64] conducted research on surface modification on Ti−6Al−4V by coating it with calcium phosphate using Nd-YAG laser operated in continuous mode. The influence of varying laser power densities (25−50 W/mm2) on the resulting property of the titanium alloy in simulated body condition, Ringer's physiological solution, was investigated. The results showed an improvement in the corrosion and mechanical behavior of the alloy when the power density was initially increased up to the value of 35 W/mm^2. Further increase in the power density resulted in decrease in the corrosion resistance property. The use of laser metal deposition process for the production of porous medical implant of titanium alloy was studied by Mahamood and Akinlabi [50]. The study revealed that laser metal deposition can be used to produce medical implant with the desired porosity and controllable pore size and shape [33,50]. Porosity in medical implant may be required to aid osseointegration as well as to reduce the mismatch between the modulus of elasticity of the implant and the human bone and additive manufacturing has been well studied in this area [65]. The ability to produce dental implant with functionally graded porosity has been investigated by Traini et al. [66]. In their study, titanium dental implants were produced with graded porosity from the core of the implant of the outer structure of the implant. The modulus of the porous implant produced was found to be similar to that of the bone tissue. Creating functionally graded porous structure on the implant or as coating on the implant is also useful to prevent stress concentration between the interface layers where the elastic modulus changes abruptly and functionally graded structures in porous medical implant were the subject of investigation by Grunsven et al [67]. The use of laser metal deposition for the production of implant with the required surface property was demonstrated by Mangano et al. [68] and they conclude that direct laser fabrication of medical implant is an economical manufacturing process and with the capability of producing implants with complex and interconnecting crevices using titanium alloy. Bandyopadhyay et al. [69] studied the fabrication of porous implant of Ti6Al4V using laser engineered net shaping (LENS) [34]. The porosity was designed to reduce the elastic modulus of the implant which was found to be very close to those of human bone [33]. The LENS was used to produce the Ti6Al4V with porosity of between 23% and 32%. The result of this study revealed that the modulus of the porous implant was equivalent to that of human bone. The *in vivo* behavior of the porous Ti6Al4V samples in male Sprague−Dawley rats was studied for 16 weeks [69]. The results showed a significant increase in calcium within the implants which indicates an excellent biological tissue ingrowth through the interconnected porosity and that the total amount of porosity plays a critical role in the tissue ingrowth [69].

Mokgalaka et al. [70] investigated the surface modification of Ti6Al4V with NiTi intermetallic coating using laser metal deposition process. The effect of varying the Ti percentage in the NiTi intermetallic coating on the microstructure and wear resistance property was studied. The results show that the coatings displayed significant improvement in wear resistance up to 80% compared to the substrate. It was concluded that the Ti dendrites formed during solidification as seen in the microstructure played a big role in improving the wear properties of the coating. Sisti et al. [71] investigated the surface modification of titanium implant and carried out biological analysis. Surfaces modification was performed by depositing hydroxyapatite using laser metal deposition process. The results obtained from the coated implant were compared to uncoated implants by inserting the samples with 3 different surfaces into the tibias of 30 rabbits. The group I samples were implants with just machined surface. The group I sample was used as a control experiment. The second group were implants that were just laser irradiated and in the third group, the implants were coated with hydroxyapatite using laser metal deposition process. The result show that the group with three implants performed better in the rabbits. The study concluded that surface modification of titanium implants using LMD can increase osseointegration. Laser metal deposition of tantalum coating on titanium in order to improve the osseointegration properties has also been studied by Balla et al. [72]. They conducted *in vitro* biocompatibility study on the coated implant using human osteoblast cell line. The study showed that an excellent cellular adherence and growth on the coated surface when compared to the uncoated surface. They also observed six times higher of living cell density on the coated surface than on the uncoated surface. High surface energy and wettability of tantalum surface were observed to contribute to its significantly better cell materials. Balla et al. [73] conducted laser deposition of Zr on titanium to improve the surface property of the titanium. After the deposition, the coated surface was oxidized using Nd-YAG laser; the wear resistance behavior of the oxidized coated surface was found to be twice of the as deposited coated surface. The oxidized coatings showed a similar in vitro biocompatibility when compared to that of the pure titanium [73]. Justin et al. [74] invented a method of performing surface modification on medical implant by coating the implant with wear resistance coating using LENS. It was proved that the hard-wear resistant surface will extend the life of the implant especially when applied to a load bearing surfaces such as artificial joint bearing surfaces or a dental implant. The invention was claimed to provide economical fabrication method of implants. Rajesh et al. [75] attempted to produce micropatterned surface structure on titanium substrate by coating hydroxyapatite using LMD process with the laser operated in pulse wave mode. The result showed that surface with an improved mechanical property when compared with the uncoated implant was achieved. Roy et al. [76] used LMD process to perform surface modification on titanium by coating it with tricalcium phosphate (TCP) ceramics to improve the bone cell—materials interactions. The laser power, scanning speed, and powder flow rate were varied to study their effect on the evolving properties. The result showed that as the laser power and powder feed rate were increased, the thickness of the coating was also increased, while the coating thickness was found to decreased as the scanning speed was increased. Also as the scanning speed was reduced from 15 to 10 mm s^{-1}, the hardness of the coating was increased from 882 $\pm$ 67 to 1049 $\pm$ 112 Hv as a result of an increase in the volume fraction of TCP in the coating which is attributed to the higher laser material interaction time. The TCP coating showed a good biocompatibility property. There has been significant research interest in the area of manufacturing of medical implants using the LMD process because of the potential advantages this process can provide [34]. Apart from the fact that the LMD process can be used to produce a customized implant directly from the digital image of the implant, the LMD process can also be used to perform surface modification on an existing implant [33,51]. The LMD process can also be used for repair on failed implant and performed surface modification to further extend the life of the implant. The fabrication of porous implant using the LMD process has also been explored by researchers [77,78]. It was recognized that an approach to promote the cellular and tissue integration of a metal implants is to design the implants with some inherent porosity so that the implant can closely match the properties of human bone. A comprehensive review was conducted by Das et al. [77] on the use of LMD process for the development of materials for load bearing implant in order to improve their properties and increase their in vivo service lifetime. The review showed that LMD process is capable of fabricating near net shaped of metallic implants with designed porosity and it has the potential of producing implants with compositionally graded structure. A number of studies have been conducted on the use of LMD process to produce functionally graded hard coatings on implant (such as functionally graded TiB−TiN coating on porous Ti6Al4V) to minimize wear-induced osteolysis. Another study conducted by Tian et al. [80] demonstrated the use of LMD for surface modification of titanium alloys in order to improve their wear and corrosion resistance properties. The influence of the processing parameters on the resulting surface properties of titanium alloys was also analyzed. The tribological behavior of titanium can effectively be improved through surface modification through addition of TiC using additive manufacturing technology [57]. The wear resistance behavior of Ti6Al4V as reported by Mahamood et al. [57] was shown to be improved when TiC was deposited on the surface using LMD. Fig. 9.2 shows the microstructure of Ti6Al4V before (2a) and

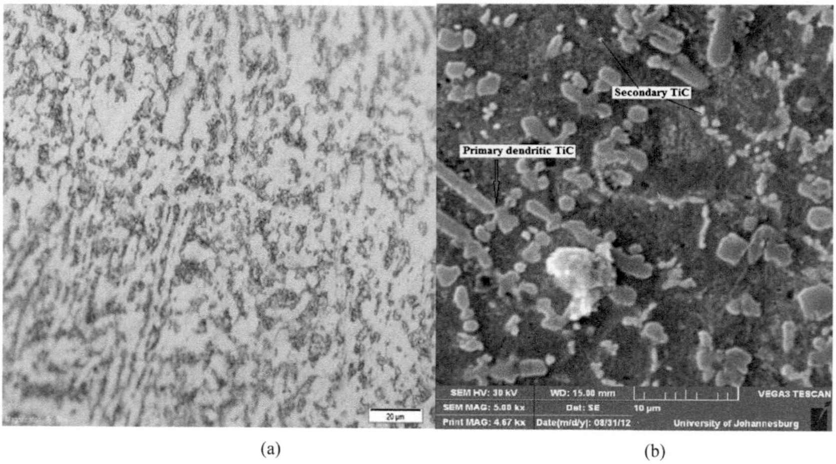

Figure 9.2 Micrograph of Ti6Al4V (A) before surface modification and (B) after surface modification [57].

after surface modification (2b). The presence of primary dendritic TiC and secondary TiC as shown in Fig. 9.2B helps to improve the wear resistance characteristics of Ti6Al4V. During the sliding wear process, the dendritic TiC and the secondary TiC are ground into fine powder that acts as lubricant which inhibits the wear process. The processing parameter (scanning speed) was also found to influence the wear resistance behavior of Ti6Al4V. Fig. 9.3A shows the wear track of sample processed at low scanning speed of 0.015 m/s and Fig. 9.3B shows the sample processed at high scanning speed of 0.065 m/s. The wear track of Ti6Al4V sample not processed is shown in

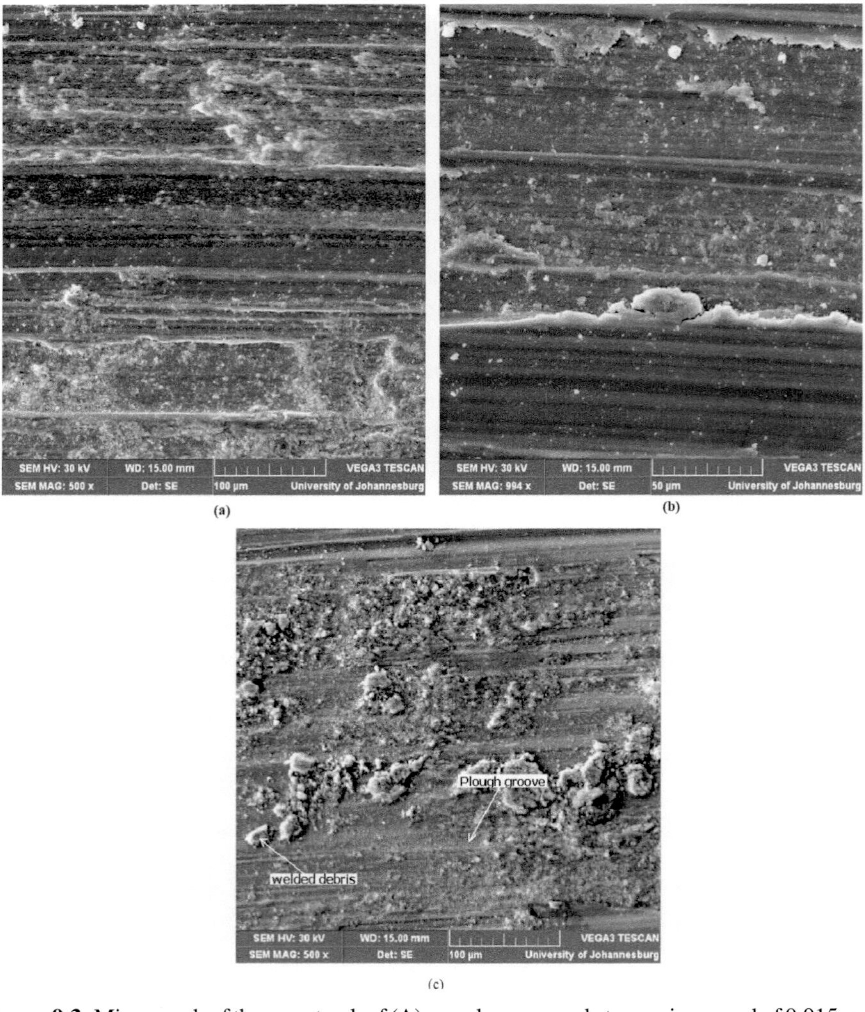

Figure 9.3 Micrograph of the wear track of (A) sample processed at scanning speed of 0.015 m/s, (B) sample processed at scanning speed of 0.065 m/s, and (C) Ti6Al4V not laser processed [57].

Fig. 9.3. The sample processed at a higher scanning speed showed a better wear resistance as can be seen in Fig. 9.3B. This was as a result of the addition of TiC and the right processing parameter. The high scanning speed ensures that the interaction time of the laser energy and the material being processed is limited, that enables some of the TiC to be retained as unmelted powders that also assist in improving the wear resistance property. Unlike at low scanning speed where all the TiC powder is completely melted because of the longer laser material interaction time and hence does not greatly improve the wear resistance property. However, the sample processed the lowest scanning speed considered in that study still has a better wear resistance property than the unprocessed Ti6Al4V as shown in Fig. 9.3C.

9.6 Summary

This chapter presents the tribological behavior of additively manufactured components with a particular reference to titanium and its alloy for biomedical application. Laser metal deposition process is an additive manufacturing technology that is an ideal manufacturing process for difficult to machine materials such as titanium and its alloys. LMD is used in applications such as prototyping and production of medical implants, such as hip and knee implants and with the desired surface properties. LMD can also be used for surface coatings on medical implants to improve the wear resistance and the corrosion properties of the implant as well as for the production of implant with designed porosity that helps in improving osseointegration. The excellent properties of titanium and its alloy as well as their biocompatibilities make them to be favored as medical implant materials. In spite of these excellent properties, the wear resistance properties of this noble material are poor which makes surface modification to be necessary where the implant will come in contact with other surfaces such as kneel implant. Also, the required surface modification on some titanium implant may be to have surface porosity that will help to promote the integration of the surrounding tissues with the implant which is important for the healing process and the stability of the implant in the host environments. Whatever the reason is for the surface modification of titanium implants, LMD process provides the needed flexibility that helps to produce the required surface property at high manufacturing speed and at a lower cost. The use of LMD for surface modification of titanium and its alloys for medical implants to improve their tribological behaviour has been presented in this chapter. A recent development in the fabrication of porous implant using the LMD process was also analyzed. The LMD process is also capable of producing the titanium-based implant with the needed surface properties in a single manufacturing process. Hard phase ceramics coatings such as TiC can readily be applied on the surface of the implant because the LMD process can handle more than one material at the same time. In summary, the capability of LMD process will have a significant influence on the production and the economy of the fabrication of medical titanium-based implants as well as other biomedical materials with tailored mechanical, biological, and the needed surface properties.

References

[1] S. Ramesh, L. Karunamoorthy, K. Palanikumar, Surface roughness analysis in machining of titanium alloy, Materials and Manufacturing Processes 23 (2) (2008) 174−181.

[2] Z.D. Cui, S.L. Zhu, H.C. Man, X.J. Yang, Microstructure and wear performance of gradient Ti/TiN metal matrix composite coating synthesized using a gas nitriding technology, Surface and Coating Technology 190 (2−3) (2005) 309−313.

[3] J.D. Destefani, Introduction to Titanium and Titanium Alloys, vol. 2, Metals Handbook ASM International, Materials Park, 1998, pp. 586−591.

[4] M.J. Donachi, Titanium—A Technical Guide, second ed., ASM International, Metals Park, OH, 2000.

[5] J. Barksdale, Titanium, its Occurrence, Chemistry, and Technology, Ronald Press, New York, 1968.

[6] D.R. Askeland, P.P. Fulay, W.J. Wright, The Science and Engineering of Materials, sixth ed., Global Engineering, Canada, 2011.

[7] R.R. Boyer, Attributes, characteristics, and applications of titanium and its alloys, JOM: The journal of the Minerals, Metals & Materials Society 62 (5) (2010) 35−43.

[8] R.R. Boyer, An overview on the use of Titanium in the aerospace industry, Materials Science and Engineering A 213 (1−2) (1996) 103−114.

[9] J. Emsley, Titanium, in: Nature's Building Blocks: An A-Z Guide to the Elements, Oxford University Press, Oxford, England, UK, 2001. ISBN 0-19-850340-7.

[10] P. Quadbeck, Titanium foams replace injured bones, in: Research News (Fraunhofer-Gesellschaft), September 2010. Retrieved 27 September 2010. Available at: http://www.fraunhofer.de/en/press/research-news/2010/09/titanium-foams-replace-injured-bones.html. (Accessed 5 July 2016).

[11] A. Pinsino, R. Russo, R. Bonaventura, A. Brunelli, A. Marcomini, V. Matranga, Titanium dioxide nanoparticles stimulate sea urchin immune cell phagocytic activity involving TLR/p38 MAPK-mediated signalling pathway, Scientific Reports 5 (2015), https://doi.org/10.1038/srep14492.

[12] R.M. Mahamood, E.T. Akinlabi, M. Shukla, S. Pityana, Characterizing the effect of processing parameters on the porosity properties of laser deposited titanium alloy, in: International Multi-conference of Engineering and Computer Science (IMECS 2014), 2013.

[13] X. Liu, P.K. Chu, C. Ding, Surface modification of titanium, titanium alloys, and related materials for biomedical applications, Materials Science and Engineering R 47 (2005) 49−121.

[14] Vasanthan, H. Kim, S. Drukteinis, W. Lacefield, Implant surface modification using laser guided coatings: in vitro comparison of mechanical properties, Journal of Prosthodontics 17 (2008) 357−364.

[15] I.M. Ghayad, N.N. Girgis, W.A. Ghanem, Laser surface treatment of metal implants: a review article, Journal of Metallurgical Engineering (ME) 4 (2015), https://doi.org/10.14355/me.2015.04.006.

[16] J.J. Polmear, Titanium alloys, in: Light Alloys, Edward Arnold Publications, London, 1981.

[17] R.W. Schutz, Beta titanium alloys in the 1990's, in: D. Eylon, R.R. Boyer, D.A. Koss (Eds.), The Mineral, Metals & Materials Society, Warrendale, PA, 1993, pp. 75−91.

[18] R.R. Boyer, International Conference on Processing and Manufacturing of Advanced Materials (Zurich: Trans Tech Publications, 2003), 2003.

[19] D.R. Lide (Ed.), CRC Handbook of Chemistry and Physics, 2005.
[20] Z.M. Wang, E.O. Ezugwu, Titanium alloys and their machinability a reviewJournal of, Material Processing Technology 68 (1997) 262–270.
[21] R.L. Williams, S.A. Brown, K. Merritt, Electrochemical studies on the influence of proteins on the corrosion of implant alloys, Biomaterials 9 (2) (1988) 181–186.
[22] L.J. Knob, D.L. Olson, Metals Handbook: Corrosion, ninth ed., vol. 13, 1987, p. 669.
[23] A. Rosa, M.M. Beloti, Effect of cpTi surface foughness on human bone marrow cell attachment, proliferation and differentiation, Brazilian Dental Journal 14 (2003) 16–21.
[24] Y. Yang, R. Glover, J.L. Ong, Fibronectin adsorption on titanium surface and its effect on osteoblast precusor cell attachment, Colloids Surface B: Interfaces 30 (2003) 291–297.
[25] S.K. Nishimoto, M. Nishimoto, S.W. Park, K.M. Lee, H.S. Kim, J.T. Koh, J.L. Ong, Y. Liu, Y. Yang, The effect of titanium surface roughening on protein absorption, cell attachment, and cell spreading, The International Journal of Oral & Maxillofacial Implants 23 (4) (2008) 675–680.
[26] M.N. Ahsana, A.J. Pinkerton, R.J. Moatb, J.A. Shackleton, A comparative study of laser direct metal deposition characteristics using gas and plasma-atomized Ti–6Al–4V powders, Materials Science and Engineering: A 528 (2011) 7648–7657.
[27] D. Williams, On the nature of biomaterials, Biomaterials 30 (2009) 5897–5909.
[28] D. Williams, in: D.M. Brunette, P. Tengvall, M. Textor, P. Thompson (Eds.), Titanium in Medicine: Material Science, Surface Science, Engineering, Biological Responses and Medical Applications, Springer-Verlag, Berlin and Heidelberg, 2001, pp. 13–24.
[29] M. Geetha, A. Singh, R. Asokamani, A. Gogia, Ti based biomaterials, the ultimate choice fororthopaedic implants – a review, Prog. Mater. Sci. 54 (2009) 397–425.
[30] K. Vasilev, Z. Poh, K. Kant, J. Chan, A. Michelmore, D. Losic, Tailoring the surface functionalities of titania nanotube arrays, Biomaterials 31 (2010) 532–540.
[31] M.A. Gepreel, M. Niinomi, Biocompatibility of Ti-alloys for long-term implantation, Journal of the Mechanical Behavior of Biomedical Materials 20 (2013) 407–415.
[32] E. Gongadze, D. Kabaso, S. Bauer, J. Park, P. Schmuki, A. Iglič, Adhesion of osteoblasts to a vertically aligned TiO_2 nanotube surface, Mini-Reviews in Medicinal Chemistry 13 (2012) 94–200.
[33] R.M. Mahamood, Laser Metal Deposition Process of Metals, Alloys, and Composite Materials, Springer, 2016.
[34] R.M. Mahamood, E.T. Akinlabi, Functionally Graded Materials, Springer, Switzerland, 2017. http://www.springer.com/gp/book/9783319537559.
[35] H. Haken, Laser Theory, Springer Berlin Heidelberg, 1983.
[36] K. Yamashita, H. Taniguchi, S. Yuyama, K. Oe, J. Sun, H. Mataki, Continuous-wave simulated emission and optical amplification in europium (III)-aluminum nanocluster-doped polymeric waveguide, Applied Physics Letters 91 (8) (2007) 081115–081117.
[37] A.E. Siegman, Lasers, University Science Books. Maple-vail Group Manufacturing Group, USA, 1986.
[38] W.T. Silfvast, Laser Fundamentals, Cambridge University Press, 1996.
[39] Y.S. Tian, C.Z. Chen, S.T. Li, Q.H. Huo, Research progress on laser surface modification of titanium alloys, Applied Surface Science 242 (2005) 177–184.
[40] F.J.C. Braga, R.F.C. Marques, E. de A. Filho, A.C. Guastaldi, Surface modification of Ti dental implants by Nd:YVO4 laser irradiation, Applied Surface Science 253 (2007) 9203–9208.
[41] R. Brånemark, L. Emanuelsson, A. Palmquist, P. Thomsen, Bone response to laser-induced micro- and nano-size titanium surface features, Nanomedicine 7 (2011) 220–227.

[42] Á. Györgyey, K. Ungvári, G. Kecskeméti, J. Kopniczky, B. Hopp, A. Oszkó, I. Pelsöczi, Z. Rakonczay, K. vNagy, K. Turzó, Attachment and proliferation of human osteoblast-like cells (MG-63) on laser-ablated titanium implant material, Materials Science and Engineering C 33 (2013) 4251–4259.
[43] F. Weng, C. Chen, H. Yu, Research status of laser cladding on titanium and its alloys: a review, Materials & Design 58 (0) (2014) 412–425.
[44] Y.S. Tian, C.Z. Chen, S.T. Li, Q.H. Huo, Research progress on laser surface modification of titanium alloys, Applied Surface Science 242 (1–2) (2005) 177–184.
[45] X.-B. Liu, X.-J. Meng, H.-Q. Liu, G.-L. Shi, S.-H. Wu, C.-F. Sun, M.-D. Wang, L.-H. Qi, Development and characterization of laser clad high temperature self-lubricating wear resistant composite coatings on Ti–6Al–4V alloy, Materials & Design 55 (0) (2014) 404–409.
[46] J. Dutta Majumdar, I. Manna, 21 - Laser surface engineering of titanium and its alloys for improved wear, corrosion and high-temperature oxidation resistance, in: J.L.G. Waugh (Ed.), Laser Surface Engineering, Woodhead Publishing, UK, 2015, pp. 483–521.
[47] A.G. Arlt, R. Muller, Technology for wear resistant inside diameter cladding of tubes, Proe. ECLAT 94 (1994) 203–212.
[48] R.M. Mahamood, E.T. Akinlabi, M. Shukla, S. Pityana, Revolutionary additive manufacturing: an overview, Lasers in Engineering 27 (2014).
[49] R.M. Mahamood, E.T. Akinlabi, Effect of laser power on surface finish during laser metal deposition process, WCECS 2 (2014) 965–969.
[50] R.M. Mahamood, E.T. Akinlabi, Influence on degree of porosity in laser metal deposition process, in: G.-C. Yang, et al. (Eds.), Transactions on Engineering Technologies, Springer, 2015, pp. 31–42.
[51] J. Schroers, G. Kumar, T.M. Hodges, S. Chan, T.R. Kyriakides, Bulk metallic glasses for biomedical applications, JOM: The Journal of the Minerals, Metals & Materials Society 61 (2009) 21–29.
[52] R.M. Mahamood, E.T. Akinlabi, Achieving mass customization through additive manufacturing, in: Advances in Ergonomics of Manufacturing: Managing the Enterprise of the Future, Springer, Cham, 2016, pp. 385–390.
[53] R.M. Mahamood, E.T. Akinlabi, Laser-assisted additive fabrication of micro-sized coatings, in: Advances in Laser Materials Processing, Woodhead Publishing, 2018, pp. 635–664.
[54] E.T. Akinlabi, R.M. Mahamood, S.A. Akinlabi, Advanced Manufacturing Techniques Using Laser Material Processing, IGI Global, Hershey, PA, 2016.
[55] R.M. Mahamood, Y. Okamoto, M.R. Maina, S.A. Akinlabi, S. Pityana, M. Tlotleng, E.T. Akinlabi, Wear resistance behaviour of laser additive manufacture materials: an overview, in: 2019 International Conference on Engineering, Science, and Industrial Applications (ICESI), IEEE, 2019, pp. 1–6.
[56] R.M. Mahamood, E.T. Akinlabi, Modelling and optimization of laser additive manufacturing process of Ti alloy composite, in: Optimization of Manufacturing Processes, Springer, Cham, 2020, pp. 91–109.
[57] R.M. Mahamood, E.T. Akinlabi, M. Shukla, S. Pityana, Scanning velocity influence on microstructure, microhardness and wear resistance performance of laser deposited Ti6Al4V/TiC composite, Materials & Design 50 (2013) 656–666.
[58] S. Amin Yavari, S. Ahmadi, R. Wauthle, B. Pouran, J. Schrooten, H. Weinans, A. Zadpoor, Relationship between unit cell type and porosity and the fatigue behavior of selective laser melted meta-biomaterials, Journal of the Mechanical Behavior of Biomedical Materials 43 (2015) 91–100.

[59] S. Kumar Malyala, Y. Ravi Kumar, C.S.P. Rao, Organ printing with life cells: a review, Materials Today: Proceedings 4 (2, Part A) (2017) 1074–1083.
[60] A.J. Pinkerton, W. Wang, L. Li, Component repair using laser direct metal deposition, Journal of Engineering Manufacture 222 (2008) 827–836.
[61] A. Sidambe, Biocompatibility of advanced manufactured titanium implants—a review, Materials 7 (2014) 8168–8188.
[62] J. Scott, N. Gupta, C. Wember, S. Newsom, T. Wohlers, T. Caffrey, Additive Manufacturing: Status and Opportunities, Science and Technology Policy Institute, 2012. Available from, https://www.ida.org/stpi/occasionalpapers/papers/AM3D_33012_Final.pdf. (Accessed 11 July 2016).
[63] R.M. Mahamood, E.T. Akinlabi, M. Shukla, S. Pityana, Improving surface integrity using laser metal deposition process, in: L. Santo, J.P. Davim (Eds.), Surface Engineering Techniques and Applications: Research Advancements, IGI Global, Pennsylvania (USA), 2014, pp. 146–176, https://doi.org/10.4018/978-1-4666-5141-8.ch005.
[64] R. Singh, S.K. Tiwari, K. Suman Mishra, B. Narendra Dahotre, Electrochemical and mechanical behavior of laser processed Ti-6Al- 4V surface in ringer's physiological solution, Journal of Materials Science: Materials in Medicine 22 (2011) 1787–1796.
[65] T. Laoui, E. Santos, K. Osakada, M. Shiomi, M. Morita, S.K. Shaik, N.K. Tolochko, F. Abe, M. Takahashi, Properties of titanium dental implant models made by laser processing, Journal of Mechanical Engineering Science 220 (2006) 857–863.
[66] T. Traini, C. Mangano, R.L. Sammons, F. Mangano, A. Macchi, A. Piattelli, Direct laser metal sintering as a new approach to fabrication of an isoelastic functionally graded material for manufacture of porous titanium dental implants. Direct laser metal sintering as a new approach to fabrication of an isoelastic functionally grad, Dental Materials 24 (2008) 1525–1533.
[67] W. Van Grunsven, E. Hernandez-Nava, G. Reilly, R. Goodall, Fabrication and mechanical characterisation of titanium lattices with graded porosity, Metals 4 (2014) 401–409.
[68] C. Mangano, M. Raspanti, T. Traini, A. Piattelli, R. Sammons, Stereo imaging and cytocompatibility of a model dental implant surface formed by direct laser fabrication, Journal of Biomedical Materials Research Part A 88 (2009) 823–831.
[69] A. Bandyopadhyay, F. Espana, V.K. Balla, S. Bose, Y. Ohgami, N.M. Davies, Influence of porosity on mechanical properties and in vivo response of Ti6Al4V implants, Acta Biomaterialia 6 (2010) 1640–1648.
[70] M.N. Mokgalaka, S.L. Pityana, P.A.I. Popoola, T. Mathebula, NiTi intermetallic surface coatings by laser metal deposition for improving wear properties of Ti-6Al-4V substrates, Advances in Materials Science and Engineering 2014 (2014), https://doi.org/10.1155/2014/363917. Article ID 363917, 8 pages.
[71] K.E. Sisti, R. de Rossi, A.M.B. Antoniolli, R.D. Aydos, A.C. Guastaldi, T.P. Queiroz, I.R. Garcia Jr., A. Piattelli, H.S. Tavares, Surface and biomechanical study of titanium implants modified by laser with and without hydroxyapatite coating, in rabbits, Journal of Oral Implantology 38 (3) (2012) 231–237, https://doi.org/10.1563/AAID-JOI-D-10-00030.
[72] V.K. Balla, S. Banerjee, S. Bose, A. Bandyopadhyay, Direct laser processing of tantalum coating on titanium for bone replacement structures, Acta Biomaterialia 6 (6) (2010) 2329–2334.
[73] V.K. Balla, W. Xue, S. Bose, A. Bandyopadhyay, Laser assisted Zr/ZrO_2 coating on Ti for load-bearing implants, Acta Biomaterialia 5 (2009) 2800–2809.
[74] D. Justin, B. Stucker, T. Fallin, D.J. Gabbita, Based Metal Deposition (LBMD) of Implant Structures, US Patent-US 20070202351 A1, 2007.

[75] P. Rajesh, C.V. Muraleedharan, M. Komath, H. Varma, Laser surface modification of titanium substrate for pulsed laser deposition of highly adherent hydroxyapatite, Journal of Materials Science: Materials in Medicine 22 (7) (2011) 1671−1679.

[76] M. Roy, B. Vamsi Krishna, A. Bandyopadhyay, S. Bose, Laser processing of bioactive tricalcium phosphate coating on titanium for load-bearing implants, Acta Biomaterialia 4 (2) (2008) 324−333.

[77] M. Das, V. Krishna Balla, T.S. Sampath Kumar, I. Manna, Fabrication of biomedical implants using laser engineered net shaping (LENS™), Transactions of the Indian Ceramic Society 72 (3) (2013) 169−174.

[78] Y.S. Tian, C.Z. Chen, D.Y. Wang, T.Q. Lei, Laser surface modification of titanium alloys — a review, Surface Review and Letters 12 (01) (2005) 123−130.

Corrosion in additively manufactured cold spray metallic deposits

10

Mohammadreza Daroonparvar [1,2] *and Charles M. Kay* [2]
[1]Department of Mechanical Engineering, University of Nevada, Reno, NV, United States;
[2]Research and Development Department, ASB Industries Inc., Barberton, OH, United States

10.1 Introduction

The electrochemical corrosion process is normally considered for metallic materials (e.g., coatings [1−7], deposits, etc.) during corrosion. In this process (as a chemical reaction), electrons from one chemical species to another are transferred. In fact, electrons (during oxidation reaction) are released from metal atoms (as their characteristic during corrosion) [8].

Anode is a place where oxidation (is sometimes called anodic reaction) occurs. The electrons produced from each metal atom (due to oxidation reaction) would be transferred and become a part of another chemical species (because of reduction reaction). Cathode is a place where reduction reaction would take place. Simultaneous occurrence of two or more of the preceding reduction reactions has been reported during corrosion [8].

The sum of one oxidation reaction (also known as half-cell reaction) and one reduction reaction (also known as half-cell reaction) can form an overall electrochemical reaction. So, the sum of the two half-cell reactions can lead to the overall chemical reaction [8]. Moreover, the four following conditions are necessary for the corrosion occurrence [9]:

(1) Anodic reaction.
(2) Cathodic reaction.
(3) A metallic path of contact between anodic and cathodic sites.
(4) Electrolyte existence.

A solution that contains dissolved ions (with capability of conducting a current) is known as an electrolyte. Aqueous solution is considered as the most common electrolyte such as water containing dissolved ions; nevertheless, other liquids, such as liquid ammonia, can act as electrolytes. Fig. 10.1 demonstrates the half-cell reactions and overall electrochemical reaction for a Fe surface soaked in a neutral or a basic aqueous solution (as electrolyte) [9].

It is interesting to note that when Mg and its alloys are exposed to the chloride containing solutions, Mg elements quickly dissolved (going into the electrolyte) and corrosion surface film is formed per following reactions [10−17]:

Tribology of Additively Manufactured Materials. https://doi.org/10.1016/B978-0-12-821328-5.00010-X

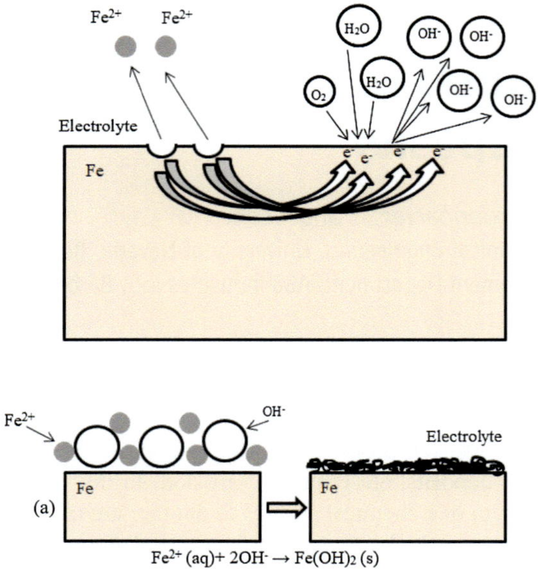

$Fe\ (s) \rightarrow Fe^{2+}(aq) + 2e^-$ at anodic sites (half-cell reaction)

$O_2 + H_2O + 4e^- \rightarrow 4OH^-$ at cathodic sites (half-cell reaction)

Overall reaction $2Fe\ (s) + O_2\ (g) + 2\ H_2O\ (l) \rightarrow 2Fe^{2+}(aq) + 4OH^-\ (aq) \rightarrow Fe(OH)_2\ (s)$

By continuation of reactions: $2Fe(OH)_2\ (s) + \frac{1}{2}\ O_2\ (g) + H_2O\ (l) \rightarrow 2Fe(OH)_3\ (s)$ (known as rust as corrosion layer)

Figure 10.1 The half-cell reactions and overall electrochemical reaction for a Fe surface soaked in a neutral or a basic aqueous solution (as electrolyte) (a) ferrous hydroxide precipitation on the surface of iron.
Modified from E. McCafferty, Introduction to Corrosion Science, Springer, 2010, e-ISBN 978-1-4419-0455-3.

$$Mg(s) \rightarrow Mg^{2+} + 2e^- \quad \text{Anodic reaction} \quad (1)$$

$$2H_2O + 2e^- \rightarrow H_2 \uparrow + 2OH^- \quad \text{Cathodic reaction} \quad (2)$$

$$Mg + 2H_2O \rightarrow Mg(OH)_2 + H_2 \uparrow \quad \text{Overall reaction} \quad (3)$$

$$Mg(OH)_2 + 2Cl^- \rightarrow MgCl_2 + 2OH^- \quad \text{reaction} \quad (4)$$

The reactions 1 and 2 take place on the alloy surface during the corrosion process. Hence, the formation of Mg $(OH)_2$ film (as brucite phase, reaction 3) on the alloy surface started. This could result in the reduction of the corrosion rate of Mg alloy (at the onset of corrosion). However, the presence of Cl^- in the chloride containing solution could destroy the protective performance of Mg $(OH)_2$ corrosion film by the following reaction (4). Severe Cl^- adsorption and Mg dissolution (simultaneously) can be accelerated, when concentration of Cl^- (more than 30 mmol/L) is very high in the corrosive electrolyte. This phenomenon could lead to the formation of pits on the Mg alloy surface and followed by formation of $MgCl_2$ or Mg $(OH)_2$ (as corrosion products) at the vicinity of pits [18,19].

Hence, production of corrosion resistant materials (notable reduction of metallic atoms tendency from going into the electrolyte during corrosion) and protective coatings with high reliability, quality, low costs, etc., is very difficult to be achieved and vital for many industries, e.g., chemical and process equipment, energy production systems, automobile, etc.

The corrosion phenomenon can take place in variant forms: e.g., uniform, pitting, crevice, and galvanic corrosion which are common forms of corrosion for coatings. Pitting and crevice corrosions are known as most common forms of localized corrosion in which corrosive electrolyte can relatively quickly infiltrate [20]. Pits form on the metal surface, and underlying fresh metal is exposed to the corrosive electrolyte for the attack, if the passive layer (as corrosion surface film) on the protecting metal is locally damaged (known as pitting corrosion phenomenon) [21]. This phenomenon leads to highly localized damages [22]. Pitting corrosion quite easily occurs for coatings whose structures are not totally adhering, uniform, and contain considerable number of microdefects and imperfections. Porosities in the protective coatings (anodically) can open the way for penetration of corrosive electrolyte into the inner regions of coating structure and could accelerate the pitting corrosion [23]. Crevice corrosion as one of the most damaging kinds of corrosion causes localized corrosion [24]. Two dissimilar metals (with different standard reduction potentials) form an electrical galvanic couple in the same electrolyte. This phenomenon is known as galvanic corrosion. In this type of corrosion, corrosion happens in more active (as anode) material (with lower standard reduction potential), while the less active (nobler) material (with higher standard reduction potential) would act as the cathode [25]. It has been reported that metallographic structure and microstructural properties have considerable effect on the corrosion behavior of cold sprayed coatings [20,26].

X. Meng et al. reported that both the plastic deformation of deposited powder particles and the porosities in the coating microstructure can considerably affect the corrosion rates of the coatings (cold sprayed deposits). Cold sprayed 304 stainless steel coating (sprayed at 450°C and 3.0 MPa for N_2 as propellant gas) showed the highest corrosion current density compared to the other 304 stainless steel coatings. This indicates that coating's microdefects (prevention of a uniform passive film formation) [27] can easily conduct the electrolyte into the inner regions of the cold sprayed coatings and increase the corrosion current density during corrosion. Nevertheless, cold sprayed 304 stainless steel coating (sprayed at 550°C and 3.0 MPa for N_2 as propellant gas) indicated higher corrosion rate and corrosion current density than cold sprayed 304 stainless steel coating (sprayed at 500°C and 3.0 MPa for N_2 as propellant gas). This observation was mainly related to the higher process gas temperature which generated a higher degree of plastic deformation in the deposited powder particles. This severe plastic deformation might change the chemical potential of the metallic atoms in the deformed regions (particularly at the interparticle boundaries). These regions provide the more active sites for the corrosion [26].

It is interesting to note that an array of site energies because of the presence of different crystal faces (i.e., grains), grain boundaries, and the other defects such as edges, steps, kink sites, screw dislocations, and point defects are existent on polycrystalline metal surfaces. Furthermore, surface energy of the underlying metal atoms around the adsorbate can also be changed by the surface contaminants due to the existence of impurity metal atoms or to the adsorption of ions from solution (electrolyte). Fig. 10.2 demonstrates some of these defects on the metal surface [9].

At the highest energy regions, metal atoms most probably go into solution (electrolyte). Atoms situated at the edges and corners of crystal planes as well as reactive atoms (on the stressed surfaces) which have a less stable crystalline environment could be considered as high-energy sites. Basically, atoms in nonstrained regions have much lower propensity to go into electrolyte in comparison with atoms located in the strained regions of the metal lattice due to the cold working or shaping (considerable plastic deformation) [9].

Severe plastic deformation of Ni powder particles upon impact led to the significant duplication of dislocations, induction of lattice microstrain, refined crystals, and residual stresses in the coating microstructure [28–31]. These could influence the unique corrosion behavior of the cold sprayed Ni coating and are also responsible for the slightly lower corrosion resistance of the CS Ni coating in comparison with the annealed Ni bulk [32].

The lattice microstrain as well as considerable dislocation multiplication is a result of the particle impact-induced deformation which is typically observed in the cold sprayed coatings/deposits [28,33–37]. In this regard, the dislocation density in the cold sprayed cBN/Ni–CrAl coating was reported to be $10^{16}\ m^{-2}$ compared with annealed alloys ($10^{12}\ m^{-2}$) [29]. It was noticed that the energy barrier for the electrochemical reactions could be lowered by the reduction of electron work function due to the presence of the defects/dislocations [30]. Laleh et al. [31] demonstrated that

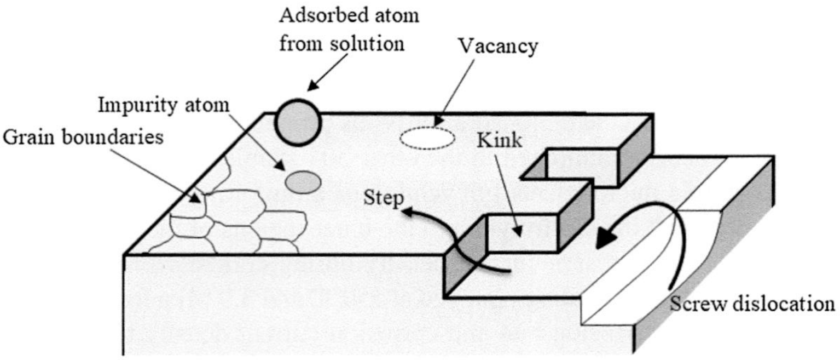

Figure 10.2 Different types of imperfections on a metal surface (the heterogeneous nature of a metal surface).
Modified from E. McCafferty, Introduction to Corrosion Science, Springer, 2010, e-ISBN 978-1-4419-0455-3.

increment of the dislocation density (a large number of active sites) on the Mg alloy surface could considerably increase the corrosion rate. Moreover, other researchers observed that higher plastic deformation of impinged pure Al powder particles can substantially raise the dislocation density and active sites on the metal surface and thus increase the corrosion rate of the cold sprayed pure Al deposit during corrosion [1,38,39]. Likewise, Balani et al. [38] reported that the higher strain (plastic deformation) rate at the interparticle interfaces in the cold sprayed coatings could be considered as preferential reaction zones on the coating surface during corrosion.

It is interesting to note that the ceramic particle reinforced composite Al-based coatings (as MMCs) showed higher corrosion current density and lower corrosion potential (less nobler potential) than those of unreinforced 7075 Al coating. As previously mentioned, the existence of more active sites for corrosion (as a result of high degree of plastic deformation) can substantially increase the corrosion current densities of the cold sprayed coatings/deposits [38]. In fact, composite Al7075-based coatings have experienced higher degree of plastic deformation than unreinforced 7075 Al coatings. This would result in the higher corrosion current density values for the composite Al7075-based coatings [39]. This was also observed for the Al-20vol% Al_2O_3 composite coating which showed lower corrosion resistance than unreinforced pure Al coating [40]. This behavior was also related to the interfaces between reinforcements (Al_2O_3) and Al matrix particles which are also considered as active sites that can simply be attacked by the corrosive electrolyte [40]. Also, in another research, the TiB2/7075Al composite coatings showed higher corrosion rates than cold sprayed 7075Al coatings and 7075Al bulk in chloride containing solutions (Fig. 10.3). This was ascribed to the galvanic cell formation between the noble TiB_2 reinforcement particles and the more active Al-based matrix in comparison with the pure 7075Al coatings. Furthermore, compared to air (as propellant gas in the CS process) processed samples, the lower corrosion resistance for He-processed samples was related to the greater plastic deformation in the He-processed samples during CS process. This was attributed to the larger generation of precipitates and crystal defects like dislocations within the CS samples. Hence, it can be said that dislocations with high density in cold sprayed metallic coatings (Fig. 10.4) have an undesirable influence on the corrosion resistance [32].

In situ grain refinement of impinged powder particles (during cold spray process, Chapter 4, Fig. 10.4) mainly due to rotational dynamic recrystallization has been frequently reported for the cold sprayed metallic coatings [43,44]. On the other hand, effect of grain size on the corrosion behavior of the passive metals has been investigated by many researchers. In this regard, smaller crystals can accelerate the formation of the uniform and compact passive films on the surface and thus improve the corrosion resistance of passive metal in alkaline or neutral aqueous corrosive solutions [45,46]. Likewise, nanocrystalline materials (with more grain boundaries) have more suitable nucleation sites for the formation of a passivation film on the surface during corrosion [47,48]. This behavior was also seen for the pure Ti (after sandblasting) [49]. Nevertheless, the regions with refined grains are very restricted in the cold sprayed coatings/deposits. In fact, high velocity impact leads to the plastic deformation and consequential rotational dynamic recrystallization mainly focuses at the interparticle

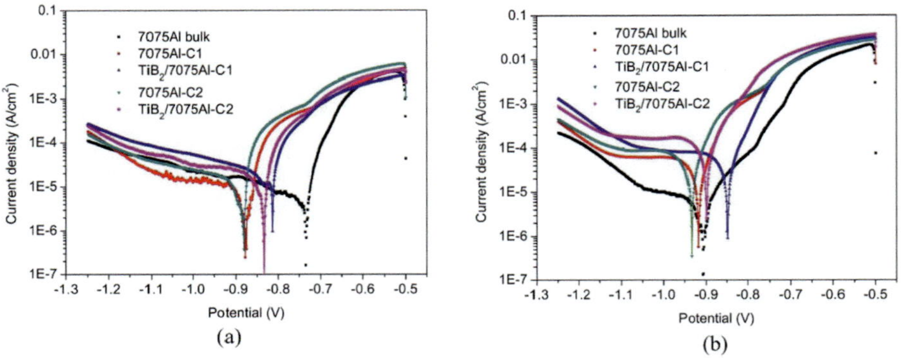

(c) Material	E_{corr} (V)		I_{corr} (E-5 A/cm^2)	
	0.1 M	0.6 M	0.1 M	0.6 M
7075Al Bulk	−0.74 ± 0.02	−0.91 ± 0.02	0.23 ± 0.05	1.6 ± 0.1
7075Al-C1	−0.88 ± 0.02	−0.92 ± 0.03	2.6 ± 0.1	4.0 ± 0.5
TiB_2/7075Al-C1	−0.81 ± 0.02	−0.85 ± 0.02	4.2 ± 0.2	8.1 ± 0.7
7075Al-C2	−0.88 ± 0.02	−0.94 ± 0.02	3.1 ± 0.3	7.9 ± 0.7
TiB_2/7075Al-C2	−0.83 ± 0.02	−0.90 ± 0.03	4.2 ± 0.3	18 ± 2

C1: air processed samples; C2: He processed samples

Figure 10.3 Electrochemical corrosion behavior (potentiodynamic polarization curves) of 7075Al bulk, and CS coatings in (A) 0.1 M NaCl solution and (B) 0.6 M NaCl solution; (C) potentiodynamic polarization tests results in 0.1 M and 0.6 M NaCl solutions (electrolytes) [41].

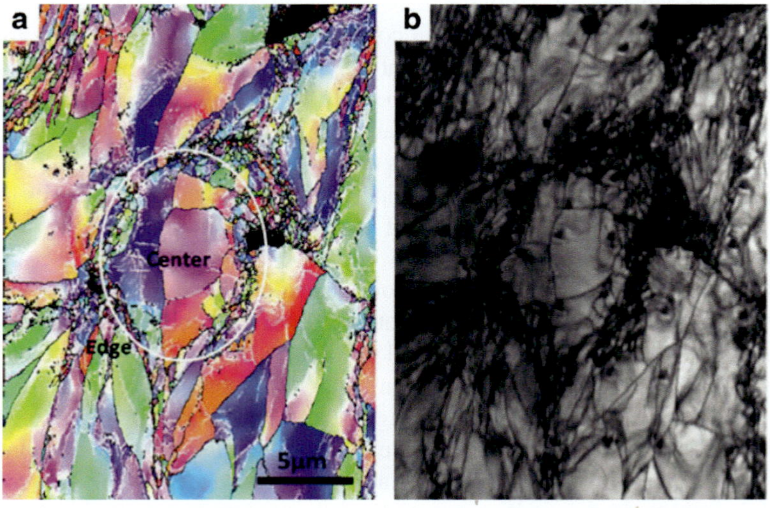

Figure 10.4 Cross-sectional of as-sprayed nickel coating; IPF map (A) and IQ map (B); A roughly bimodal grain-size microstructure including 100−200 nm sized grain/subgrains with a large number of LABs (<15 degrees) is present at interparticle boundaries and micron-sized grains with a small number of LABs in the particle interiors with the same microstructure as the as-received powder particles. Darker color is lower IQ value (average IQ = 126) which implies higher crystal strain (more strain in the particle/particle bonding regions (a large number of dislocations)); brighter color is higher IQ value (average IQ = 225) which signifies lower crystal strain in this region [42].

boundaries during cold spray process. Hence, the refined grains are mostly situated at the interparticle boundaries (Fig. 10.4). The grain size of the interior regions of the impacted particles is largely retained in the cold sprayed coatings/deposits (Fig. 10.4). Hence, corrosion resistance improvement in the cold sprayed coatings due to the grain refinement may be very restricted as compared to the nanocrystalline coatings prepared by other methods such as electrodeposition technique [50].

The following Eq. (10.1) displays the correlation between the corrosion current density and plastic deformation in the materials [26]:

$$\Delta i_a = i_a \left(\exp \frac{n \Delta \tau}{\alpha R' T} - 1 \right) \tag{10.1}$$

In Eq. (10.1), i_a is the corrosion current density of nondeformed (strained) material, n is the number of dislocations in the dislocation pile-up, $\Delta \tau$ is degree of plastic deformation, α is the density of dislocations, $R' = KN\text{max}$, K is the Boltzmann constant, and Nmax is the max number of dislocations in the unit volume and T is the temperature. The amount of n and $\Delta \tau$ could considerably increase when severe plastic deformation of powder particles occurs during cold spray process. This can lead to the sizable increment of anodic corrosion current density (Δi_a) of the cold sprayed deposit. Moreover, increment of corrosion rate (which is in direct relationship with the corrosion current density) of the deformed metal materials due to the cold working has been reported by the other researchers [51,52]. Regarding cold sprayed deposits, it was proven that both the degree of plastic deformation of powder particles and the porosity percentage of the cold sprayed coatings/deposits could influence the corrosion kinetics [26]. In fact, the larger degree of plastic deformation and the higher porosity percentage can considerably raise the corrosion rate of the cold sprayed coatings/deposits. Hence, a more protective coating could be achieved by conformity between these two aspects [26].

Motivated by the above mentioned concepts of the corrosion behavior of the cold sprayed deposites, mostly heat treatments pre- and post-cold spray treatments should be done to modify the corrosion resistance of the cold sprayed deposits.

10.2 Effect of heat treatments on the CS deposites

The microstructural details of the cold sprayed coatings can be affected by heat treatments under several mechanisms which are dependent to the temperature of heat treatments. Fig. 10.5 depicts the microstructural changes (including recovery, recrystallization, and grain growth [53]) at different temperature regions [54]. Several studies indicated the refinement of grain structures after heat treatment. This was attributed to the recrystallization during heat treatment. It is interesting to note that the fracture behavior of cold sprayed deposits/coatings obviously changed to ductile (heat-treated state) from the brittle (as-sprayed state) [55] (and also refer to Chapter 4).

The recrystallization temperature (T_r) typically is between 1/3 and ½ of T_m (absolute melting temperature) of a metal or alloy. The amount of prior cold work and also the

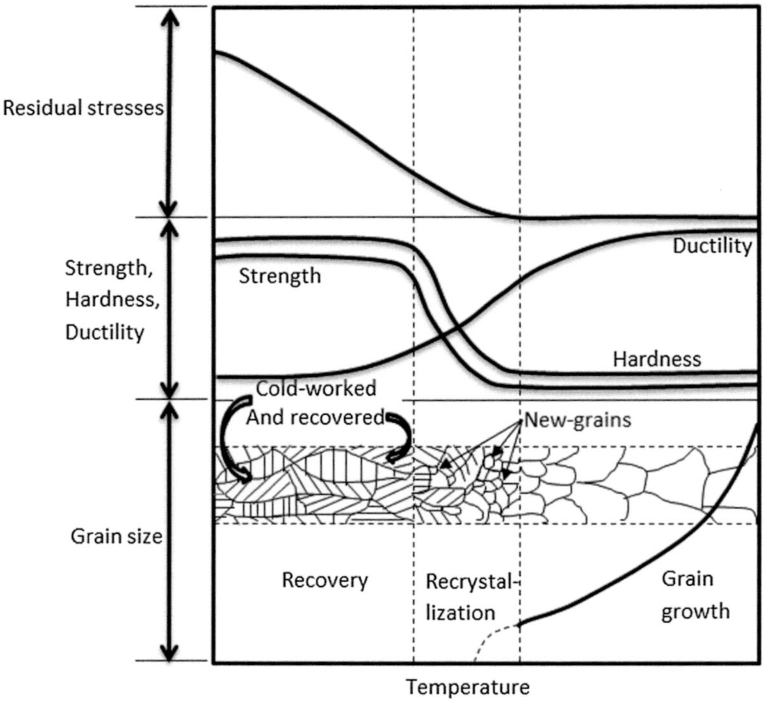

Figure 10.5 Properties and microstructure of cold worked samples; annealing temperature influence on the recovery, recrystallization, and grain growth phenomena.
Modified from S. Semiatin, Recovery, Recrystallization, and Grain-Growth Structures, Metalworking: Bulk Forming, Volume 14A, ASM Handbook, ASM International, 2005, pp. 552–562.

purity of the alloy influence the recrystallization temperature. The recrystallization rate could be enhanced by increasing the cold work percentage. Thus, the recrystallization temperature is lowered to a constant or limiting value (at high deformations), as shown in Fig. 10.6. It is apparently seen that the recrystallization cannot be made to take place below some critical degree of cold work (Fig. 10.6) [8].

One of the most common methods which is employed to modify the microstructure and mechanical properties of the cold sprayed deposits is heat treatment (as postcold spray treatment). This could be mostly achieved by the furnace or eddy current heating [56]. Table 10.1 summarizes a number of preformed studies on this topic.

The materials' properties have an important role in the determination of HT (heat treatment) temperature and duration. In this regard, HT temperature is reported to be lower than 400C [57] and 700C [58] for Al and copper deposits, respectively. On the contrary, heat treatment was performed on niobium deposits (with melting point: ~2500°C) at about 1500°C. So, good ductility and tensile strength [67] could be achieved. Vacuum or inert gas (such as Argon) protection rather than exposure to

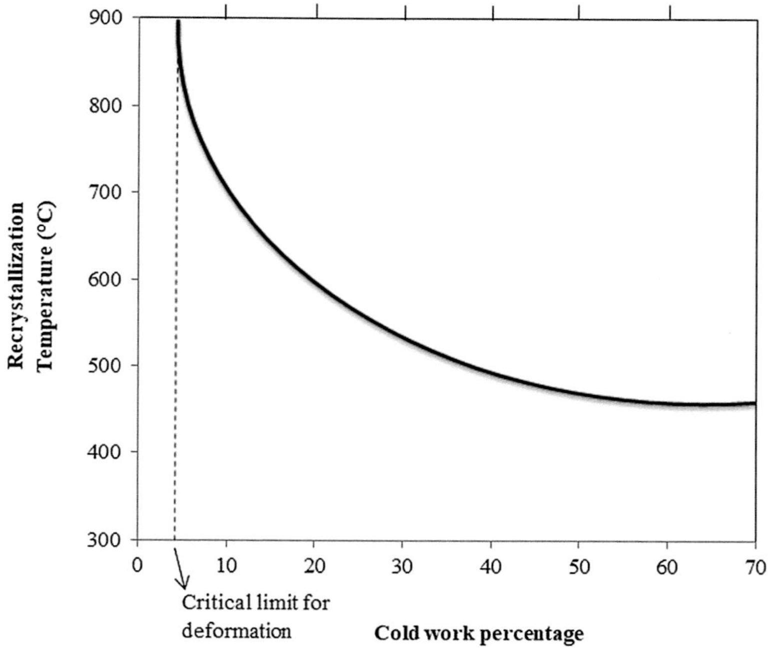

Figure 10.6 Recrystallization temperature versus percent cold work for iron metal; recrystallization will not take place if deformations are lower than the critical (about 5% cold work).
Modified from D. William, Callister Jr, et al., Materials Science and Engineering, An Introduction, Wiley, ninth ed.

air is recommended for heat treatment environment. This could substantially prevent oxidation and lower porosity level of the coatings/deposits [58,59,68].

The effects of HT on cold sprayed dense (Al and Cu) and porous (Ti and SS316) coatings were reported by Huang et al. [65]. High dislocation density and some micro-sized and nanosized grains were identified in the as-sprayed coating. This observation was ascribed to the high kinetic energy impact and severe plastic deformation during cold spray process. The porosity level slightly dropped in the relatively dense coatings at a lower HT temperature [65]. This was due to the diffusion phenomenon across the interparticle boundaries which could close up the interparticle interfaces and tiny pores as well [59].

In contrast, diffusion and porosity decrease were not so noticeable and sizable for relatively porous coatings compared to the dense coatings. This was related to the presence of gaps between weak-bonded particles (Fig. 10.7A and B) [65]. More severe diffusion led to the further elimination of interparticle boundaries and grain size got more uniform (as a result of the recrystallization) (Fig. 10.7C) when heat treatment temperature is further increased. Moreover, enlargement of grains due to grain growth

Table 10.1 The influence of postcold spray heat treatment (PCSHT) on the microstructure and mechanical properties of cold sprayed coatings [57,58,59,60–64,65].

Type of coatings	Heat treatment processes	Main results
Stainless steel 304	Temprature: 600–950°C Duration: 1 h Atmosphere: Vacuum	The weakly bonded interfaces were healed up by annealing heat treatment. This treatment (under certain annealing conditions) could also alter the particle bonding mechanism from mechanical interlocking to metallurgical bonding [61].
Stainless steel 316L	Temprature: 1000°C Duration: 4–8 h Atmosphere: Air, vacuum	Tensile strength and ductility of SS316L deposits were substantially improved by vacuum annealing. However, the tensile strength and ductility of SS316L deposits were somewhat improved by air annealing. Improved interparticle boundaries and particle grain structure had major role in the enhancement of the mechanical properties of the coatings compared to porosity percentage reduction in the coating's microstructure [59].
Fe–40Al	Temprature: 650–1100°C Duration: 5 h Atmosphere: Ar	Fe–Al intermetallic compounds formed within the coating microstructure after heat treatment. The conversion of erosion mechanism from spalling-off to microcutting and also ploughing corrosive particles led to the improvement of erosion resistance of the coating [62].
Al	Temprature: 400°C Duration: 20 h Atmosphere: Ar	Formation of relatively hard Al_3Mg_2 and $Al_{17}Mg_{12}$ intermetallic compound layers (compared to aged AZ91 alloy) at coating/substrate interface considerably increased the corrosion resistance of Mg alloy (comparable to that of Al alloys) [57].
Al-25-Ni, Al–25Ti	Temprature: 450–630°C Duration: 4 h Atmosphere: N_2	Coatings hardness was greatly improved due to the well distribution of intermetallic compounds in the coating microstructure after heat treatment [63].
Al6061	Temprature: 176°C Duration: 1 h, 8 h Atmosphere: Air	The formation of strengthening precipitates and the localized improvement in the metallurgical bonding caused the considerable enhancement of ultimate tensile strength of coatings after heat treatment [64].
Cu–4Cr–2Nb	Temprature: 250–950°C Duration: 2 h Atmosphere: Vacuum	After HT at 350°C, coating microhardness was the highest. This was attributed to the formation and homogeneous distribution of Cr_2Nb phase in the coating microstructure. However, coating microhardness reduced at higher temperatures of HT. This was related to the Cr_2Nb phase coarsening and softening of the copper matrix as well [66].
Cu	Temprature: 300°C Duration: 1 h Atmosphere: Vacuum, air	Coatings showed considerable electrical conductivity (after HT) which was similar to that Cu bulk. However, higher conductivity and lower porosity percentage were observed in the heat-treated Cu coatings in vacuum compared to Cu coatings (heat-treated in air) [58].

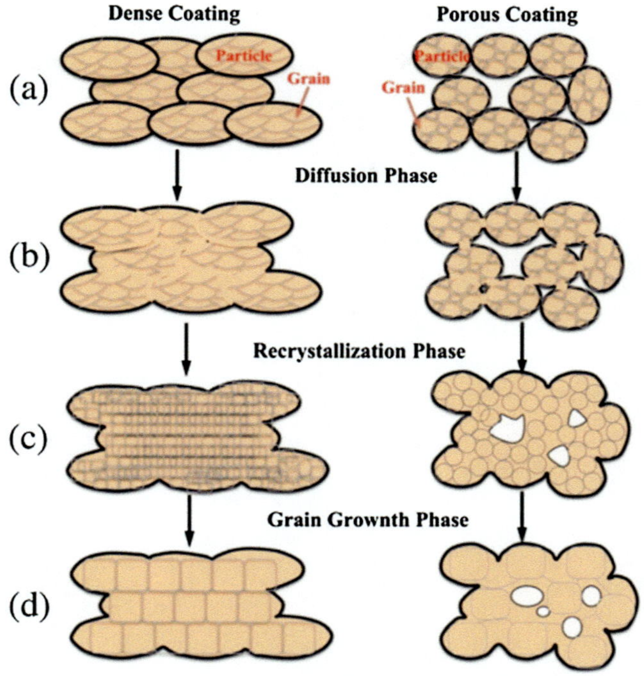

Figure 10.7 The effect of postcold spray heat treatment temperature on the microstructural evolution of cold sprayed dense and porous deposits [60,65].

and further healing some defects in the cold sprayed coatings are anticipated with even higher heat treatment temperatures (Fig. 10.7D).

Nevertheless, all the defects or pores inside the coatings cannot be fixed by using heat treatment even with higher HT temperatures [60]. This should be taken into account for production of cold sprayed deposits with lower and smaller porosities and microdefects at interparticle boundaries.

Electron backscatter diffraction as well as transmission electron microscopy (TEM) analyses [56,69] proved that the dislocation density in the coating microstructure declines and grain size gets larger because of the recovery phenomenon in the coating microstructure during HT. In this regard, considerable reduction of Kernel Average Misorientation (KAM) values was observed in cold sprayed Inconel 718 coatings after HT. As a matter of fact, a decrease of dislocation density can lower the KAM values [69], Fig. 10.8. High-density dislocations and narrow twin bands due to severe plastic deformation were observed in the deformed splats (in the cold sprayed Inconel 718 coatings [56]) by using TEM. Interestingly, the dislocations were reorganized into cell networks and formed some recrystallized subgrains (nonstrained grains) after HT. Moreover, heat-treated Inconel 718 samples showed grain growth and twin-band thickening as well [56]. The hardness of cold sprayed deposits decreased after HT. This could be an obvious sign of dislocation curing (dislocation density

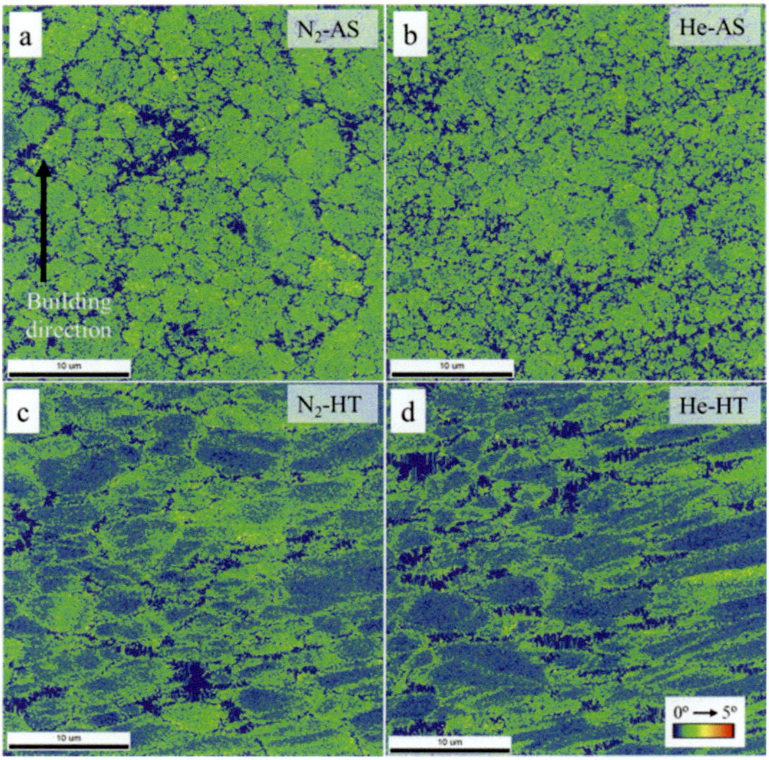

Figure 10.8 EBSD characterization**:** Kernel average misorientation maps of the CS Inconel 718 coatings at different conditions (A) as-sprayed (N_2 at 7 MPa, 1000°C), (B) as-sprayed (He at 3 MPa, 1000°C), (C) N_2-heat-treated, and (D) He-heat-treated [69].

reduction), grain growth, and stress relief [58,60,61,69]. It is expected that these phenomena could also apparently affect the cotorsion behavior of cold sprayed deposits (after HT) in the corrosive environments.

As shown in Table 10.2, it is clearly seen that higher process N_2 gas temperature can produce Ti deposit with lower porosity than Ti deposit (produced at lower process N_2 gas temperature). Postspray heat treatment significantly affected the porosity volume percentage and followed by the corrosion rate of cold sprayed Ti deposits [70,71]. Vacuum annealing heat treatment (at 1050 °C for 60 min) on Ti deposits substantially decreased the coating interconnected porosities and thus declined the corrosion current density as compared to the as-sprayed Ti deposits [70,71].

S. Kumar et al. [72] produced relatively thick free-standing coatings of Ta using compressed air (as propellant gas in the cold spray process). These deposits were then subjected to postspray heat treatments at 750, 1000, and 1500°C in a vacuum furnace for 2h. Vacuum heat treatments could lower the porosity level in the cold sprayed deposits. This was due to the elimination of interparticle boundaries. Likewise, the ductility of the Ta deposit enhanced after vacuum HT. In this regard, fracture

Table 10.2 Interconnected porosities (volume fraction) in the free standing (FS)1 and (FS)2 deposits before and after PCSHT [70].

Process conditions	**Total intrusion, mL/g**	**Total volume percentage of interconnected porosity, vol.%**	**Volume percentage of porosity with pore size >1 μm, vol.%**	**Volume percentage of porosity with pore size <1 μm, vol.%**
As-sprayed FS1 deposit	0.025	11.3	9.9	1.4
Heat-treated FS1 deposit	0.010	4.5	4.0	0.5
As-sprayed FS2 deposit	0.013	5.9	1.4	4.5
Heat-treated FS2 deposit	0.004	1.8	0.2	1.6

Table 10.3 PDP data for Ta [72].

Sample	$R_p (\Omega/cm^2)$	I_0 (μA/cm^2)	E_0 (mV)	E_{RP} (mV)	E_b (mV)	I_{pass} (A × 10^{-5})
Ta-bulk	49,322	0.528	−631	−1149	510	1.16
As coated	3630.8	7.184	−632	−385	412	4.81
750°C	3957.9	6.591	−561	−211	509	4.39
1000°C	24,476	1.065	−706	−433	462	1.28
1500°C	37,058	0.703	−516	−383	483	1.02

surfaces of Ta deposits disclosed the ductile fracture rather than brittle fracture which is mostly observed in cold sprayed coatings/deposits. Annealed Ta deposits at 1500°C (very close to the recrystallization temperature of tantalum) were noticed to act almost like Ta bulk. It is worth mentioning that the corrosion current density gradually declines with increment of HT temperature [72] (Table 10.3).

Warburg impedance (WI) due to the diffusion process was observed for as-sprayed samples and those heat treated at 750°C. However, this phenomenon was less appreciated for the samples heat treated at 750°C which showed lower porosity level than as-sprayed samples (Figs. 10.9 and 10.10). Moreover, heat-treated samples at 750°C showed higher R_{ct} (charge transfer resistance) value (i.e. lower corrosion rate [73,74]) compared to as-sprayed samples. No diffusion process was observed for the samples heat treated at 1000 and 1500°C (Figs. 10.9 and 10.10). Moreover, pore resistance (R_{pore}) was seen for both coatings. Total resistance ($R_{pore} + R_{ct}$) is raised by increasing heat treatment temperature from 1000°C to 1500 °C. In general,

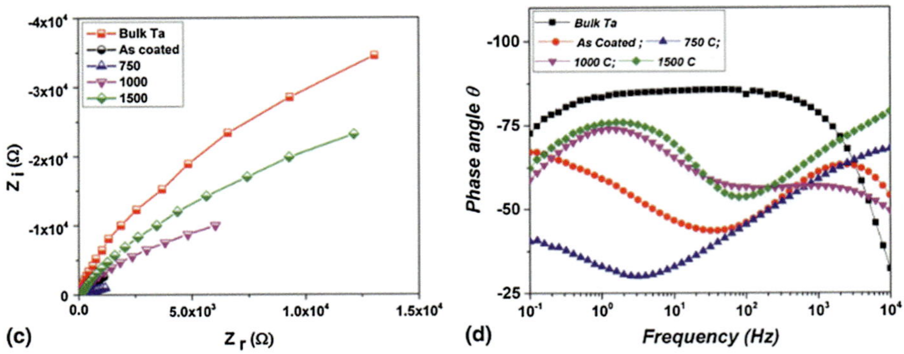

Figure 10.9 EIS curves of AS and heat-treated Ta deposits at different temperatures in KOH solution; Nyquist (c) and (d) Bode phase angle plots [72].

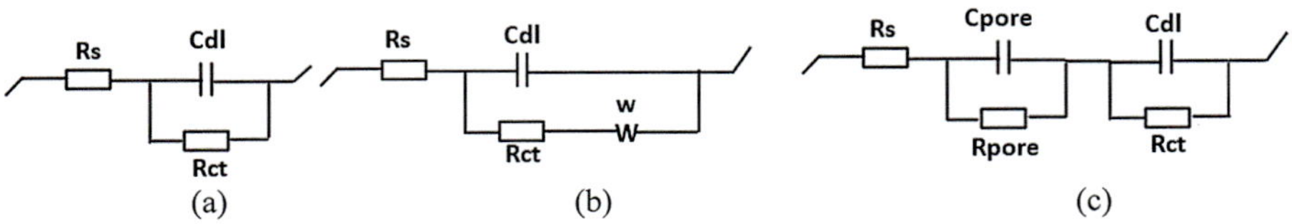

Figure 10.10 EEC employed to fit EIS spectra of (A) bulk Ta, (B) AS and heat-treated Ta deposits at 750°C, and (C) heat treated at Ta deposits at 1000°C and 1500°C.
Modified from S. Kumar, V. Vidyasagar, A. Jyothirmayi, S.V. Joshi, Effect of heat treatment on mechanical properties and corrosion performance of cold-sprayed tantalum coatings, Journal of Thermal Spray Technology 25(4) (2016) 745–756.

samples heat treated at 1500°C exhibited better results, comparable to those of bulk Ta [72].

Cyclic potentiodynamic polarization (CPP) tests were performed to determine the resistance of cold sprayed Nb deposits to pitting corrosion [67], Fig. 10.11. Fig. 10.11A and B present cyclic polarization curves of bulk Nb and as-sprayed niobium deposits, respectively. Arrows indicate the forward and reverse scans. At breakdown or pitting potential (E_b or E_p) where pits initiate and grow, the passive oxide film would be damaged during forward scan. On the contrary, pits could be repassivated using reverse scan.

The pits would not grow if E_{Corr} is less than E_{rp} (repassivation potential or protection potential) during reverse scan. However, the pits keep growing if E_{Corr} is more than E_{rp}. In contrast to bulk Nb where $E_{Corr} > E_{rp}$, as-sprayed and heat-treated CS Nb deposits where $E_{rp} > E_{Corr}$ (irreversible growth of pits) showed obvious negative hysteresis loop at lower current densities.

Lower passivation current and positive repassivation potential are characteristics of the repassivation behavior of some CS deposits (e.g., Nb, Ti, Al, Al 5083) during corrosion in an electrolyte [67,75,76]. Stored energy (due to sever plastic deformation) in the cold sprayed deposits can help these coatings (passive metals) to get easily passivated. It is clearly seen that the slope of the hysteresis loop declines as the heat

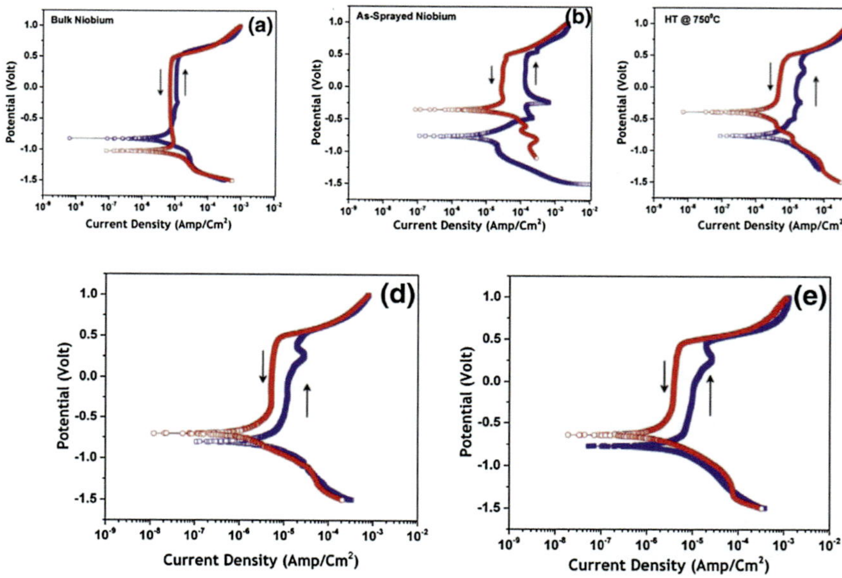

Figure 10.11 CPP curves of (A) Nb bulk; (B) As-sprayed Nb and heat-treated Nb deposits at (C) 750°C, (D) 1000°C, and (E) 1500°C in 1 M KOH solution [67].

Table 10.4 PDP data for Bluk Nb, cold sprayed Nb and heat treated cold sprayed Nb deposits [67].

Sample	Corrosion rate (MPY)	R_p (Ohms/cm^2)	I_o (A/cm^2)	E_o (volts)
Bulk	0.498	14,656	1.78E−6	−0.773
As coated	1.326	5513	4.71E−6	−0.755
750°C	0.524	13,954	1.86E−6	−0.761
1000°C	0.459	15,910	1.63E−6	−0.800
1500°C	0.443	16,490	1.58E−6	−0.769

treatment temperature increases. The stored energy (sever plastic deformation, residual stresses) decreases as the heat treatment temperature is increased. So, heat-treated Nb deposits would gradually behave (decreasing E_{rp} and narrowing the hysteresis loop) analogous to the Nb bulk as HT temperature goes up. Table 10.4 has tabulated the data obtained from polarization test results. Higher corrosion rate due to higher stored energy [77] was observed for the as-sprayed coatings. Samples heat treated at 1000 °C and 1500 °C showed the lower corrosion current densities (a measure of corrosion rate) than Nb bulk. Moreover, sample heat treated at 750 °C displayed low corrosion current density which was close to that of the Nb bulk. It indicates that heat treatment (almost irrespective of the heat treatment temperature) can considerably lower the corrosion current density and increases the resistance to pitting corrosion for the cold sprayed Nb deposits. This could be attributed to the closing up the pores and defects in the cold sprayed deposits, with decrease in interparticle boundaries, recovery phenomenon, reduction of dislocation density, stress relief, and recrystallization of non-strained equiaxed grains by heat treatment [67].

In summary, considering that both the powder particles plastic deformation degree (upon impact) and the presence of microdefects at the interparticle boundaries can influence the corrosion resistance of as-sprayed coatings/deposits, proper postcold spray heat treatments should play a role in reducing corrosion susceptibility of as-sprayed samples, because of stress relief, dislocations elimination, and thus reducing the number of active sites on the surface of sample.

10.3 Effect of heat treatments on feedstcok powder particles (before CS process)

Recently, pre-cold spray heat treatment was performed on the feedstock powders to modify the microstructure of the deposits as well as the deposition efficiency (DE) [78,79]. Large stress-free grains were seen in the solution heat-treated Al 6061 powder [80], while as-atomized feedstock powder showed smaller grains with the presence of dislocations [78]. A certain amount of LAGBs (low angle grain boundaries) located mostly toward the edge of the particle were detected in the feedstock powder [78]. It is assumed that LAGBs are mostly composed of an array of dislocations. This is a potential indicator of residual stresses in the droplet postatomization [78]. In contrast, a negligible amount of LAGB coherent with the formation of the stress-free grains postheat treatment were identified in the solution heat-treated powder. In fact, recrystallization occurring during the heat treatment can substantially reduce the LAGB in the heat-treated powder. Solution heat-treated powder produced a cold sprayed coating with a more intimate bonding between the powder particles [78,79]. However, the feedstock powder produced a cold sprayed coating with gaps and voids at the interparticle boundaries [78,79]. Likewise, more developed misorientation was identified in the coating deposited (using the feedstock powder) (Figs. 10.12 and 10.13) [78]. These recent results are promising for the subsequent studies on the corrosion behavior of cold sprayed coatings produced with heat-treated powders (as pre-cold spray heat treatments (Pre-CSHT) on feedstock powders).

10.4 Pre-cold spray treatments for reducing the corrosion rate of the additively manufactured cold sprayed metallic coatings/deposits

Apart from Section 10.3, this part will cover some pre-cold spray treatments (Pre-CST) which include varying cold spray parameters, changing type of propellant gases and morphology of feedstock powder particles, and their effects on the corrosion behavior of cold sprayed deposits/coatings.

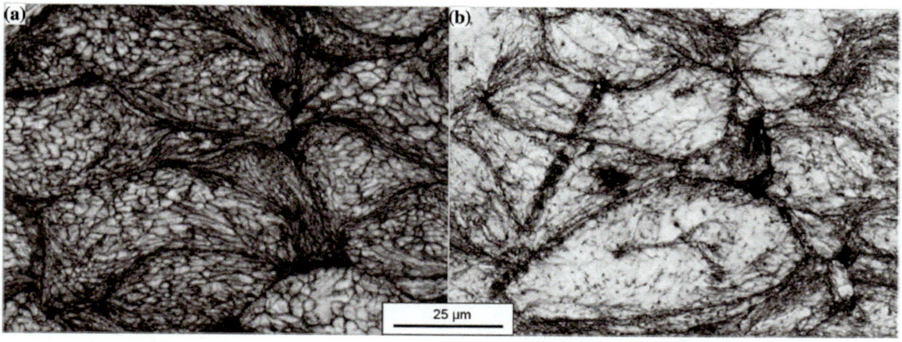

Figure 10.12 IQ patterns of cross-section of cold spray deposits using (A) feedstock powder and (B) solution heat-treated powder. IQ pattern with high quality was observed for (B) implying a lower amount of lattice defects compared to the coating (A) which displays a darker contrast because of a high number of grain boundaries [78].

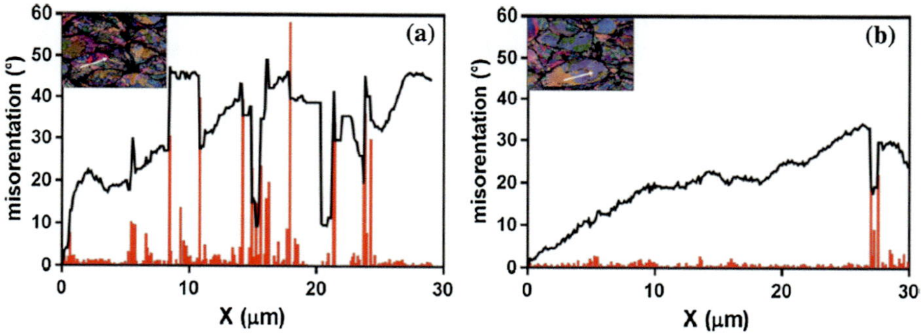

Figure 10.13 MP of the interior of the particle in the coating cross-section deposited with (A) feedstock Al 6061 powder and (B) solution heat-treated Al 6061 powder. The interior of the deformed feedstock particle contains a large number of grains indicated by the HAGB as well as misorientation and LAGB across the entire deformed particle. However, the deformed heat-treated Al 6061 particle displays a limited misorientation signifying a low lattice strain and dislocation density as well [78].

It is obviously (Fig. 10.14) seen that Ti coating sprayed at 350°C shows higher corrosion current density than coating sprayed at 500 and 650°C. This could be attributed to the possible presence (10.3%) of porosities in the coating microstructure (sprayed at 350°C). Ngai et al. [81] reported a high corrosion rate for the cold sprayed coating due to the severe microcrevice corrosion formation and irregular pitting network on the corroded surface. This occurrence was ascribed to the limited particle deformation and a high degree of resultant porosity in the coating microstructure. Hence, it can be said that the corrosion resistance of the CS coatings [82] can negatively be affected by a porous microstructure (Table 10.5).

It is worth mentioning that the corrosion current density of cold sprayed Ti coatings (processed with compressed air as propellant gas at 500°C) decreases as propellant gas

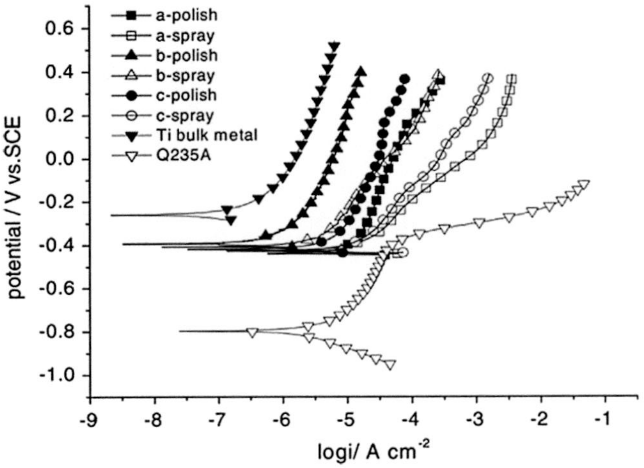

Figure 10.14 PDP curves of cold sprayed Ti deposits, sprayed at different temperatures (using nitrogen at a pressure of 2 MPa) in a 3.5 wt% NaCl solution, (A) 350°C, (B) 500°C, and (C) 650°C. TA2 is Ti bulk metal and Q235A substrate represents a low carbon steel [71].

pressure increases, and the corrosion current density of as-sprayed Ti coating surface is larger than that of as-polished Ti coating surface [83]. As previously stated, the rough outer layer of coating including small pores could be peeled, and the *actual surface area* of the coating could also be decreased using polishing treatment which ultimately caused the considerable enhancement of the corrosion resistance (reduction of the corrosion rate). However, it was reported that the polarization current of both polished and as-sprayed Ti coatings (processed with compressed air) is mostly higher than that of both polished and as-sprayed Ti coatings (processed with N_2). This behavior is mainly ascribed to higher level of porosity in the processed Ti coatings with compressed air than processed coatings with N_2 gas (21).

Ti coating sprayed (processed with N_2 as propellant gas) at 500°C (with porosity of 5.8%) shows lower corrosion current density than coating sprayed at 650°C (with porosity of 1.6%), Fig. 10.14 [71]. This phenomenon could be related to the severe plastic deformation of powder particles and the possible presence of more active sites in the coating (particularly at interparticle boundaries). High corrosion current densities due to the presence of active sites as a result of a high degree of plastic deformation were also reported by Balani et al. [38] and Meydanoglu et al. [39].

As previously mentioned [38], both the stress generation (due to severe plastic deformation) during cold spray process and the porosity introduction because of the cold spray processing nature can increase the corrosion rate. So, conformity (balance) between these two factors is crucial to attain coatings with high corrosion protection performance (especially when the postspray heat treatment (PCSHT) is not recommended for some cold sprayed deposits/coatings).

In this regard, the Al coating processed with a mixture of He and N_2 gases showed lower corrosion rate than the coating, where only He was employed as processing

Table 10.5 PDP results (E_{corr} and i_{corr}) of bare Ti64 substrate and Ti64 coatings deposited with different propellant gases. Porosity level of 1.3 ± 0.4% was obtained for the Ti64 coating deposited with He as propellant gas, while higher porosity level of 11.5 ± 0.7% was obtained for the Ti64 coating deposited with N_2 as propellant gas [82].

	Measured time (hr)							
	3	6	9	12	15	18	21	24
Samples	E_{corr} (V vs. Ag/AgCl)							
Ti64 substrate	−0.29	−0.263	−0.259	−0.246	−0.238	−0.221	−0.216	−0.198
Ti64 coating with N_2 gas	−0.077	−0.087	−0.074	−0.062	−0.055	−0.053	−0.055	−0.051
Ti64 coating with He gas	−0.268	−0.265	−0.264	−0.258	−0.251	−0.243	−0.24	−0.235
	i_{corr} (A/cm^2)							
Ti64 substrate	6.78×10^{-9}	6.97×10^{-9}	6.92×10^{-9}	6.8×10^{-9}	6.69×10^{-9}	6.51×10^{-9}	7.31×10^{-9}	6.7×10^{-9}
Ti64 coating with N_2 gas	1.84×10^{-5}	3.24×10^{-5}	5.02×10^{-5}	4.75×10^{-5}	5.13×10^{-5}	4.9×10^{-5}	3.94×10^{-5}	3.81×10^{-5}
Ti64 coating with He gas	1.06×10^{-8}	1.11×10^{-8}	1.09×10^{-8}	1.12×10^{-8}	1.12×10^{-8}	1.14×10^{-8}	1.16×10^{-8}	1.18×10^{-8}

gas. It was reported that powder particles reach much higher impact velocity and thus greater particle deformation when He (due to higher specific heat ratios and lower mass density of He) uses as the propellant gas compared to N_2+He processed coatings. Although a slightly denser structure was obtained using He compared to N_2+He processed coatings, highly stressed regions in the 100%He processed coatings led to the promotion of number of active sites (as preferential sites to undergo rapid corrosion kinetics) for corrosion. This could be a main reason for the higher corrosion rate of 100%He processed coatings in comparison with N_2+He processed coatings which showed inferior hardness [38]. It can be said the processing gas (as propellant gas) could have a determining role in the corrosion resistance of cold sprayed coatings/deposits. So, it could be inferred that the degree of plastic deformation of powder particles may play a more important role than porosity (to a certain extent or when the amount of porosity is low) in determining the corrosion behavior of cold sprayed coatings/deposits. However, this should be more investigated on the different metal materials.

As previously mentioned, one of criterions for good corrosion resistance of coatings is their denseness, or in other words their impermeability. The compactness of the coatings on corrodible substrates (anodic protection) can be evaluated by open circuit potential (OCP) measurements (Fig. 10.15, [84]) and salt spray (fog) tests as wet corrosion tests (relevant methods). It is also reported that the quality of various coatings could be assessed by the salt spray (fog) test (a commonly used test method). In a controlled test condition, the use of different corrosive solutions and different test temperatures will also be enabled by the salt spray (fog) test [85].

The potential behavior of coating, substrate (in anodic protection condition), and porous coating is demonstrated in Fig. 10.16. Mixed potential which is comprised of potentials of both coating and substrate is basically observed for the coatings that contain interconnected porosities (through thickness porosities). So, the existence of interconnected porosities, in the other words, through-thickness porosities in the sprayed coatings could be proven by the open circuit potential measurements during long immersion test.

Fig. 10.17A shows OCP curves of the cold sprayed samples in the 3.5 wt.% NaCl aqueous solution. For comparison, OCP curves of bulk Cu and bare Al substrate were also employed. OCP values of all samples get mostly stable over immersion time. Highest OCP values were observed for the coating deposited with mixed powders, the (using commercial dendritic and porous Cu (E) powder) E Cu coating, and (using spherical and dense gas atomized Cu powder (GA)) GA Cu coating, respectively [87]. It was deduced that the coating sprayed with the mixed powders may contain the least interconnected porosities compared to the other Cu coatings. Fig. 10.17B depicts the polarization curves of the cold sprayed coatings as well as Cu plate and Al substrate for comparison. The highest E_{Corr} and lowest I_{corr} which correspond to the best corrosion resistance were also observed for the Cu coating deposited with mixed powders [87]. However, performing longer immersion tests (more than 3600 S) along with OCP test are recommended due to possible further penetration of the corrosive electrolyte into the inner regions of the sprayed coatings in the course of immersion time which may change the results.

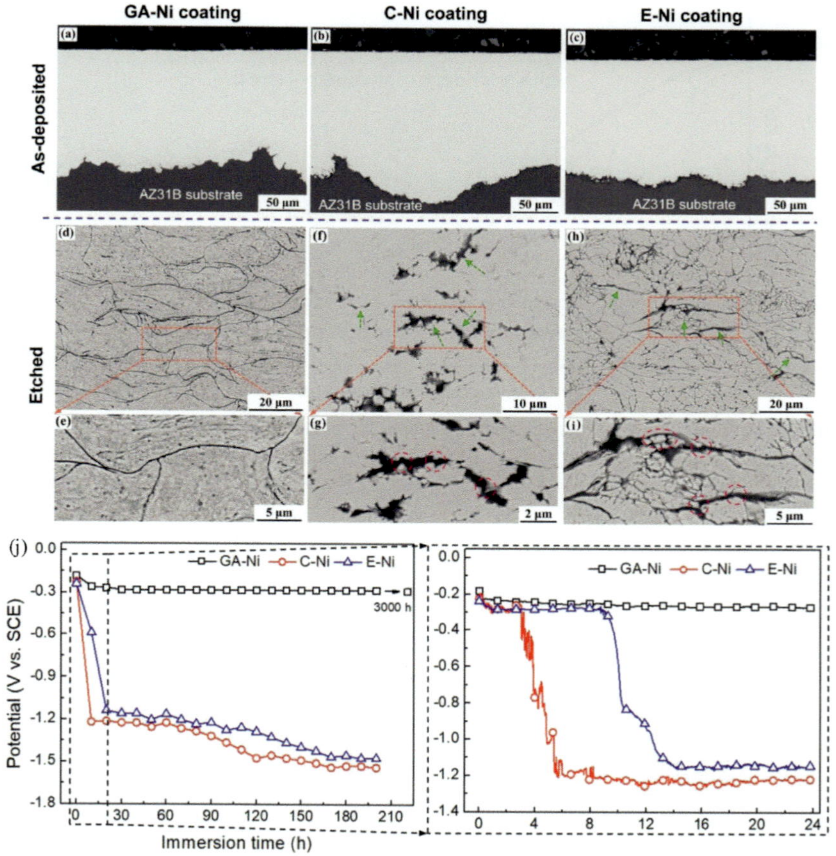

Figure 10.15 Microstructure of cross sectional polished of as-cold sprayed and etched GA-Ni (gas-atomized spherical Ni (GA-Ni) (A and D), carbonyl irregular Ni (C-Ni) (B and F) and electrolytic dendritic porous Ni (E-Ni) coatings. (E), (G) and (I) are the higher magnifications of (D), (F) and (H), respectively; (J) E_{OCP} vs. time curves of in-situ micro-forging assisted cold sprayed Ni coatings in 0.6 mol L^{-1} NaCl electrolyte [84].

10.5 In-situ cold spray treatments for reducing the corrosion rate of the additively manufactured cold sprayed metallic coatings/deposits

This part will cover some in situ cold spray treatments (ISCSTs) which include laser-assisted cold spraying (LACS), in situ SP (shot peening)-assisted cold spraying methods, and the influence of these methods on the corrosion behavior of cold sprayed deposits/coatings.

LACS method could be considered as one of the ISCSTs in which the coating properties could be improved by simultaneous interaction between the laser irradiation and the spraying spot on the substrate or deposited layer surfaces. It has been reported that

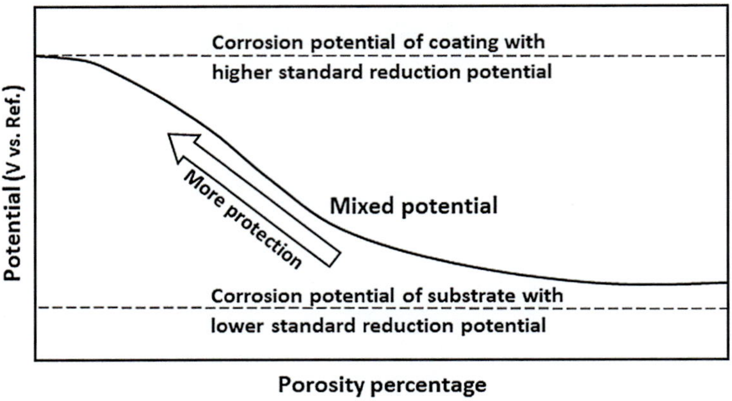

Figure 10.16 Potential behavior of coating, substrate (in anodic protection condition), and porous coating is demonstrated. Mixed potential which is comprised of potentials of both coating and substrate is basically observed for the coatings that contain interconnected porosities.
Modified from M. Vreijling, Electrochemical Characterization of Metallic Thermally Sprayed Coatings, Printed in the Netherlands, 1998, p. 143.

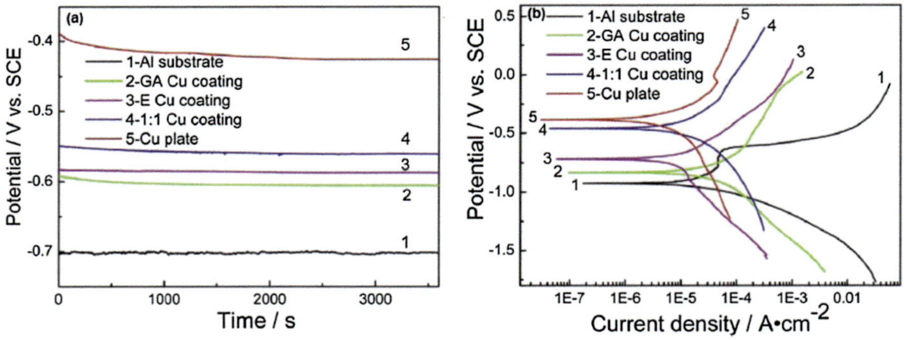

Figure 10.17 Electrochemical corrosion test results of the Cu coatings, Al substrate, and Cu plate for the comparison; (A) OCP curves (B) PDP curves [87].

the deposition at powder particle velocities lower than those normally required for the cold spray process is expected by using laser-assisted cold spray technology in which no melting of the deposited materials and also heat-affected zone in the substrate were observed [88].

In this regard, corrosion behavior of laser-assisted low-pressure cold sprayed (LALPCS) copper coatings (irrespective of the applied laser power) was investigated by Kulmala and Vuoristo [88]. LACS technology increased the copper coating denseness and also improved DE. Moreover, OCP of LACS Cu coatings with denser surface was comparable to that of bulk Cu (Fig. 10.18A). In their research, the functional performance of LACS Cu coatings was also compared to that of LACS Ni coatings. Although LACS Ni coatings showed denser surface than CS Ni coatings, OCP test

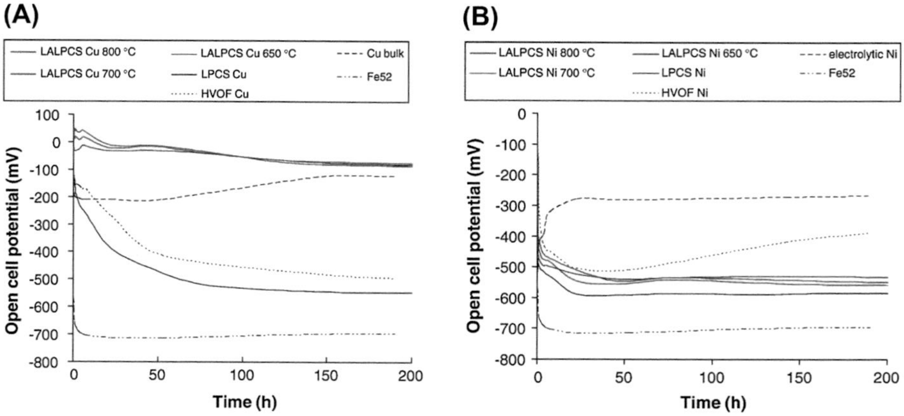

Figure 10.18 (A) OCP curves of LALPCS Cu coatings (pyrometer temperatures 650, 700, and 800°C) and LPCS Cu coating (without laser assistance) and (B) OCP curves of the LALPCS Ni coatings (pyrometer temperatures 650, 700, and 800°C), and LPCS Ni coating (without laser assistance); HVOF sprayed Cu coating, bulk Cu, and substrate material Fe52 (unalloyed low carbon steel) were used for the comparison [88].

proved the existence of open porosities in both LACS Ni coatings and CS Ni coatings (regardless of the applied laser power) (Fig. 10.18B).

This observation was attributed to the inadequate laser energy power that could not influence sufficient thermal softening of Ni particles (lowering the critical velocity of Ni particles) prior to their impact (consolidation) on the substrate surface [88]. So, it is speculated that the materials with higher T_m such as Ni require higher energy density (higher laser power) for effective softening of powder particles prior to the deposition on the surface during LACS process.

In situ SP (shot peening)-assisted cold spraying (SPACS) method by intentionally introducing SP particles (large spherical stainless-steel powder particles) into the spray powders (Al6061 powder) was presented by [89] Ying-Kang Wei et al. They presented that this method could substantially enhance the plastic deformation and consequently cohesion of deposited Al606l powder particles (by enhanced accumulative tamping effect). They reported that a completely dense microstructure with a porosity of 0.4% compared to as sprayed Al6061 coating with porosity of 14.40% could be obtained if the SP particle content rises to 60 vol.%. However, with increasing the ratio of SP particles, deposition efficiency of Al6061 powder reduced. The electrochemical corrosion tests disclosed that the open circuit potential and a dynamic polarization behavior of the fully dense Al6061 coated AZ31B alloy could be comparable to those of Al6061 bulk. Hence, in situ SP-assisted cold spraying method could be used to produce the protective and fully dense Al6061 coatings on Mg alloys [89].

Z. Zhang et al. also used the in situ shot-peening-assisted low-pressure cold spray method to spray dense Al-based coatings (with good mechanical and corrosion protection properties) on AA 2024-T3 alloy (as substrate) [75].

Average corrosion parameter values for the three coatings (derived from three CPP curves) are tabulated in Table 10.6. Al 2024 coating indicates noticeably higher Icorr

Table 10.6 CPP curves results for different samples after immersion in 3.5 wt% NaCl solution for 6 h [75].

Coatings	I_{corr} (μA/cm^2)	E_{corr} (V/SCE)	E_{pit} (V/SCE)	E_{rp} (V/SCE)	$E_{pit}-E_{corr}$ (V)	$E_{rp}-E_{corr}$ (V)
Al	0.5 ± 0.1	−0.88 ± 0.04	−0.74 ± 0.02	−0.78 ± 0.03	0.15 ± 0.03	0.10 ± 0.03
Al 2024	3.2 ± 2.1	−0.71 ± 0.01	−0.71 ± 0.01	−0.76 ± 0.01	0	−0.05 ± 0.01
Al 5083	0.6 ± 0.2	−0.87 ± 0.01	−0.73 ± 0.01	−0.84 ± 0.01	0.15 ± 0.03	0.03 ± 0.01

than the commercially pure (CP) Al and Al 5083 coatings which showed similar and relatively lower Icorr. This indicates that the corrosion reactions on CP Al and Al 5083 coatings surface are less severe than the Al 2024 coating surface.

From ($E_{pit}-E_{Corr}$) value, it is postulated that the CP Al and Al 5083 coatings are less susceptible to pitting than Al 2024 coating with the lowest ($E_{pit}-E_{Corr}$) value [89]. Better resistance to pitting corrosion could be related to the higher pitting potential, and so, Epit may be a good choice to show the corrosion resistance [90]. Nevertheless, when the value of ($E_{pit}-E_{corr}$) is very low, for instance, for aluminum alloys in chloride containing electrolyte, a small increase in potential could lead to the pitting commencement. Under this condition (although E_{pit} is slightly higher than E_{corr}), initiated pitting would keep growing, but new pitting would not take place. Hence, it can be said that Epit cannot be considered as a proper parameter for the corrosion resistance determination.

G. E Kiourtsidis et al. reported that the [91] ($E_{pit}-E_{Corr}$) value may be a better guide to pitting corrosion resistance. Hence, the larger ($E_{pit}-E_{Corr}$) value indicates lower susceptibility to the pitting during corrosion [91,92]. In CPP, all the coatings displayed a positive hysteresis loop (on the reverse scan). In this condition, further development of pits is expected. It has been reported that [93] pitting corrosion would not be further developed, if the reversed (anodic) curve is moved to lower anodic currents (as negative hysteresis loop: HL), or the reversed and the forward curves overlap (neutral hysteresis loop). On the contrary, additional pitting development is anticipated, if the reversed anodic curve is moved to the higher current densities compared to the forward anodic curves (as positive-hysteresis loop). In positive-hysteresis loop, pits would get repassivated, when the potential keeps reducing, and the reversed anodic curve intersects the forward scan at protection potential Erp (or repassivation potential). Present pits would not keep growing if Erp to be more than E_{Corr}. Likewise, the (E_{rp}-E_{Corr}) values show that CP Al coating has high ability to repassivation compared to the Al 5083 and Al 2024 coatings [75], Fig. 10.19.

The anodic and cathodic current density peaks for the CP Al coating reduce as immersion time increases. This is indicated by SVET (Scanning Vibrating Electrode Technique permits the identification of anodic zones (positive currents) and cathodic zones (negative currents) above the corroding surface in the corrosive electrolyte) current maps, as shown in Fig. 10.20. This resulted in a decrease in pitting activity. The Al 2024 and Al 5083 coatings showed several anodic and cathodic peaks with low current density on the surface (after 24 h of immersion), Figs. 10.21 and 10.22. This was attributed to the formation of microlocal cell between the precipitates and the α-Al matrix which led to the increment of the average current density for the Al 2024 and Al 5083 coatings. Moreover, uniform corrosion on the surface may be anticipated if the expansion of the microlocal cells happens and when the pits get interconnected. Nonetheless, some problems are associated with the SVET on certain conditions. Since the probe in SVET vibrates about 100 μm above the sample surface (in this experiment), the only ionic currents which can reach the probe could be detected in SVET. So, it can be said that the values attained by SVET are underestimated and could only be employed as a semiquantitative measurement for determining the corrosion mechanism of the CS coatings/deposits.

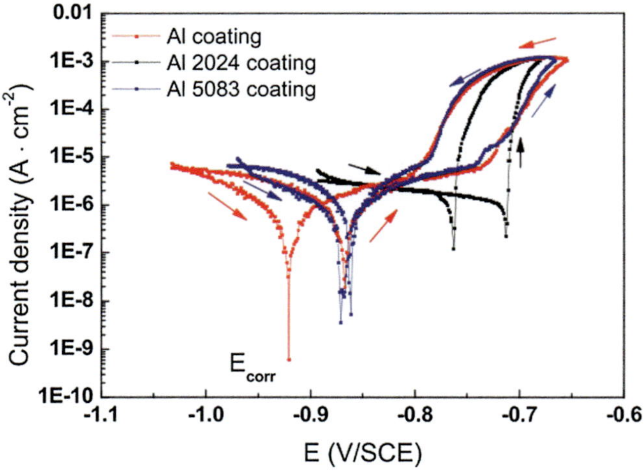

Figure 10.19 CPP curves of coated samples in 3.5 wt% NaCl solution (at 1 mV/s, after 6 h of immersion) [75].

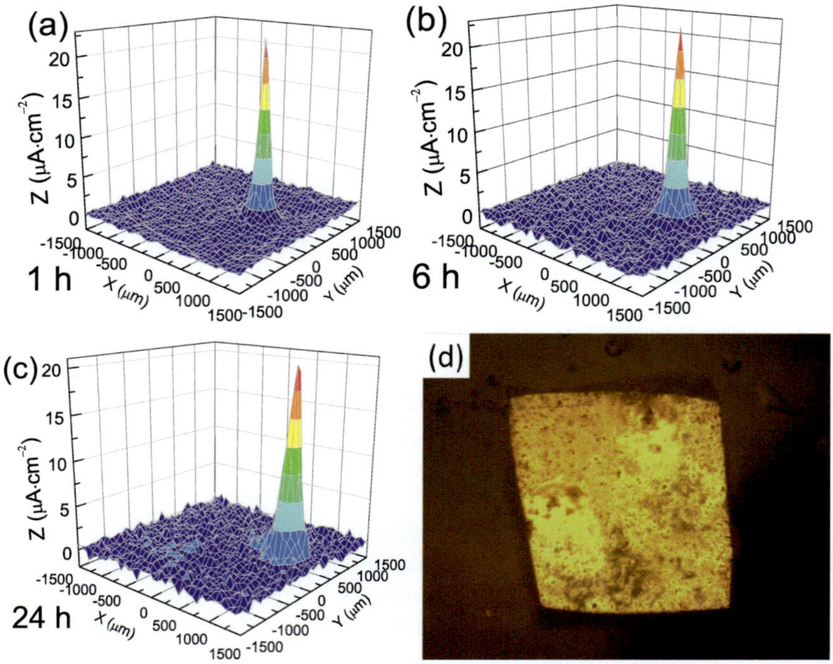

Figure 10.20 SVET maps of CP Al coating after 1 h (A), 6 h (B), and 24 h (C) exposure in 3.5wt% NaCl solution, and optical image of the corroded surface after 24 h of immersion (D) [75].

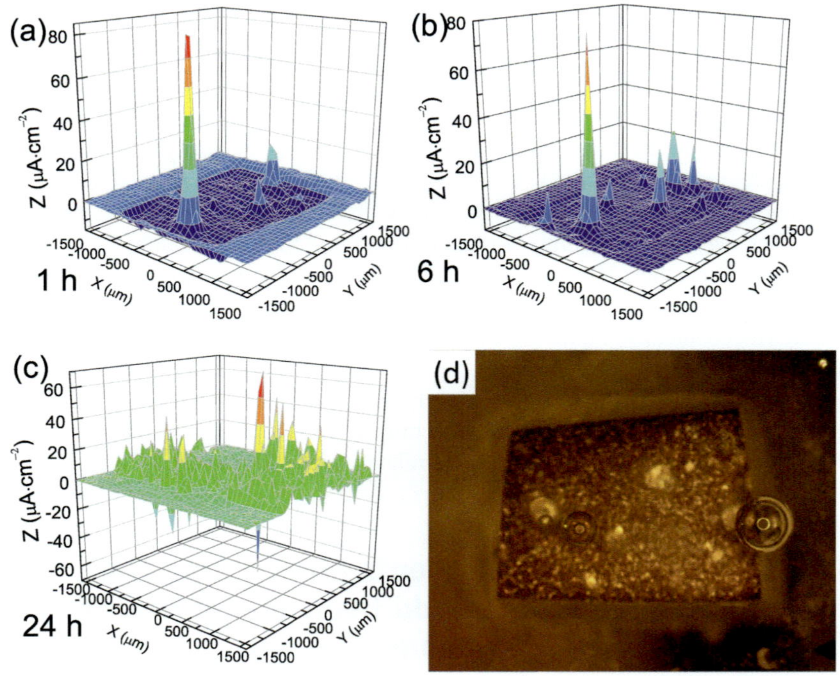

Figure 10.21 SVET maps of Al 2024 coating after 1 h (A), 6 h (B), and 24 h (C) exposure in 3.5wt% NaCl solution, and optical image of the corroded surface after 24 h of immersion (D) [75].

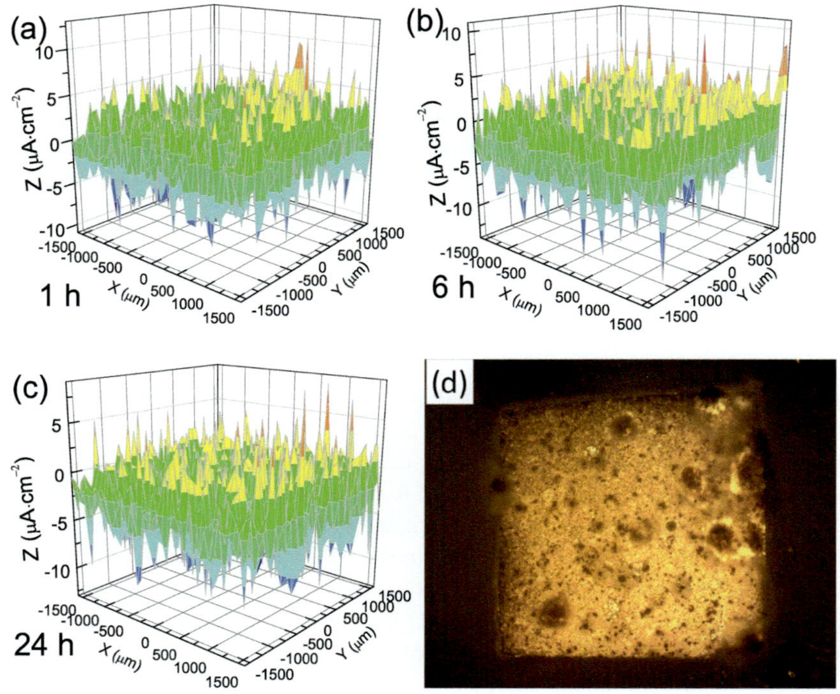

Figure 10.22 SVET maps of Al 5083 coating after 1 h (A), 6 h (B), and 24 h (C) exposure in 3.5wt% NaCl solution, and optical image of the corroded surface after 24 h of immersion (D) [75].

Ying-Kang Wei et al. [89] also reported that SP particles could remain in the coating structure when they chose 350°C and 3.5 MPa for N_2 process gas temperature and pressure, respectively, though the SP powder particle (PSD: 100−200 μm) velocity is completely lower than its critical velocity. In fact, the inclusion mechanism of the SP particles may not dominantly originate from the ASI phenomenon. As a matter of fact, several researchers have observed the inclusion of the hard particles into the soft metallic coatings due to the deep penetration when ceramic and metallic powder particles are (mechanically) mixed for composite coatings/deposits preparation [94−98]. This matter was sorted out when the N_2 process gas temperature and pressure were lowered to 300°C and 2.5 MPa, respectively. However, higher volume of SP particles was used to produce coatings with higher compactness. This considerably lowered the DE of Al powder and increased the costs related to the SP particles.

10.6 Post-cold spray treatments for reducing the corrosion rate of the additively manufactured cold sprayed metallic coatings/deposits

It was reported that the influence of shot peening treatment (tamping effect) on the bulk materials is less significant than cold sprayed coatings. This treatment has been reported to be appropriate for the production of cold sprayed coatings with high compactness and corrosion resistance as well [99].

Recently, dense Al coatings on LA43M substrate were produced using conventional cold spraying process and postshot peening treatment. In fact, the kinetic energy of martensitic stainless-steel particles can be absorbed by deposited Al particles. This could considerably densify the higher fraction of the deposited Al particles. The level of porosity in the shot-peened Al coating was significantly lowered to 0.2% from 12.4% for as-sprayed Al coating, Figs. 10.23 and 10.24. Moreover, inclusion of the hard SP particles into the soft metallic coatings like Al was not reported in this research [99].

Higher open circuit potential, lower corrosion current density, and better impedance values were observed for shot peened Al-coated LA43M sample compared to the as-sprayed Al-coated LA43M sample. Likewise, surface morphology of samples after fog salt spray test, surface analysis (EDS), and impedance behavior of the samples during long-term immersion tests (Fig. 10.25) all proved that the postshot peening treatment could significantly raise the pure Al coating compactness and substantially enhance the corrosion protective performance of the pure Al-coated Mg LA43M alloy during corrosion. This could be probably ascribed to higher level of porosity (12.4% in this research) in the as-sprayed Al coating compared with the shot-peened pure Al costing with only 0.2% porosity level [99]. Nevertheless, the microhardness and wear resistance of the shot peened Al-based coatings were not reported in this research.

Newly, M. Daroonparvar et al. [100,101] developed multilayered coatings on Mg-based alloys using high-pressure cold spray process. The tamping effect caused by powder particles (with high kinetic energy and microhardness) significantly enhanced the

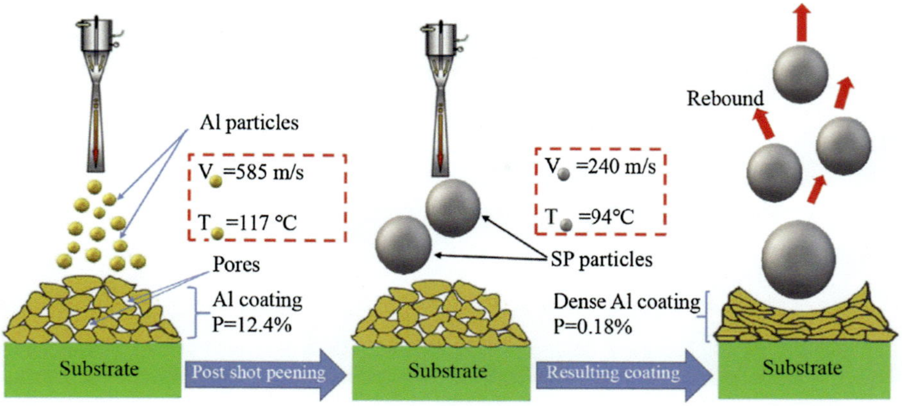

Figure 10.23 The schematic illustration of obtaining dense Al coatings on LA43M substrates using conventional cold spray and postcold spray shot peening processes [99].

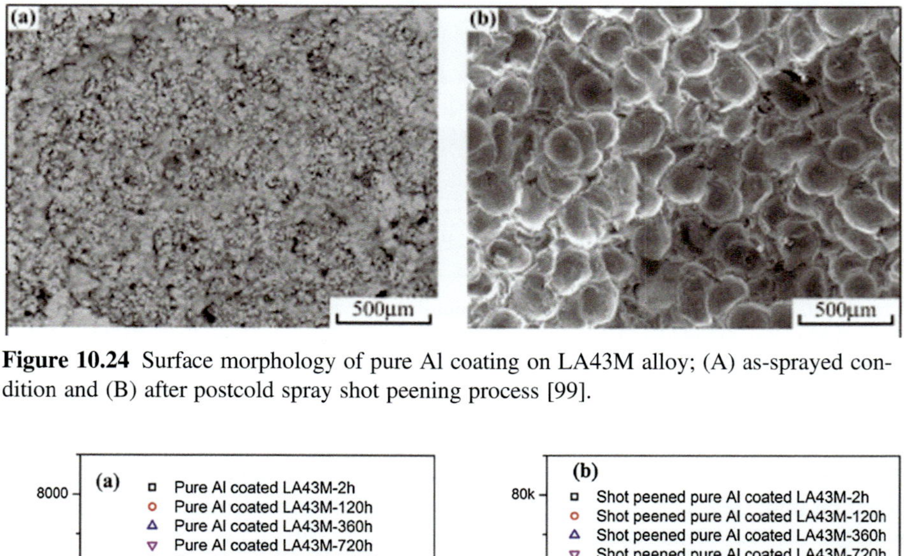

Figure 10.24 Surface morphology of pure Al coating on LA43M alloy; (A) as-sprayed condition and (B) after postcold spray shot peening process [99].

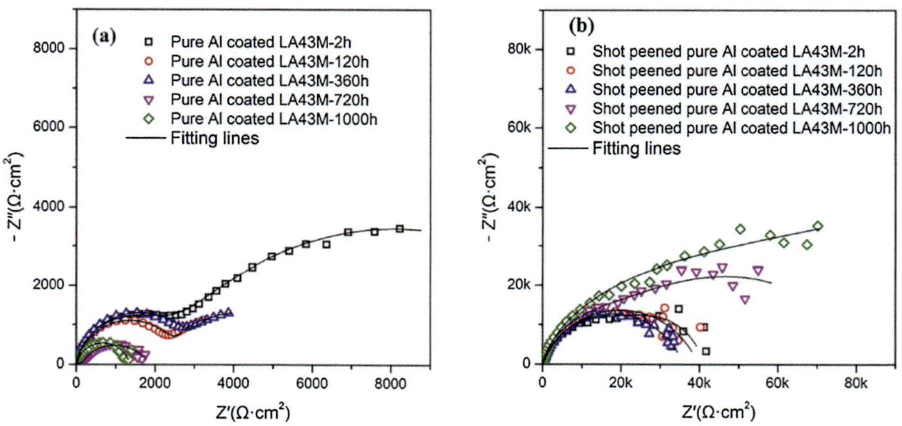

Figure 10.25 The Nyquist plots (A) (as-sprayed) pure Al-coated LA43M alloys (B) (shot peened) pure Al-coated LA43M alloys after different immersion periods of 2, 120, 360, 720, and 1000 h in 3.5 wt% NaCl [99].

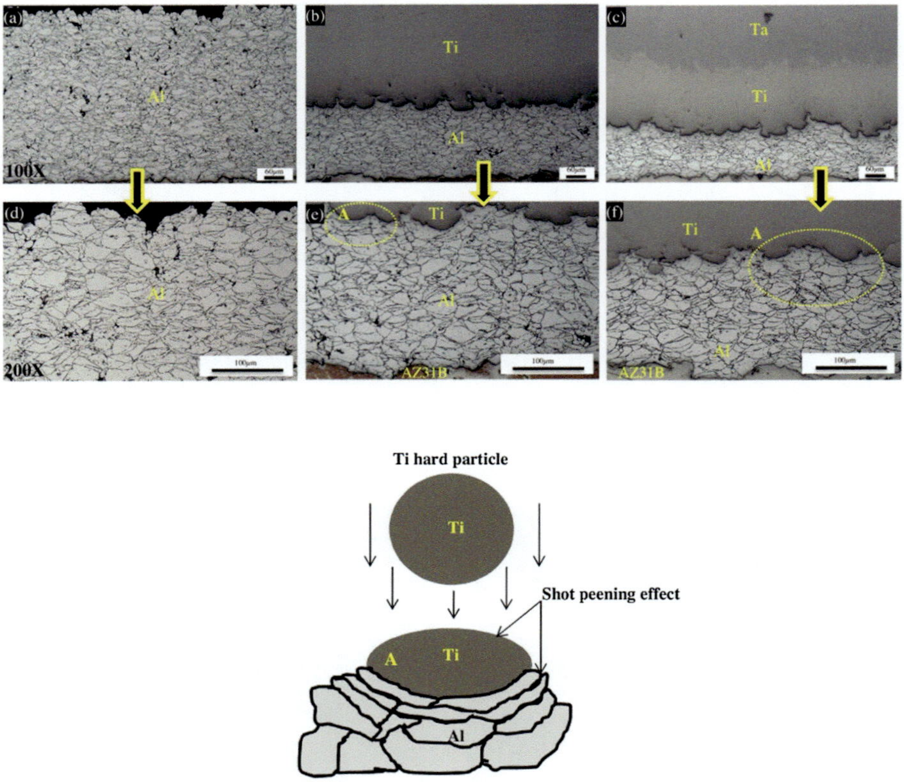

Figure 10.26 Densification of Al coating in multilayer systems, (A and D) Al coating on AZ31B Mg alloy, (B and E) Ti/Al coating on AZ31B Mg alloy, and (C and F) Ta/Ti/Al coating on AZ31B Mg alloy [101].

denseness of the cold sprayed Al coating (Fig. 10.26) and outstandingly protected the Mg alloy substrate in corrosive solutions. The drawbacks associated with CP-Al coatings could be mitigated with high-pressure cold sprayed top layers (mostly passive metals with low standard reduction potential (SRP) mismatch with underneath coating). Actually, a part of cold sprayed Al coating thickness could be replaced by high-pressure cold sprayed top layers (mostly passive metals such as Ti, Zr, Ta, etc.) for modifying the surface properties of the cold sprayed Al-based coatings on Mg alloys. This method also significantly enhanced the microhardness and wear resistance of the Al-coated Mg alloy surface. In fact, a balance between the cost reduction and greater surface properties of passive metals could be achieved by using multilayered materials, Fig. 10.27 [101].

F. S. da Silva et al. [102] found out that the top layer of Al coating (as-prepared) did not contribute to increase in the corrosion resistance in corrosive solution. This was related to the porous nature of the top layer of the coating and also the existence of irregular and small microdefects along the coating cross-section. Thus, the chemical attacks could be increased by the initial accelerated corrosion through the pores.

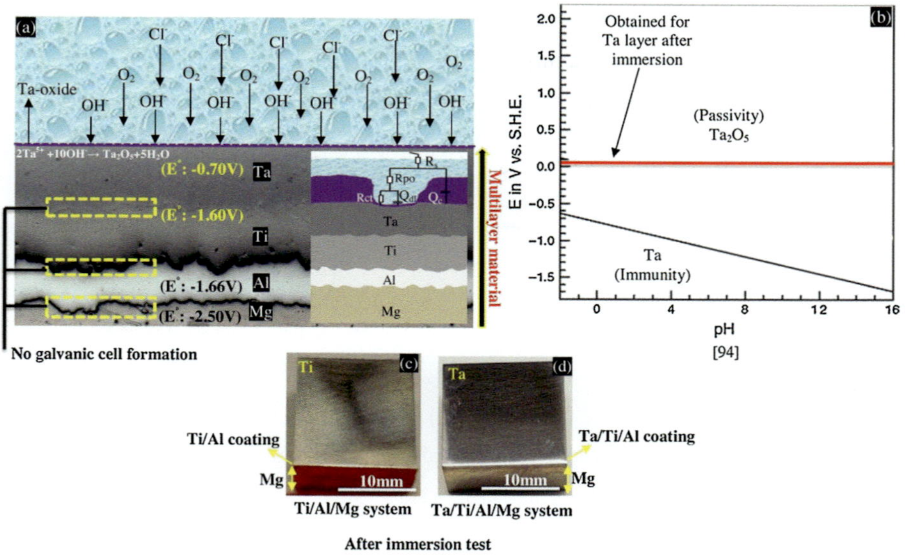

Figure 10.27 (A) Schematic drawing of Ta/Ti/Al-coated Mg alloy in 3.5 wt.% NaCl solution and electrical equivalent circuit for fitting EIS spectra (inset), (B) Pourbaix (E-pH) diagram of Ta, (C) photos of Ti/Al coated Mg alloy, and (D) Ta/Ti/Al coated Mg alloy after immersion in corrosive solution [101].

This also could lead to the small defects enlargement and significant attack at the interface between coating and the substrate as well. It was reported that the initial corrosion rate of the coating surface could be reduced by the removal of the coating top layer. This led to the preventing strong attack on the small microdefects and permitting the corrosive electrolyte to reach only some regions at the interface of Al coating/7075-T6 Al alloy substrate. In general, it was found that the coating top layer can be very harmful to the coating lifetime (reduction of corrosion resistance or impedence modulous at low frequency 0.01Hz [103]) and should be removed [102].

A porosity gradient is often observed in the cold sprayed coatings including an inner dense region, a more porous outer region and a final rough surface finish (high true surface area). This observation could be related to the incoming powder particles on densifying (shot peening effect) the previously deposited particles [104]. Removal of the coating top layer resulted in a flattened coating surface with low surface roughness average (R_a) [102]. In this regard, as-sprayed Ti surfaces [71] showed lower corrosion resistance than Ti deposits with polished surfaces [71,102], Fig. 10.14.

Above-mentioned behavior was also reported by Hong-Ren Wang et al. [105]. It was noticed from the polarization curves (Fig. 10.28) that the as-sprayed Ti coating has higher corrosion rate than TA2 (bulk Ti). This observation originates from the higher active surface of the as-sprayed Ti coating and also its higher surface area due to the higher surface rough average of the cold sprayed coating surface. In contrast, lower corrosion current density and higher corrosion potential were observed for the

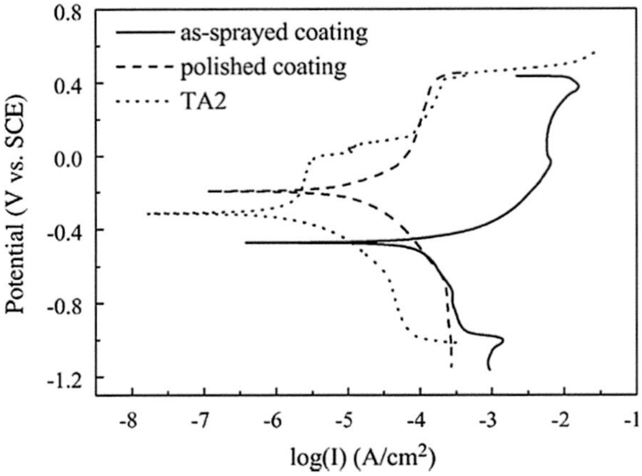

Figure 10.28 PDP curves of as-sprayed, polished Ti coatings and TA2 (Ti bulk) in seawater [105].

Table 10.7 PDP results of (as-sprayed) Ti coating on carbon steel (laser treated), Ti coating on carbon steel, Bulk Ti, and bulk carbon steel in aerated 3.5wt% NaCl solution [106].

Specimens	OCP, mV[a]	E_{corr}, mV[a]	I_{corr}, mA/cm^2	I_{pp}, mA/cm^2
Bulk titanium	−285	−281	1×10^{-4}	0.004
As-sprayed coating (onto carbon steel)	−500	−507	5×10^{-3}	0.6
Laser-treated coating	−285	−294	5×10^{-5}	0.005
Carbon steel [a:]vs Ag/AgCl	−590	−567	2×10^{-2}	

polished coating in comparison with the as-sprayed Ti coating. This could be ascribed to the elimination of the porous surface layer of the cold sprayed Ti coating which had more active surface than the polished Ti coating [105].

Laser remelting (as a local thermal treatment on the top of the coating) is an interesting method which is employed to reduce the surface roughness and seal open porosities. Sufficiently high heat energy from the laser beam is able to melt the coatings surface during laser remelting (LR) process. However, rapid cooling off the sample is expected at the end of process [60]. It was noticed that LR process can vary the corrosion resistance of cold sprayed Ti coatings [71]. A shift in the open circuit potential (OCP) and corrosion potential (E_{corr}) values toward the nobler potentials (lower tendency to corrosion) was observed for LR-treated coatings [60].

A significant decrease in the corrosion current density (I_{corr}) (Table 10.7) was observed in the LRed coating [106]. This was ascribed to the dense LR-treated layer and the presence of protective surface oxide film (formed during corrosion) on the coating surface. Nevertheless, LRed Ti coating showed the different corrosion behavior compared to the bulk Ti. So, the unclear corrosion mechanism of the

LRed coating layer, the unique microstructural features, and the role of surface oxide film (corrosion surface film) on the corrosion behavior of LRed layer should be further clarified in the future research studies.

The lifetime of CS fabricated alloys (especially alloys with relatively low corrosion resistance such as Al alloys in aggressive electrolytes) with acceptable mechanical properties could be prolonged with protective coatings [107].

Among the other beneficial surface treatment methods [108−110,111,112], Plasma Electrolytic Oxidation (PEO) method has been widely used to deposit hard ceramic coatings for protection of Al, nonferrous alloys and cold sprayed coatings as well [113−118]. In a research work, the surface of CS carbon nanotube-aluminum (CNT-Al) and pure Al coatings were post-cold spray treated using PEO process to acquire CNT-Al/PEO and Al/PEO duplex coatings on Mg-base alloy, respectively (Fig. 10.29A-D). The corrosion current density of the CS-Al coating was reduced using PEO-treatment by two orders of magnitude. However, the PEO treatment had little influence on the enhancement of the corrosion resistance of the CS CNT-Al coating (Fig. 10.29E−H). This was related to the considerable presence of cracks and holes

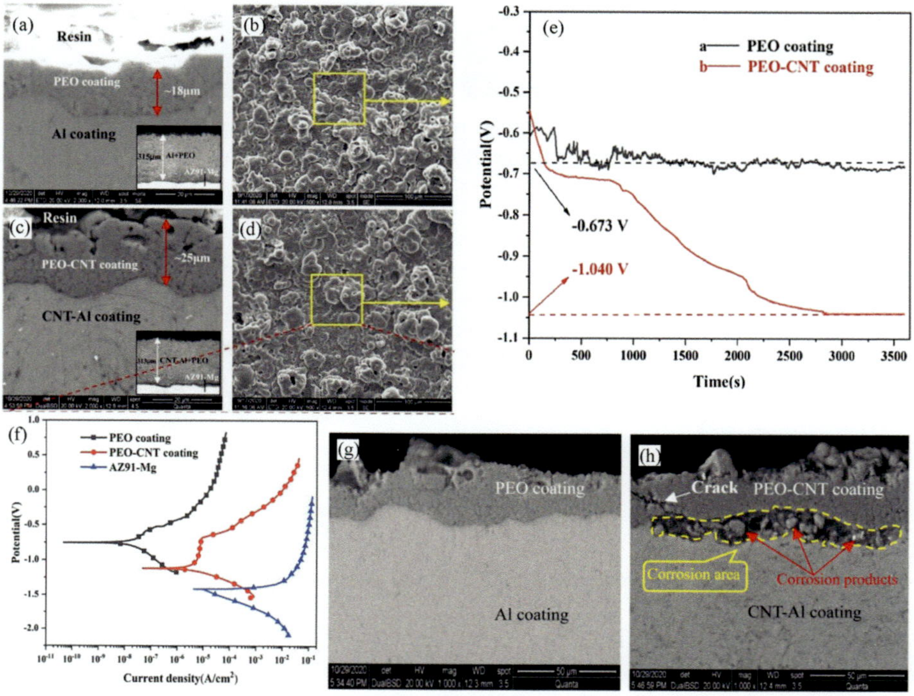

Figure 10.29 (A−D) Cross-sectional and surface SEM images of CS carbon nanotube-aluminum (CNT-Al) and pure Al coatings post-cold spray treated by the PEO process, (E) OCP Vs time, (F) PDP curves of samples in 3.5 wt% NaCl solution, (G, H) cross-sectional SEM images of CS Al/PEO and CS CNT-Al/PEO duplex coatings after electrochemical corrosion test [119].

in the PEO-CNT coating [119]. Recently, an alumina coating on a cold sprayed 7075 aluminum alloy deposit (with 10 mm thickness) was deposited by the PEO process to improve the surface properties (Fig. 10.30). PEO treatment improved the electrochemical corrosion resistance of the cold sprayed 7075 aluminum alloy (Fig. 10.31 and Table 10.8). It was noticed that the coating microstructure could influence the coating ability for preventing a corrosive attack during corrosion [107].

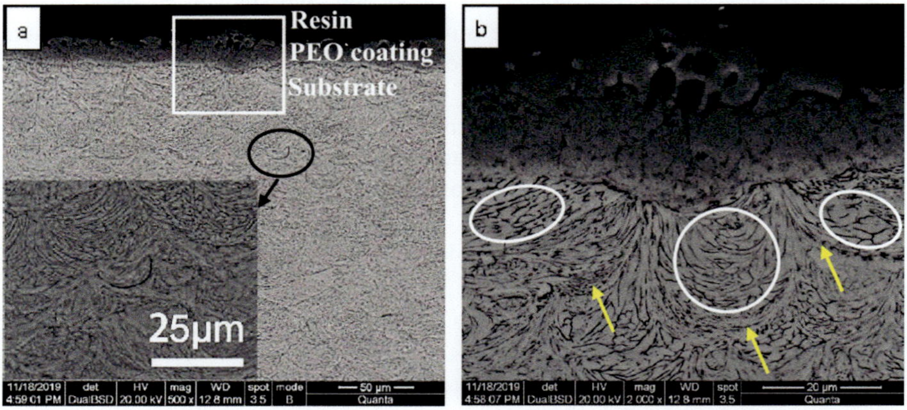

Figure 10.30 Cross-section of PEO coatings on cold sprayed 7075 aluminum deposit (SEM images (BSE: backscattered electron images)); (A and B) at different magnifications [107].

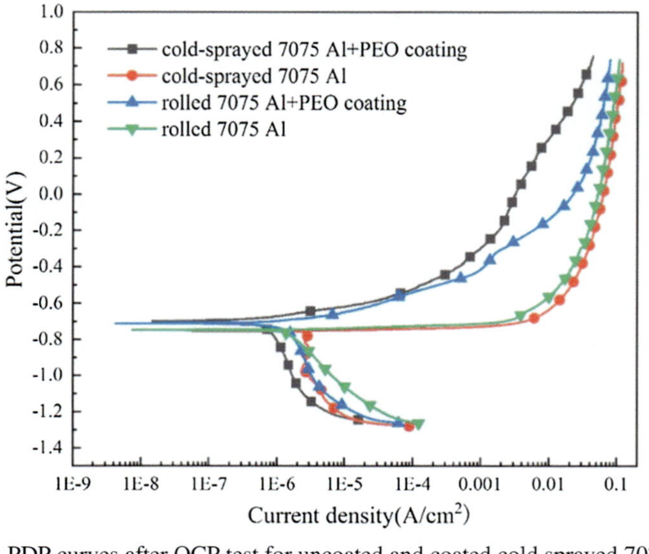

Figure 10.31 PDP curves after OCP test for uncoated and coated cold sprayed 7075 aluminum alloy deposits in 3.5 wt% NaCl solution [107].

Table 10.8 Results of PDP test from Fig. 10.7 for uncoated and coated cold sprayed 7075 aluminum alloy deposits in 3.5 wt% NaCl solution [107].

Sample	E_{corr} (V)	I_{corr} (A/cm^2)	R_p (Ω-cm^2)
Cold sprayed 7075 Al + PEO	−0.699	6.27×10^{-7}	3.90×10^{4}
Cold sprayed 7075 AI	−0.781	1.16×10^{-4}	6.45×10^{2}
Rolled 7075 Al (T6) + PEO	−0.711	2.33×10^{-6}	1.55×10^{4}
Rolled 7075 Al (T6)	−0.747	3.19×10^{-5}	1.87×10^{3}

Cold spray technology which is mostly used to improve the surface properties of the bulk materials could also be considered as another surface treatment on the CSAM metallic components (as substrate) and as potential future studies.

Abbreviations

AA	Air Annealing
AM	Additive manufacturing
BSE	Backscattered Electron
CP	Commercially pure
CPP	Cyclic Potentiodynamic Polarization
CS	Cold spraying
CSAM	Cold spraying additive manufacturing
DE	Deposition efficiency
EBSD	Electron Backscatter Diffraction
Ecorr	Corrosion Potential
EDS	Energy Dispersive X-ray Spectroscopy
EEC	Electrical Equivalent Circuit
EIS	Electrochemical Impedance Spectroscopy
Epitt	Pitting Potential
Erp	Repassivation Potential (Protection Potential)
GRCop-42	Cu−4%Cr−2%Nb
HAGB	High Angle Grain Boundary
HL	Hysteresis Loop
HPCS	High-Pressure Cold Spray
HT	Heat Treatment
Icorr	Corrosion Current Density
IN718	Inconel 718
IPF	Inverse Pole Figure
IQ	Image Quality
ISCST	In Situ Cold Spray Treatment
KAM	Kernel Average Misorientation
LACS	Laser-Assisted Cold Spray
LAGB	Low Angle Grain Boundary
LALPCS	Laser-Assisted Low-Pressure Cold Spray

LAM	Local Average Misorientation
LPCS	Low-Pressure Cold Spray
LR	Laser Remelting
LRed	Laser Remelting Treated
MMC	Metal Matrix Composite
MP	Misorientation Profiles
OCP	Open Circuit Potential
PCSHT	Post-Cold Spray Heat Treatment
PCST	Post-Cold Spray Treatment
PDP	Potentiodynamic Polarization
PEO	Plasma Electrolytic Oxidation
PreCSHT	Pre-Cold Spray Heat Treatment
PreCST	Pre-Cold Spray Treatment
PSD	Powder Particle Size Distribution
Rct	Charge Transfer Resistance
Rpo	Pore Resistance
SEM	Scanning Electron Microscopy
SP	Shot Peening
SPACS	Shot Peening-Assisted Cold Spray
SVET	Scanning Vibrating Electrode Technique
TEM	Transmission Electron Microscopy
Tm	Absolute Melting Temperature
Tr	Recrystallization Temperature
UTS	Ultimate Tensile Strength
VA	Vacuum Annealing
WI	Warburg Impedance

References

[1] Y. Ta, T. Xiong, C. Sun, et al., Microstructure and corrosion performance of a cold sprayed aluminum coating on AZ91D magnesium alloy, Corrosion Science 52 (2010) 3191−3197.

[2] B.S.D. Force, T.J. Eden, J.K. Potter, Cold spray Al-5% Mg coatings for the corrosion protection of magnesium alloys, Journal of Thermal Spray Technology 20 (6) (2011) 1352−1358.

[3] M. Diab, X. Pang, H. Jahed, The effect of pure aluminum cold spray coating on corrosion and corrosion fatigue of magnesium (3% Al-1% Zn) extrusion, Surface & Coatings Technology 309 (2017) 423−435.

[4] Z. Jing, K. Dejun, Effect of laser remelting on microstructure and immersion corrosion of cold sprayed aluminum coating on S355 structural steel, Optics and Laser Technology 106 (2018) 348−356.

[5] Z. Monette, A.K. Kasar, M. Daroonparvar, P. L Menezes, Supersonic particle deposition as an additive technology: methods, challenges, and applications, The International Journal of Advanced Manufacturing Technology 106 (2020) 2079−2099.

[6] C.M. Kay, J. Karthikeyan, High Pressure Cold Spray: Principles and Applications, ASM International, 2016. ISBN: 978-1-62708-096-5.

[7] W. Sun, A.W.Y. Tan, Adhesion, tribological and corrosion properties of cold-sprayed CoCrMo and Ti6Al4V coatings on 6061-T651 Al alloy, Surface & Coatings Technology 326 (2017) 291–298.

[8] D. William, Callister Jr, et.al, Materials Science and Engineering, An Introduction, ninth ed., Wiley (October 8, 2012).

[9] E. McCafferty, Introduction to Corrosion Science, Springer, 2010 e-ISBN 978-1-4419-0455-3.

[10] M. Daroonparvar, M.A.M. Yajid, N.M. Yusof, H.R. Bakhsheshi-Rad, Preparation and corrosion resistance of a nanocomposite plasma electrolytic oxidation coating on Mg-1%Ca alloy formed in aluminate electrolyte containing titania nano-additives, Alloys and Compounds 688 (2016) 841–857.

[11] M. Daroonparvar, M.A.M. Yajid, N.M. Yusof, H.R. Bakhsheshi-Rad, E. Hamzah, H.A. Kamali, Microstructural characterization and corrosion resistance evaluation of nanostructured Al and Al/AlCr coated Mg–Zn–Ce–La alloy, Alloys and Compounds 615 (2014) 657–671.

[12] M. Daroonparvar, M.A. M Yajid, N.M. Yusof, H.R. Bakhsheshi-Rad, Fabrication and properties of triplex NiCrAlY/nano $Al_2O_3 \cdot$ 13% TiO_2/nano TiO_2 coatings on a magnesium alloy by atmospheric plasma spraying method, Alloys and Compounds 645 (2015) 450–466.

[13] M. Carboneras, M.D. López, P. Rodrigo, M. Campo, B. Torres, E. Otero, J. Rams, Corrosion behaviour of thermally sprayed Al and Al/SiCp composite coatings on ZE41 magnesium alloy in chloride medium, Corrosion Science 52 (2010) 761–768.

[14] M. Daroonparvar, M.A.M. Yajid, H.R. Bakhsheshi-Rad, N.M. Yusof, S. Izman, E. Hamzah, M.R. Abdul Kadir, Corrosion resistance investigation of nanostructured Si- and Si/TiO_2-coated Mg alloy in 3.5% NaCl solution, Vacuum 108 (2014) 61–65.

[15] H.R. Bakhsheshi-Rad, E. Hamzah, R. Ebrahimi-Kahrizsangi, M. Daroonparvar, M. Medraj, Fabrication and characterization of hydrophobic micro arc oxidation/poly-lactic acid duplex coating on biodegradable Mg–Ca alloy for corrosion protection, Vacuum 125 (2016) 185–188.

[16] M. Esmaily, J.E. Svensson, S. Fajardo, N. Birbilis, G.S. Frankel, S. Virtanen, R. Arrabal, S. Thomas, L.G. Johansson, Fundamentals and advances in magnesium alloy corrosion, Progress in Materials Science 89 (2017) 92–193.

[17] M. Daroonparvar, M.A. Mat Yajid, N.M. Yusof, H.R. Bakhsheshi-Rad, E. Hamzah, Microstructural characterisation of air plasma sprayed nanostructure ceramic coatings on Mg–1% Ca alloys (bonded by NiCoCrAlYTa alloy), Ceramics International 42 (2016) 357–371.

[18] H.S. Ryu, D.S. Park, S.H. Hong, Improved corrosion protection of AZ31 magnesium alloy through plasma electrolytic oxidation and aerosol deposition duplex treatment, Surface and Coatings Technology 219 (2013) 82–87.

[19] H.R. Bakhsheshi-Rad, E. Hamzah, M. Daroonparvar, M. Kasiri-Asgarani, M. Medraj, Synthesis and biodegradation evaluation of nano-Si and nano-Si/TiO_2 coatings on biodegradable Mg–Ca alloy in simulated body fluid, Ceramics International 40 (9) (2014) 14009–14018.

[20] D. Jones, Principles and Prevention of Corrosion, second ed., Prentice-Hall, Upper Saddle River, NJ, Page, 1996, p. 572.

[21] G. Frankel, Pitting Corrosion, Corrosion Fundamentals, Testing, and Protection, ASM Handbook, Volume 13A, ASM International, 2003, pp. 236–241.

[22] P.A. Schweitzer (Ed.), Corrosion Engineering Handbook, Marcel Dekker, 1996, 736 pp.

[23] U. Chatterjee, S. Bose, S. Roy, Environmental Degradation of Metals, Marcel Dekker Inc., New York, USA, 2001, 498 pp.
[24] R. Kelly, Crevice corrosion, Corrosion Fundamentals, Testing, and Protection, ASM Handbook, Volume 13A, ASM International, 2003, pp. 242−247.
[25] R. Baboian, Galvanic Corrosion, Corrosion Fundamentals, Testing, and Protection, ASM Handbook, Volume 13A, ASM International, 2003, pp. 210−213.
[26] X. Meng, J. Zhang, J. Zhao, Y. Liang, Y. Zhang, Influence of gas temperature on microstructure and properties of cold spray 304SS coating, Journal of Materials Science and Technology 27 (9) (2011) 809−815.
[27] H. Koivuluoto, G. Bolelli, et al., Corrosion resistance of cold-sprayed Ta coatings in very aggressive conditions, Surface & Coatings Technology 205 (2010) 1103−1107.
[28] H. Koivuluoto, G. Bolelli, L. Lusvarghi, F. Casadei, P. Vuorist, Cold-sprayed copper and tantalum coatings — detailed FESEM and TEM analysis, Surface and Coatings Technology 204 (2010) 2353−2361.
[29] X.-T. Luo, G.-J. Yang, C.-J. Li, Multiple strengthening mechanisms of cold-sprayed cBNp/NiCrAl composite coating, Surface and Coatings Technology 205 (2011) 4808−4813.
[30] W. Li, D.-Y. Li, Variations of work function and corrosion behaviors of deformed copper surfaces, Applied Surface Science 240 (2005) 388−395.
[31] M. Laleh, F. Kargar, Effect of surface nanocrystallization on the microstructural and corrosion characteristics of AZ91D magnesium alloy, Journal of Alloys and Compounds 509 (2011) 9150−9156.
[32] Y. Kang Wei, Y. Juan Li, Y. Zhang, et al., Corrosion resistant nickel coating with strong adhesion on AZ31B magnesium alloy prepared by an in-situ shot-peening-assisted cold spray, Corrosion Science 138 (2018) 105−115.
[33] Z. Arabgol, H. Assadi, et al., Analysis of thermal history and residual stress in cold-sprayed coatings, Journal of Thermal Spray Technology 23 (2014) 84−90.
[34] T. Suhonen, et al., Residual stress development in cold sprayed Al, Cu and Ti coatings, Acta Materrilia 61 (2013) 6329−6337.
[35] D. Goldbaum, et al., Mechanical behavior of Ti cold spray coatings determined by a multi-scale indentation method, Materials Science and Engineering A 530 (2011) 253−265.
[36] A.M. Ralls, A.K. Kasar, M. Daroonparvar, A. Siddaiah, P. Kumar, C.M. Kay, M. Misra, P.L. Menezes, Effect of gas propellant temperature on the microstructure, friction, and wear resistance of high-pressure cold sprayed Zr702 coatings on Al6061 alloy, Coatings 12 (2022) 263, https://doi.org/10.3390/coatings12020263.
[37] A.M. Ralls, M. Daroonparvar, S. Sikdar, M.H. Rahman, M. Monwar, K. Watson, C.M. Kay, P.L. Menezes, Tribological and corrosion behavior of high pressure cold sprayed duplex 316 L stainless steel, Tribology International 169 (2022) 107471.
[38] K. Balani, et al., Effect of carrier gases on microstructural and electrochemical behavior of cold-sprayed 1100 aluminum coating, Surface and Coatings Technology 195 (2005) 272−279.
[39] O. Meydanoglu, et al., Microstructure, mechanical properties and corrosion performance of 7075 Al matrix ceramic particle reinforced composite coatings produced by the cold gas dynamic spraying process, Surface and Coatings Technology 235 (2013) 108−116.
[40] N. Lia, W. Lia, X. Yang, Y. Xu, A. Vairisa, Corrosion characteristics and wear performance of cold sprayed coatings of reinforced Al deposited onto friction stir welded AA2024-T3 joints, Surface and Coatings Technology 349 (2018) 1069−1076.

[41] X. Xie, B. Hosni, C. Chen, H. Wu, Y. Li, Z. Chen, C. Verdy, O.E.I. Kedim, Q. Zhong, A. Addad, C. Coddet, G. Ji, H. Liao, Corrosion behavior of cold sprayed 7075Al composite coating reinforced with TiB_2 nanoparticles, Surface and Coatings Technology 404 (2020) 126460.
[42] Y. Zou, D. Goldbaum, A. Jerzy, J.A. Szpunar, Y. Stephen, Microstructure and nanohardness of cold-sprayed coatings: electron backscattered diffraction and nanoindentation studies, Scripta Materialia 62 (2010) 395–398.
[43] C.K.S. Moy, et al., Investigating the microstructure and composition of cold gas-dynamic spray (CGDS) Ti powder deposited on Al 6063 substrate, Surface and Coatings Technology 204 (2010) 3739–3749.
[44] Y. Zou, et al., Dynamic recrystallization in the particle/particle interfacial region of cold-sprayed nickel coating: electron backscatter diffraction characterization, Scripta Materialia 61 (2009) 899–902.
[45] L.-Y. Qin, et al., Effect of grain size on corrosion behavior of electrodeposited bulk nanocrystalline Ni, Transaction of Nonferrous Metals Society of China 20 (2010) 82–89.
[46] D. Sachdeva, et al., Electrochemical characterization of nanocrystalline nickel, Defense Science Journal 58 (2008) 525–530.
[47] S.S. Visw, et al., Corrosion protection and control using nanomaterials, in: A Volume in Woodhead Publishing Series in Metals and Surface Engineering, 2012.
[48] R. Mishra, et al., Effect of nanocrystalline grain size on the electrochemical and corrosion behavior of nickel, Corrosion Science 46 (2004) 3019–3029.
[49] X.-P. Jiang, et al., Enhancement of fatigue and corrosion properties of pure Ti by sandblasting, Materials Science and Engineering A 429 (2006) 30–35.
[50] A. Vakilipour Takaloo, K. Ahmadi Majlan, M. Keshavarz, S. Farahani, M. Daroonparvar, S.V. Takaloo, A mechanism of nickel deposition on titanium substrate by high speed electroplating, in: Proceedings of the 12th World Conference on Titanium, 1920–1923.
[51] W.Y. Guo, J. Sun, J.S. Wu, Effect of deformation on corrosion behavior of Ti–23 Nb–0.7Ta–2Zr–O alloy, Materials Characterization 60 (2009) 173–177.
[52] C.G.a.F. Martın, P. DeTiedra, J. AHeredero, M. L Aparicio, Effects of prior cold work and sensitization heat treatment on chloride stress corrosion cracking in type 304 stainless steels, Corrosion Science 43 (2001) 1519–1539.
[53] F. Gärtner, T. Stoltenhoff, J. Voyer, H. Kreye, S. Riekehr, M. Kocak, Mechanical properties of cold-sprayed and thermally sprayed copper coatings, Surface and Coatings Technology 200 (2006) 6770–6782.
[54] S. Semiatin, Recovery, Recrystallization, and Grain-Growth Structures, Metalworking: Bulk Forming, Volume 14A, ASM Handbook, ASM International, 2005, pp. 552–562.
[55] M. Decker, R. Neiser, D. Gilmore, H. Tran, Microstructure and properties of cold spray nickel, in: C. Berndt, K. Khor, E. Lugscheider (Eds.), Thermal Spray 2001: New Surfaces for a New Millennium, ASM International, Singapore, May 28–30, 2001, pp. 433–439.
[56] W. Sun, A. Bhowmik, A. Wei-Yee Tan, R. Li, F. Xue, I. Marinescu, E. Liu, Improving microstructural and mechanical characteristics of cold sprayed Inconel 718 deposits via local induction heat treatment, Journal of Alloys and Compounds 797 (2019) 1268–1279.
[57] K. Spencer, et al., Heat treatment of cold spray coatings to form protective intermetallic layers, Scripta Materialia 61 (2009) 44–47.
[58] S. Phani, et al., Effect of process parameters and heat treatments on properties of cold sprayed copper coatings, Journal of Thermal Spray Technology 16 (2007) 425–434.

[59] S. Yin, et al., Annealing strategies for enhancing mechanical properties of additively manufactured 316L stainless steel deposited by cold spray, Surface and Coatings Technology 370 (2019) 353–361.
[60] W. Sun, et al., Post-process treatments on supersonic cold sprayed coatings: a review, Coatings 10 (123) (2020) 1–35.
[61] X.-M. Meng, et al., Influence of annealing treatment on the microstructure and mechanical performance of cold sprayed 304 stainless steel coating, Applied Surface Science 258 (2011) 700–704.
[62] G.-J. Yang, et al., Effect of annealing on the microstructure and erosion performance of cold-sprayed FeAl intermetallic coatings, Surface and Coatings Technology 205 (2011) 5502–5509.
[63] H.Y. Lee, et al., Fabrication of cold sprayed Al-intermetallic compounds coatings by post annealing, Materials Science and Engineering A 433 (2006) 139–143.
[64] M.R. Rokni, et al., Microstructure and mechanical properties of cold sprayed 7075 deposition during non-isothermal annealing, Surface and Coatings Technology 276 (2015) 305–315.
[65] R. Huang, et al., The effects of heat treatment on the mechanical properties of cold-sprayed coatings, Surface and Coatings Technology 261 (2015) 278–288.
[66] W.Y. Li, et al., Improvement of microstructure and property of cold-sprayed Cu–4 at.%Cr–2at.%Nb alloy by heat treatment, Scripta Materialia 55 (2006) 327–330.
[67] S. Kumar, et al., Influence of annealing on mechanical and electrochemical properties of cold sprayed niobium coatings, Surface and Coatings Technology 296 (2016) 124–135.
[68] J.S. Yu, et al., Densification and purification of cold sprayed Ti coating layer by using annealing in different heat treatment environments, Advanced Materials Research 602–604 (2012) 1604–1608.
[69] W. Ma, et al., Microstructural and mechanical properties of high-performance Inconel 718 alloy by cold spraying, Journal of Alloys and Compounds 792 (2019) 456–467.
[70] T. Hussain, D.G. McCartney, P.H. Shipway, T. Marrocco, Corrosion behavior of cold sprayed titanium coatings and free-standing deposits, Journal of Thermal Spray Technology 20 (1–2) (2011) 260–274.
[71] W. Li, C. Cao, S. Yin, Solid-state cold spraying of Ti and its alloys: a literature review, Progress in Materials Science 110 (2020) 100633.
[72] S. Kumar, V. Vidyasagar, A. Jyothirmayi, S.V. Joshi, Effect of heat treatment on mechanical properties and corrosion performance of cold-sprayed tantalum coatings, Journal of Thermal Spray Technology 25 (4) (2016) 745–756.
[73] A.V. Takaloo, M.R. Daroonparvar, M.M. Atabaki, K. Mokhtar, Corrosion behavior of heat treated nickel-aluminum bronze alloy in artificial seawater, Materials Sciences and Applications 2 (11) (2011) 1542.
[74] H.R. Bakhsheshi-Rad, E. Hamzah, M.R. Abdul-Kadir, M. Daroonparvar, M. Medraj, Corrosion and mechanical performance of double-layered nano-Al/PCL coating on Mg–Ca–Bi alloy, Vacuum 119 (2015) 95–98.
[75] Z. Zhang, F. Liu, E.H. Han, L. Xu, Mechanical and corrosion properties in 3.5% NaCl solution of cold sprayed Al-based coatings, Surface and Coatings Technology 385 (2020) 125372.
[76] M. Daroonparvar, A.K. Kasar, M.U. Farooq Khan, P.L. Menezes, C.M. Kay, M. Misra, R.K. Gupta, Improvement of wear, pitting corrosion resistance and repassivation ability of Mg-based alloys using high pressure cold sprayed (HPCS) commercially pure-titanium coatings, Coatings 11 (2021) 57, https://doi.org/10.3390/coatings11010057.

[77] X. Meng, J. Zhang, J. Zhao, W. Han, Y. Liang, Influence of annealing treatment on microstructure and properties of cold sprayed stainless steel coatings, Acta Metallurgica Sinica 24 (2) (2011) 92–100.
[78] A. Sabard, T. Hussain, Inter-particle bonding in cold spray deposition of a gas atomized and a solution heat-treated Al 6061 powder, Journal of Materials Science 54 (2019) 12061–12078.
[79] A. Sabard, P. McNutt, H. Begg, T. Hussain, Cold spray deposition of solution heat treated, artificially aged and naturally aged Al 7075 powder, Surface and Coatings Technology 385 (2020) 125367.
[80] T. Liu, W.A. Story, L.N. Brewer, Effect of heat treatment on the Al-Cu feedstock powders for cold spray deposition, Metallurgical and Materials Transaction A 50A (2019) 3373.
[81] S. Ngai, T. Ngai, F. Vogel, W. Story, G.B. Thompson, L.N. Brewer, Saltwater corrosion behavior of cold sprayed AA7075 aluminum alloy coatings, Corros. Sci. 130 (2018) 231–240.
[82] N.W. Khun, A.W.Y. Tan, K.J.W. Bi, E. Liu, Effects of working gas on wear and corrosion resistances of cold sprayed Ti-6Al-4V coatings, Surface and Coatings Technology 302 (2016) 1–12.
[83] H.R. Wang, B.R. Hou, J. Wang, Q. Wing, W.Y. Li, Effect of process conditions on microstructure and corrosion resistance of cold-sprayed Ti coatings, Journal of Thermal Spray Technology 17 (2008) 736–741.
[84] Y.K. Wei, X.T. Luo, X. Chu, Y. Ge, G.S. Huang, Y.C. Xie, R.Z. Huang, C.J. Li, Ni coatings for corrosion protection of Mg alloys prepared by an in-situ micro-forging assisted cold spray: Effect of powder feedstock characteristics, Corrosion Science 184 (2021) 109397.
[85] "Standard Test Method of Salt Spray (Fog) Testing," B117-90, Annual Book of ASTM Standards, ASTM, 19–25.
[86] M. Vreijling, Electrochemical Characterization of Metallic Thermally Sprayed Coatings, Printed in the Netherlands, 1998, p. 143.
[87] Y. Juan Li, X. Tao Luo, H. Rashid, C.-J. Li, A new approach to prepare fully dense Cu with high conductivities and anti-corrosion performance by cold spray, Journal of Alloys and Compounds 740 (2018) 406–413.
[88] M. Kulmala, P. Vuoristo, Influence of process conditions in laser-assisted low-pressure cold spraying, Surface and Coatings Technology 202 (2008) 4503–4508.
[89] Y. Kang Wei, X. Tao Luo, C. Xin Li, C.-J. Li, Optimization of in-situ shot-peening-assisted cold spraying parameters for full corrosion protection of Mg alloy by fully dense Al-based alloy coating, Journal of Thermal Spray Technology 26 (2017) 173–183.
[90] M.A. Amin, A newly synthesized glycine derivative to control uniform and pitting corrosion processes of Al induced by SCN anions – chemical, electrochemical and morphological studies, Corrosion Science 52 (10) (2010) 3243–3257.
[91] G.E. Kiourtsidis, et al., A study on pitting behavior of AA2024/SiC p composites using the double cycle polarization technique, Corrosion Science 41 (6) (1999) 1185–1203.
[92] J. Ma, et al., Electrochemical polarization and corrosion behavior of Al–Zn–In based alloy in acidity and alkalinity solutions, International Journal of Hydrogen Energy 38 (34) (2013) 14896–14902.
[93] Y. Liu, et al., Electronic structure and pitting behavior of 3003 aluminum alloy passivated under various conditions, Electrochimca Acta 54 (17) (2009) 4155–4163.
[94] X.-T. Luo, C.-J. Li, Large sized cubic BN reinforced nanocomposite with improved abrasive wear resistance deposited by cold spray, Materials and Design 83 (2015) 249–256.

[95] E. Irissou, J.G. Legoux, B. Arsenault, C. Moreau, Investigation of Al-Al_2O_3 cold spray coating formation and properties, Journal of Thermal Spray Technology 16 (5) (2007) 661−668.
[96] E. Sansoucy, P. Marcoux, L. Ajdelsztajn, B. Jodoin, Properties of SiC-reinforced aluminum alloy coatings produced by the cold gas dynamic spraying process, Surface and Coatings Technology 202 (16) (2008) 3988−3996.
[97] Y.-S. Tao, T.-Y. Xiong, C. Sun, H. Jin, H. Du, T.-F. Li, Effect of α-Al_2O_3 on the properties of cold sprayed Al/a-Al_2O_3 composite coatings on AZ91D magnesium alloy, Applied Surface Science 256 (1) (2009) 261−266.
[98] Y.-Y. Wang, B. Normand, N. Mary, M. Yu, H.-L. Liao, Microstructure and corrosion behavior of cold sprayed SiCp/Al 5056 composite coatings, Surface and Coatings Technology 251 (2014) 264−275.
[99] F. Fei Lua, K. Maa, C. Xin Lia, M. Yasirb, X.-T. Luoa, C.-jiu Lia, Enhanced corrosion resistance of cold-sprayed and shot-peened aluminum coatings on LA43M magnesium alloy, Surface and Coatings Technology 394 (2020) 125865.
[100] M. Daroonparvar, M.U. Farooq Khan, Y. Saadeh, C.M. Kay, R.K. Gupta, A.K. Kasar, P. Kumar, M. Misra, P.L. Menezes, H.R. Bakhsheshi-Rad, Enhanced corrosion resistance and surface bioactivity of AZ31B Mg alloy by high pressure cold sprayed monolayer Ti and bilayer Ta/Ti coatings in simulated body fluid, Materials Chemistry and Physics 256 (2020) 123627.
[101] M. Daroonparvar, M.U. Farooq Khan, Y. Saadeh, C.M. Kay, A.K. Kasar, P. Kumar, M. Misra, P.L. Menezes, P.R. Kalvala, R.K. Gupta, H.R. Bakhsheshi-Rad, L. Esteves, Modification of surface hardness, wear resistance and corrosion resistance of cold spray Al coated AZ31B Mg alloy using cold spray double layered Ta/Ti coating in 3.5 wt% NaCl solution, Corrosion Science (2020) 109029.
[102] F.S. da Silva, N. Cinca, S. Dosta, I.G. Cano, J.M. Guilemany, A.V. Benedetti, Effect of the outer layer of Al coatings deposited by cold gas spray on the microstructure, mechanical properties and corrosion resistance of the AA 7075-T6 aluminum alloy, Journal of Thermal Spray Technology (2020), https://doi.org/10.1007/s11666-020-01023-8.
[103] H.R. Bakhsheshi-Rad, E. Hamzah, A.F. Ismail, Z. Sharer, M.R. Abdul-Kadir, M. Daroonparvar, S.N. Saud, M. Medraj, Synthesis and corrosion behavior of a hybrid bioceramic-biopolymer coating on biodegradable Mg alloy for orthopaedic implants, Journal of Alloys and Compounds 648 (2015) 1067−1071.
[104] V.K. Champagne, The Cold Spray Materials Deposition Process Fundamentals and Applications, Woodhead Publisher, Sawston, 2007.
[105] H.R. Wang, W.Y. Li, L. Ma, J. Wang, Q. Wang, Corrosion behavior of cold sprayed titanium protective coating on 1Cr13 substrate in seawater, Surface & Coatings Technology 201 (2007) 5203−5206.
[106] T. Marrocco, T. Hussain, D.G. McCartney, P.H. Shipway, Corrosion performance of laser post-treated cold sprayed titanium coatings, Journal of Thermal Spray Technology 20 (4) (2011) 909−917.
[107] Y. Raoa, Q. Wanga, D. Okab, C.S. Ramachandranc, On the PEO treatment of cold sprayed 7075 aluminum alloy and its effects on mechanical, corrosion and dry sliding wear performances thereof, Surface & Coatings Technology 383 (2020) 125271.
[108] M. Daroonparvar, M.A.M. Yajid, C.M. Kay, H. B.-R., R. Kumar Gupta, N.M. Yusof, H. Ghandvar, A. Arshad, I.S.M. Zulkifli, Effects of Al_2O_3 diffusion barrier layer (including Y-containing small oxide precipitates) and nanostructured YSZ top coat on the oxidation behavior of HVOF NiCoCrAlTaY/APS YSZ coatings at 1100 °C, Corrosion Science 144 (2018) 13−34.

[109] H. R Bakhsheshi-Rad, et al., Coating biodegradable magnesium alloys with electrospun poly-L-lactic acid-åkermanite-doxycycline nanofibers for enhanced biocompatibility, antibacterial activity, and corrosion resistance, Surface and Coatings Technology 377 (2019) 124898.
[110] M. Daroonparvar, Effects of bond coat and top coat (including nano zones) structures on morphology and type of formed transient stage oxides at pre-heat treated nano NiCrAlY/nano ZrO_2-8%Y_2O_3 interface during oxidation, Journal of Rare Earths 33 (2015) 983–994.
[111] H.R. Bakhsheshi-Rad, E. Hamzah, A.F. Ismail, M. Daroonparvar, M. Kasiri-Asgarani, S. Jabbarzare, M. Medraj, Microstructural, mechanical properties and corrosion behavior of plasma sprayed NiCrAlY/nano-YSZ duplex coating on Mg–1.2 Ca–3Zn alloy, Ceramics International 41 (10) (2015) 15272–15277.
[112] M. Daroonparvar, M. Azizi Mat Yajid, N.M. Yusof, M. Sakhawat Hussain, Improved thermally grown oxide scale in air plasma sprayed NiCrAlY/Nano-YSZ coatings, Journal of Nanomaterials 2013 (2013) 520104, https://doi.org/10.1155/2013/520104.
[113] J. Martina, K. Akoda, V. Ntomprougkidis, O. Ferry, A. Maizeraya, A. Bastien, P. Brenot, G. Ezo'o, G. Henrion, Duplex surface treatment of metallic alloys combining cold spray and plasma electrolytic oxidation technologies, Surface and Coatings Technology 392 (2020) 125756.
[114] M. Daroonparvar, et al., Antibacterial activities and corrosion behavior of novel PEO/nanostructured ZrO_2 coating on Mg alloy, Transactions of Nonferrous Metals Society of China 28 (2018) 1571–1581.
[115] S. Luo, Q. Wang, R. Ye, C.S. Ramachandran, Effects of electrolyte concentration on the microstructure and properties of plasma electrolytic oxidation coatings on Ti-6Al-4V alloy, Surface and Coatings Technology 375 (2019) 864–876.
[116] M. Daroonparvar, M.A.M. Yajid, R.K. Gupta, N.M. Yusof, H.R. Bakhsheshi-Rad, H. Ghandvar, Investigation of corrosion protection performance of multiphase PEO (Mg_2SiO_4, MgO, $MgAl_2O_4$) coatings on Mg alloy formed in aluminate-silicate- based mixture electrolyte, Protection of Metals and Physical Chemistry of Surfaces 54 (2018) 425–441.
[117] M. Trevino, et al., Wear of an aluminium alloy coated by plasma electrolytic oxidation, Surface and Coatings Technology 206 (2012) 2213–2219.
[118] M. Daroonparvar, M.A.M. Yajid, N.M. Yusof, H.R. Bakhsheshi-Rad, E. Hamzah, T. Mardanikivi, Deposition of duplex MAO layer/nanostructured titanium dioxide composite coatings on Mg–1% Ca alloy using a combined technique of air plasma spraying and micro arc oxidation, Journal of Alloys and Compounds 649 (2015) 591–605.
[119] Y. Zhang, Q. Wang, R. Ye, C.S. Ramachandran, Plasma electrolytic oxidation of cold spray kinetically metallized CNT-Al coating on AZ91-Mg alloy: Evaluation of mechanical and surficial characteristics, Journal of Alloys and Compounds 892 (2022) 162094.

Index

'*Note:* Page numbers followed by "f" indicate figures and "t" indicate tables.'

G

Printed in the United States
by Baker & Taylor Publisher Services